P9-CKV-863

GEOCHEMICAL THERMODYNAMICS

GEOCHEMICAL THERMODYNAMICS

DARRELL KIRK NORDSTROM
U.S. Geological Survey

JAMES L. MUNOZ
University of Colorado

Blackwell Scientific Publications

Palo Alto   Oxford   London   Edinburgh   Boston   Victoria

Editorial Offices

667 Lytton Avenue, Palo Alto, California 94301, USA

Osney Mead, Oxford, OX2 0EL, UK

8 John Street, London, WC1N 2ES, UK

23 Ainslie Place, Edinburgh, EH3 6AJ, UK

52 Beacon Street, Boston, Massachusetts 02108, USA

107 Barry Street, Carlton, Victoria 3053, Australia

Distributors

USA and Canada
Blackwell Scientific Publications
% PBS
P.O. Box 447
Brookline, MA 02147

Australia
Blackwell Scientific Publications (Australia) Pty Ltd.
107 Barry Street
Carlton, Victoria 3053

United Kingdom
Blackwell Scientific Publications
Osney Mead
Oxford OX2 0EL

© 1986 by Blackwell Scientific Publications

All rights reserved. No part of this publication may be reproduced, stored in a retrieval system, or transmitted, in any form or by any means, electronic, mechanical, photocopying, recording or otherwise without the prior permission of the copyright owner.

Library of Congress Cataloging in Publication Data

Nordstrom, D. Kirk (Darrell Kirk)
  Geochemical thermodynamics

  Bibliography: p.
  Includes index.
    1. Geochemistry.   2. Thermodynamics.   I. Munoz,
James L.   II. Title.
QE515.N67   1985   551.9   84-24380
ISBN 0-86542-319-9

CONTENTS

PREFACE

This book had a modest beginning. We had both taught graduate courses that cover the basic principles of thermodynamics for geochemists and water chemists, and we had found no up-to-date text that presents thermodynamics in a geochemical context. Several excellent books on geochemistry have been published recently, but the thermodynamics usually takes a back seat. Authors and educators commonly assume that geologists and geochemists can acquire an adequate understanding of thermodynamics through the chemistry department. Unfortunately, our experience has shown that geochemical thermodynamics has evolved along quite different avenues than has the chemical thermodynamics taught in most chemistry departments. For example, the phase rule, chemography, solid solutions, ionic activities in mixed aqueous electrolytes, and chemical potential diagrams for multiphase systems are treated too briefly or neglected altogether in traditional chemistry courses, whereas they play a central role in geochemistry. The practice of applying thermodynamics to geology has forced the geochemical community to derive new methods of describing stability, to develop new techniques for experimental measurements, and to find new procedures for compiling, evaluating, and computing geochemical data. Furthermore, most texts leave out the important subjects of experimental measurements and evaluation of data, which are perhaps more critical to the utilization of thermodynamics than are the fundamental equations. After all, the equations originally evolved from careful measurements of pressure, volume, temperature, solubility, electromotive force, and so forth. It seems inappropriate to neglect such essential aspects of thermodynamics.

Our aims are (1) to emphasize the basic principles of thermodynamics, (2) to describe these principles within a geochemical framework, (3) to provide examples for the entire range of pressure and temperature from weathering to magmatism, and (4) to provide some information on the sources and the evaluation of geochemical thermodynamic data. Our approach is to stress the thermodynamics because we feel that it is impossible to gain enough exposure to the subject. It has been said of Arnold Sommerfeld, the famous physicist, that he was reluctant to write a monograph or treatise on thermodynamics because he wasn't sure that he understood it. The first time he studied the subject, he thought he understood it except for a few minor points. The second time he studied it, he thought he *didn't* understand it except for a few minor points. The third time, he *knew* he didn't understand it, but by then it didn't matter (because he could use it effectively).

This text is intended for first-year graduate students in geology and geochemistry, and it should be a useful prerequisite for more advanced courses in solution–mineral equilibria, petrology, and aqueous geochemistry. Mineralogy, the calculus, and basic principles of geology, chemistry, physics, and mathematics are required subjects for this text, although intensive study in these areas is not necessary. The mathematics and sciences background can be at a relatively simple level because detailed and sophisticated treatments of thermodynamic derivations have been minimized. Students not familiar with the units of physical chemistry should read Appendix A immediately. Appendix B also is important for several of the early chapters.

This book is an experimental synthesis of high-temperature with low-temperature geochemistry, of measurements and data evaluation with fundamental relations, and of principles with applications. We feel that this synthesis is a unique contribution to the teaching of geochemical thermodynamics, and we hope that others will find as much enjoyment in reading and using the text as we have found in writing it.

We wish to acknowledge the manuscript reviewers, Bruce Hemingway, Jane Selverstone, Eric Reardon, Charles Gilbert, Ingmar Grenthe, and Niel Plummer, without whom this text could not have reached the standard of quality we hoped to achieve. The friends, students, and colleagues who gave their encouragement, criticisms, and suggestions are too numerous to name, but they have had a considerable influence on this book. Ken Hon kindly contributed the perplexed penguins, and Paulina Franz graciously accomplished most of the typing. We also thank Mary Eberle for her careful reading of most of the page proofs. Finally, D. K. N. would like to express deep gratitude to his wife, Karen, whose patience and forebearance made it possible to find the time to write.

D. K. N.
J. L. M.

1

EQUILIBRIUM, DISEQUILIBRIUM, AND THE STEADY STATE

In the whole of the natural sciences there are only a dozen or two of these
sweeping and powerful generalizations, such as the Newtonian laws of motion
and of gravitation, the laws of thermodynamics . . . . if they are understood
they can be used in a wide range of diverse problems.

M. KING HUBBERT
Vetlesen Prize Acceptance Speech (1982)

Rocks, minerals, natural waters, and gases form in response to specific physical and chemical conditions. The fundamental role of geochemistry is to determine those conditions. In this sense, geochemistry is the study of the origins and transformations of the chemical elements that make up our planet (and our solar system, as samples become available). Geochemistry is the study of the distribution of elements in the oceans, of metals in ore deposits, of gases from volcanoes, of supercritical fluids in magmas, of clays in weathering profiles, and of organic compounds in sediments and sedimentary rocks. These topics can be described quantitatively only when basic knowledge from physical chemistry is applied to geologic processes. Thermodynamics and kinetics, two subdisciplines of physical chemistry, provide the working tools with which we can reconstruct the physical and chemical origins of geologic systems. *Thermodynamics is the study of energy and its transformations. Kinetics is the study of the rates and mechanisms of reactions.* Thermodynamics tells us which geochemical processes are possible, whereas kinetics tells us which processes are the fastest.

The beauty of thermodynamics lies in its capacity to synthesize numerous measured properties into a unified framework of quantitative relationships. For example, thermodynamic measurements cover a vast number of possibilities including heat capacity, solubility, electromotive force, heat of reaction, density, pressure, volume, and temperature—to name just a few. All of these measurements can be related to each other by quantitative mathematical expressions based on only three simple laws: the ideal gas equation and the first two laws of thermodynamics. Thermodynamic relationships enable us to calculate several unknown properties of mineral assemblages from a limited set of known properties. Thermodynamics also lends itself to simple yet powerful graphic representations of mineral and fluid equilibria for complex geologic systems.

1-1 THERMODYNAMICS AND PHYSICAL CHEMISTRY

Physical chemistry is a comprehensive subject that claims, with some justification, to investigate any property or process that involves transformations of matter and energy. Implicit in this statement is the chemical nature of the subject. As the name suggests, physical chemistry is the result of integrating physics with chemistry. Under this ostentatious banner, three general disciplines have evolved in the teaching of physical chemistry: thermodynamics, kinetics, and quantum mechanics. These represent three different approaches to solving the same basic problem: what is matter and how does it behave?

Classical thermodynamics, based on the equilibrium state and reversible processes, uses a totally macroscopic approach. Only macroscopic measurements such as pressure, temperature, viscosity, and electrical potential are necessary. No knowledge of the underlying molecular structure is required. Furthermore, thermodynamics predicts the energy change for any given transformation and, therefore, whether or not it is energetically possible. In geochemistry, this leads to the ability to predict which mineral assemblages should form in a given environment. Thermodynamics cannot tell us how fast a reaction will go. Although thermodynamics may indicate that a reaction should take place, some reaction rates are so slow that they do not proceed noticeably over millions of years. Diamonds that have formed in the mantle at hundreds of kilobars are a good example. They are mined in kimberlite pipes where they have existed near the earth's surface for many millions of years, yet all our best thermodynamic data predict that the stable form of carbon at surface pressures is graphite. The persistence of minerals outside their stability ranges is fortunate. It allows us, for example, to study metamorphic rocks on the earth's surface even though they are out of equilibrium with their present environment.

Kinetics deals with the important subject of reaction rates. Kinetics uses a macroscopic approach but has a microscopic objective, because rates are dependent on the interactions of individual particles. Ultimately, any physical transformation can be related to the properties of electrons, protons, neutrons, and other subatomic particles that exist in organized assemblies called ions, atoms, and molecules. This highly mathematical approach to the structure and properties of matter is the subject of quantum mechanics and may be called ultramicroscopic.

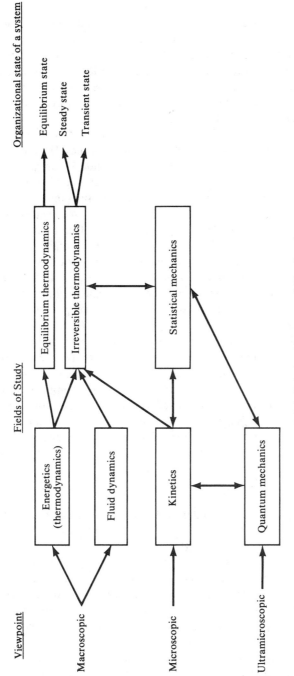

Figure 1-1. Relationships among fields of study.

As Figure 1-1 shows, all these fields of study are interrelated. The two main macroscopic disciplines are thermodynamics and fluid dynamics. Because classical thermodynamics does not treat the "dynamic" nature of matter, as does fluid dynamics, many authors have suggested that it should be called thermo-statics, or energetics. Although we heartily endorse this recommendation, we are compelled to accept the long-standing use of the word thermodynamics whose meaning is unequivocal, albeit semantically incorrect.

The younger discipline known as irreversible thermodynamics evolved from the synthesis of fluid dynamics and thermodynamics. Here, fluid dynamics is used in the broadest sense of the word to include electrical and chemical transport phenomena. The applicability of irreversible thermodynamics to quantitative treatment of geochemical problems is still subject to question; its main use is in the unification of many empirically based relationships into a formal structure. It can be shown by derivation that equilibrium thermody-namics is just a special limiting case of the larger domain of relationships embodied by irreversible thermodynamics. Furthermore, thermodynamic equalities (rather than inequalities) can be derived as criteria for the transient, steady, and equilibrium states of a system through methods of irreversible thermodynamics. Classical thermodynamics, by its very nature, is unable to make such statements.

The leap from individual molecular properties to bulk properties of matter containing millions of molecules requires a statistical approach for calculating time averages. Thus, statistical mechanics is a vital part of physical chemistry that links these disciplines together in very important ways. The same macroscopic relationships obtained from classical thermodynamics can be derived using statistical mechanics. The limitation of the statistical approach is that it can be applied only to a rather small number of substances of geochemical interest.

1-2 FUNDAMENTAL CONCEPTS

To begin the study of thermodynamics, it is necessary to define certain ordinary words that have very specific meanings here. A thermodynamic *system* is a portion of matter separated from the rest of the observable universe by well-defined boundaries. Systems are defined at the convenience of the observer, and system boundaries are chosen to permit the investigator to use thermodynamics according to his or her needs. Thermodynamics deals with the macroscopic properties of matter and of processes involving energy changes. Therefore, it is useful to describe systems as open, closed, adiabatic, or isolated. A system is *isolated* if it is entirely removed from environmental influences. Matter and energy cannot be transferred across the boundaries of an isolated system. An *open* system may exchange both matter and energy across its boundaries. Quantitative interpretations of open systems typically require a dynamic analysis based on transport theory, unless part of the system behaves in a manner that closely approximates a closed system. A *closed* system allows energy but not matter to be transferred across its boundaries. Closed systems have impermeable, diathermal, and movable boundaries that permit the transfer of heat and work between the system and its surroundings but prevent transfer of matter. The planet earth is an example of a closed system because it receives heat from the sun whereas matter is not transferred (except for the meteorites that penetrate our atmosphere, the light gases such as hydrogen and helium that escape earth's gravitational field, and spacecraft that are launched). An *adiabatic* system is a closed system in which the boundaries are thermally insulating, so that work but no heat is exchanged with the system's surroundings.

Real systems* do not behave exactly according to any of the definitions just given. Your experience, insight, and objectives play a critical role in properly defining a system and its thermodynamic properties. Howard Reiss (1965) put the problem in perspective when he stated:

> The boundaries are further blurred by the fact that the experimenter must himself make judgements as to what constitutes an acceptable thermodynamic system. Thus, he must also determine whether a given system is in equilibrium (without having at his disposal an unchallenged definition of equilibrium), and he must also determine what are the independent variables of state. The procedure for making such decisions usually consists of assuming that a given system may be treated by thermodynamic methods and that certain variables may be regarded as state variables. The experimental and theoretical consequences of these assumptions are then explored. If grave inconsistencies do not appear, and useful results are forthcoming, the methods of thermodynamics may be considered applicable. This method of proceeding inescapably possesses aspects of circular reasoning. It amounts to the assertion that thermodynamics works because it works! On the other hand, the retention of a degree of openmindedness is of considerable value in the avoidance of misunderstanding. If nothing else, it relieves the student of any compulsion to seek absolute comprehension in a domain where it does not exist.

The description of geochemical processes by thermodynamics requires a sense of space and time. Observations of the geologic environment assume, either explicitly or implicitly, a specific time period and fixed spatial dimensions to the system chosen for thermodynamic interpretation. With respect to temporal variations, a system is judged to be either time-dependent or time-independent. For example, a flowing river frequently changes its composition at any one point during a year in response to diurnal, precipitation, and seasonal perturbations. During the time interval of a few minutes, however, there is negligible change in a river's composition. The time dependency of river-water chemistry depends on the time scale chosen by the observer. Similarly, the water chemistry

---

* "Real systems" may be a contradiction in terms, because a "system" is an arbitrarily defined anthropocentric concept and "real" implies a physical reality apart from the human conceptualization of it.

varies along the width, depth, and length of a river. Over short distances (perhaps centimeters), there may be no observable chemical variation; over larger distances, however, there is always some variation. Thus, the temporal and spatial constancy of chemical gradients depends on the observer's personal judgment.

The time dependency and the boundary characteristics of a system are essential to the definition of equilibrium. Time-dependent processes taking place within a system characterize a *transient state*. As a lava flow cools, its temperature, viscosity, density, and total gas content continuously change as a function of time. During this cooling period, the lava is in a transient state. Although time-dependent processes are always in motion, time-independent processes may be either dynamic or static. When they are dynamic, the system is in a *steady state*; when they are static, the system is in an *equilibrium state*. Energy transport and/or mass transport take place at the boundaries of a steady-state system, but only so that any chosen parameter within the system does not vary with time. If a thin layer of lava were sandwiched between two larger layers of lava, each at a different temperature, a steady-state heat flow might be set up in the thin layer. The temperature gradient in the thin layer is constant with time, but heat energy is continually transferred from the lava at the higher temperature to the lava at the lower temperature. No flow of matter or energy takes place in an equilibrium-state system. As the lava flow cools, it approaches thermal equilibrium with its environment. The introduction of state variables is necessary before we can provide proper definitions for the *states* of a system.

Measurable properties of matter that describe the macroscopic state of a system are called *state variables*. Examples of state variables are pressure, volume, temperature, density, refractive index, and magnetic susceptibility. State variables are *external* or *internal* to the system. Internal variables are classified as *extensive* if they depend on the total mass of the system or *intensive* if they are independent of mass. For example, volume is an extensive variable because it varies with the mass of the system. Pressure and temperature are independent of mass and therefore are intensive variables.

More precise statements can now be given for the equilibrium and nonequilibrium states of a system. Our *equilibrium state* is the time-invariant and space-invariant state of an isolated system. All intensive variables are constant throughout the system. A *steady state* is the time-invariant (but spatially variant) state of an open or closed system. The intensive variables may vary with location in the system but are time invariant at any given point. A simple example should clarify these definitions.

If a metal bar, which easily conducts heat, is placed between reservoir 1 at temperature T_1 and reservoir 2 at temperature T_2 such that $T_2 > T_1$, then heat flows from reservoir 2 to reservoir 1 as shown in Figure 1-2. Assuming that the bar is homogeneous (i.e., that it has the same physical and chemical properties throughout its length, width, and depth), a linear decrease in temperature will be observed along the length of the bar. At each point of measurement (x_1, x_2, x_3, and x_4) there is a corresponding temperature (T_{x_1}, T_{x_2}, T_{x_3}, and T_{x_4}) that is constant and independent of time, provided that the thermal reservoirs are large enough. Under these conditions, a steady state is established because the system (metal bar) has a continuous exchange of heat energy with the surroundings.

The distinction between a steady state and an equilibrium state becomes quite clear if we perform the following thought experiment. Imagine that we can

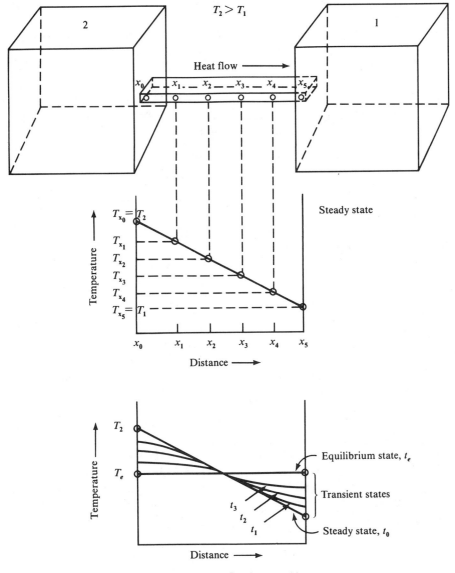

Figure 1-2. Heat flow in a metal bar.

isolate the system (i.e., separate the metal bar from the two thermal reservoirs) while continuing to monitor the temperature gradient of the bar as a function of time. Under these conditions, the steady state established previously becomes a transient state, because time-dependent changes in temperature will take place until the metal bar reaches thermal equilibrium. Figure 1-2 shows the expected variation in the temperature gradient going from the steady state at time t_0, to the equilibrium state at time t_e. The transient state is shown at instantaneous time t_1, t_2, and t_3. At t_e, the temperature measured at any point is not only independent of time but is also equal to the temperature measured at any other point in the bar.

The description and definitions for the various states of a system can also be formulated in mathematical expressions. Using temperature T as an example of an intensive variable that is a function of time t and location x, we have

$$T = T(t,x)$$

The space and time derivatives are therefore partial differential quantities. For a transient state, these quantities are real and nonzero:

$$\left|\left(\frac{\partial T}{\partial t}\right)_x\right| > 0 \quad \text{and} \quad \left|\left(\frac{\partial T}{\partial x}\right)_t\right| > 0$$

For a steady state, the temperature changes with position but not with time:

$$\left(\frac{\partial T}{\partial t}\right)_x = 0 \quad \text{and} \quad \left|\left(\frac{\partial T}{\partial x}\right)_t\right| > 0$$

And for an equilibrium state, the intensive variable is independent of time and position:

$$\left(\frac{\partial T}{\partial t}\right)_x = 0 = \left(\frac{\partial T}{\partial x}\right)_t$$

Students who have studied heat flow, fluid dynamics, or vector field theory will immediately recognize these partial derivatives as components of the continuity equation. The form of the curves in Figure 1-2 can be calculated by solving the equation of heat flow:

$$\rho c \frac{\partial T}{\partial t} = k \frac{\partial^2 T}{\partial x^2}$$

in which ρ, c, and k are density, specific heat, and thermal conductivity of the substance conducting heat (see Margenau and Murphy, 1956, or Sokolnikoff and Redheffer, 1966). This equation represents transient-state conditions. If steady-state heat flow is induced, the equation is reduced to a form of Laplace's equation:

$$\rho c \frac{\partial T}{\partial t} = 0 \quad \text{and} \quad k \frac{\partial^2 T}{\partial x^2} = 0$$

because

$$\frac{\partial T}{\partial x} = \text{constant}$$

At equilibrium, a unique limiting condition is reached where

$$\rho c \frac{\partial T}{\partial t} = k \frac{\partial^2 T}{\partial x^2} = \frac{\partial T}{\partial x} = 0$$

These same general equations are used in hydrodynamics, electromagnetics, and irreversible thermodynamics.

In natural systems, temperature or pressure gradients that may have existed at some previous time when a rock formed will not be obvious to present-day observers. Nonetheless, such "fossil" gradients can be inferred from mineralogic texture and composition in a rock that has long been isolated from its environment of formation but has not yet equilibrated with its present environment. Metamorphic and igneous rocks are commonly many millions of years removed from the original surroundings in which they formed, and we should be grateful that their rate of reequilibration with the earth's atmosphere and hydrosphere is often very slow.

1-3 REVERSIBILITY AND LOCAL EQUILIBRIUM

The development of "irreversible" or "nonequilibrium" thermodynamics by DeDonder, van Rysselberghe, Onsager, Denbigh, Prigogine, and many others has reemphasized the meaning of reversible processes and the applicability of equilibrium conditions. It is desirable to gain some understanding of the difference between a reversible process and an irreversible process.

Any natural process that proceeds at a finite rate is an *irreversible* process; it is called irreversible because it proceeds in only one direction. If the process in question is a chemical reaction, it is important to understand that the term irreversible applies only for a given set of state variables. Thus, a reaction A → B that proceeds from left to right at one fixed pressure and temperature may proceed from right to left if either pressure or temperature is changed; both reactions A → B and B → A are irreversible.

On the other hand, a *reversible* process is an abstraction that can be very closely approached but never totally realized in physical terms. In fact, reversibility is an alternative view of the equilibrium state; the impossibility of achieving a totally reversible process hinges on the paradox of achieving a *process* that is simultaneously an *equilibrium* state. One definition of a reversible process is a process that proceeds in such infinitely small stages that the system is at equilibrium for every step. The importance of this concept is that changes in thermodynamic properties of reversible processes are easily calculable and, if it can be shown that the change in the property is independent of the path taken from initial to final state, then that calculated change in the property must be identical for the same process carried out irreversibly.

One example of how one could imagine a reversible process invc.ves the expansion of a gas against a piston in a cylinder in which the pressure inside the cylinder is exactly counterbalanced by the atmospheric pressure outside the cylinder. The paradox of reversibility should be clear from this example—if the pressures inside and outside the cylinder were actually equal, the gas could not expand! A closer approach to reversibility in the real world occurs in electrochemical cells. When the emf of a galvanic cell is measured by means of a very sensitive potentiometer, the balance in the cell is so precise that changes in voltage on the order of one microvolt are sufficient to drive the electrochemical reaction in either direction away from equilibrium.

From the preceding discussion, it should be clear that the concepts of equilibrium and reversibility are closely related, especially with regard to chemical reactions. When a chemical reaction has reached equilibrium, it must be occurring in the forward direction at the same rate (on the average) as in the reverse direction. This general principle, known as the *principle of detailed balancing*, is based on statistical mechanics. For a reaction

$$A \rightarrow B$$

equilibrium is achieved when the time rate of change for the number of moles of A and the number of moles of B is zero:

$$\frac{dn_A}{dt} = 0 = \frac{dn_B}{dt}$$

However, on a microscopic scale, we recognize that molecules of A are still breaking down and molecules of B are still forming. At equilibrium, these molecular collision rates are finite and equal in magnitude but opposite in sign:

$$-\frac{dn_A}{dt} = \frac{dn_B}{dt} \neq 0$$

The principle that individual molecular collisions and the reverse of those collisions have equal probability of occurring is known as the *principle of microscopic reversibility*. Microscopic reversibility leads to the principle of detailed balancing when applied to large-scale systems.

The principle of microscopic reversibility and the postulate of *local equilibrium* are fundamental to the formulation of irreversible thermodynamics and to the application of reversible thermodynamics to natural systems. The postulate of local equilibrium is simply an extension of the equilibrium condition applied to nonequilibrium systems. It states that, in a nonequilibrium system, a small volume element can be chosen such that equilibrium is maintained, and all of the thermodynamic variables and relationships can be applied to this subsystem just as they are in a macroscopic equilibrium system. For example, if a groundwater aquifer from recharge to discharge is chosen as a thermodynamic system, it clearly is out of chemical (and hydrodynamic) equilibrium. Chemical gradients are caused by mineral dissolution and precipitation, oxidation-reduction processes, and sorption processes. However, a small portion of the same aquifer, a subsystem of defined volume, may achieve chemical equilibrium. The subsystem experiencing chemical equilibrium is an example of local equilibrium. Local equilibrium is reached when the chemical reaction rate is faster than the flow rate of the medium (in this case water), a condition common to many aquifer systems. This concept is implicit in many geochemical investigations, and it has been explicitly developed in the field of irreversible thermodynamics. The postulate of local equilibrium as applied to metamorphic reactions was introduced to geochemistry through the work of Korzhinskii (1959) and Thompson (1959, 1970). Other examples of its use include the works by Rubin and James (1973) and James and Rubin (1979) for soil-water systems, Vallochi et al. (1981) for groundwater systems, and Chapman et al. (1982) for surface-water systems.

A question inevitably arises: How do you know if the volume element is small enough to assume equilibrium? One answer is to assume that it is and then to examine the consequences. If the experimental or field measurements match the postulated equilibrium calculations, then the assumption is feasible (not necessarily correct, but at least feasible). In addition, the closer a system is to equilibrium, the more valid the local-equilibrium assumption will be. Local equilibrium is much more applicable to a groundwater system than to a river system, especially if the groundwater flow rate is slow. Similarly, local equilibrium is much more applicable to a portion of a cooling magma chamber than to a volcanic eruption, where rapid changes in temperature and pressure combine with explosive mixing of melt and crystals virtually precluding the possibility of equilibrium.

The concept of *partial equilibrium* is closely related to, yet different from, the concept of local equilibrium. Partial equilibrium is attained when some of the intensive variables in a system have reached equilibrium while others are still in a nonequilibrium state. Consider, for example, a crystal of calcite at 25°C that has just been dropped into a beaker of pure water at 25°C. This system is in apparent thermal equilibrium, but chemical equilibrium will not be reached until the calcite crystal stops dissolving and the solution reaches saturation.* Such a system is in a state of partial equilibrium. For a rock containing many different minerals dissolving in a groundwater aquifer, partial chemical equilibrium is the rule rather than the exception because different minerals have quite different dissolution rates. The minerals that dissolve most rapidly will achieve equilibrium first.

These considerations raise a further point about subtle differences between definitions of the equilibrium state. If a system is being studied from only the

---

* Actually, some heat is released due to calcite dissolution, but this heat may not be detected in the system as defined.

chemical point of view, then the equilibrium state could be defined as the time-invariant state of a closed system (Johnson, 1971; Stumm and Morgan, 1981). A closed system is not isolated, however; as long as there are energy inflows and/or outflows (such as heat absorbed or pressure–volume changes), then time invariance with respect to chemical parameters means simply that the system is in partial equilibrium. A closed system could never be in complete equilibrium with respect to all of the intensive variables. If it were, it would be an isolated system, not a closed one.

Metastable equilibrium is another important condition requiring definition. For any given closed or isolated system, each substance and group of substances can be in its most stable state, which would be the state of *stable equilibrium*. If any substance or assemblage of substances had not reached that state but showed no apparent change with time, then it would be in a state of *metastable equilibrium*. For example, liquid water can be supercooled below the freezing point of 0°C without converting it to solid ice. If the temperature is maintained at about −5°C and there are no seed nuclei added nor any physical perturbations, then the liquid water will remain in that state indefinitely. Liquid water at −5°C is a metastable substance because ice is more stable under these conditions, a fact easily demonstrated by adding some seed crystals of ice, which will induce spontaneous crystallization.

All of these definitions and their discussions are general guidelines that do not escape one of the underlying themes of this chapter: *the definition of a thermodynamic system is observer dependent.* In other words, human and subjective factors are important in the choice of the system boundaries. A system may be described as open or closed, static or dynamic, stable or metastable, and in an equilibrium state, steady state, or in a transient state; the determining factors include the length of time, the spatial dimensions, and the objectives that the observer wishes to consider, in addition to his or her experience with similar systems. "Such subjective aspects are, of course, present in all idealized models of real physical systems and can never be completely removed." (Woods, 1975).

SUMMARY

• A thermodynamic system is a portion of the universe defined by suitable boundaries.

• A system may be—
  1. open (allowing the flow of both mass and energy);
  2. closed (allowing the flow of energy but not mass);
  3. adiabatic (closed to mass and allowing the flow of energy but not heat);
  4. isolated (no flow of either mass or energy).

• The state of a system depends on the type of system and the time variation of its intensive variables:
  1. the equilibrium state is the time-invariant state of an isolated system;

    2. the steady state is the time-invariant state of an open or closed system;

    3. the transient state is the time-variant state of any system.

- Geochemical systems that are in a nonequilibrium state overall may achieve local and/or partial equilibrium.
    1. Local equilibrium describes the equilibrium state of a small-volume subsystem within the main system boundaries.
    2. Partial equilibrium describes the equilibrium state for some of the intensive variables in a system but not for all of them.
    3. Metastable equilibrium is the invariant state of substances or groups of substances in a closed or isolated system when that state is not the most stable one.

- The description of a thermodynamic system cannot be divorced from the observer's viewpoint. Every observer must consider the time and space boundaries, the objectives of his or her investigations, and all the possible state variables that may be important before making any conclusions about the state of the system being studied.

Open system

2

PROPERTIES OF PURE GASES

The leading thermodynamic properties of a fluid are determined by the relations which exist between the volume, pressure, temperature, energy and entropy of a given mass of the fluid in a state of thermodynamic equilibrium.

J. W. GIBBS (1906)

Many students have an initial opinion that gases are not too important in petrology or geochemistry—after all, problems involving the solid state (rock-forming minerals) or the liquid state (groundwater, seawater, or magmas) often come to mind as potentially more relevant or exciting than problems involving gases. Actually, nothing could be further from the truth! For one thing, gases do in fact pose many significant geochemical problems on their own such as the composition and significance of volcanic emanations, oxidation–reduction reactions, the origin and evolution of planetary atmospheres, and biogeochemical processes such as photosynthesis and microbial respiration. Another more subtle aspect of the importance of gases stems from the fact that, in any equilibria involving solids or liquids, gases may also be involved by exerting a vapor pressure over the liquid or solid phases. Variations in these vapor pressures act as sensitive barometers that reflect changes in the compositions of the condensed phases and that can be used to measure the approach of the system to equilibrium. Partly as the result of this fact, many of the early studies on the thermodynamics of gases set patterns both in approach and in specific equations that later were applied to the study of condensed phases. It is therefore both highly relevant and fundamentally important to study the thermodynamics of gases.

   Which gases are especially important in geochemistry? The overwhelming presence and significance of liquid water on our planet certainly suggests that H_2O vapor (or steam) should have the highest importance. Moreover, the gaseous components of H_2O vapor—H_2 and O_2—are almost equally important because of the control they exert over oxidation and reduction reactions, which are of critical importance in all aspects of geochemistry. After H_2O, probably the next most important species is CO_2; its significance is highlighted by the abundance of carbonate rocks in the stratigraphic record. Carbon can, of course, combine with oxygen and/or hydrogen to form a seemingly infinite array of organic molecules, many of which occur naturally. Organic geochemistry has evolved as a separate discipline that focuses on the origin, diversity, and significance of these molecules, many of which are gases. Nonetheless, averaging over all geologic environments, CO_2 is still the most important gaseous form of fixed carbon at low temperatures. In more reducing conditions, methane (CH_4) may become locally important.

Gaseous sulfur species (especially SO_2 and H_2S) also are significant in geologic environments such as hydrothermal ore deposits, volcanoes, hot springs, marine sediments, and groundwaters. Also, the halogen gases HCl and HF have a significant bearing on the evolution of magmatic and hydrothermal fluids, and they indirectly affect the halogen contents of surface waters. Other common gases such as N_2 or NH_3, and even rare species such as the noble gases helium through radon, are significant species that must be considered in certain specific geochemical problems.

2-1 THE IDEAL GAS LAW —A FOUNDATION OF CLASSICAL THERMODYNAMICS

Many of the earliest and most fundamental physicochemical measurements concentrated on the gaseous state of matter—perhaps because gases respond in more dramatic fashion to changes in their physical environment (especially changes in pressure and temperature) than do liquids or solids, and because they can be studied easily using fairly primitive measuring devices. All of these early studies focused on how variations in pressure, temperature, and volume affect a fixed amount of gas—the so-called $P-V-T$ relations of gases. In the mid-seventeenth century, the British physicist Robert Boyle became fascinated by the ability of air to be compressed under stress and to recoil with release of energy when stress is removed —in fact, he frequently referred to the springlike nature of compressed air. In his most significant experiment—one familiar to high-school physics students—he added increasing amounts of mercury to a U-tube, closed at one end, and observed that the volume of the air trapped in the closed end shrank in direct proportion to the amount of mercury added. Because the weight of the mercury column exerted a pressure on the trapped air, it was apparent that an inverse relation existed between the volume of the air and the pressure exerted on it. Stated mathematically,

$$PV = \text{constant} \qquad (2\text{-}1)$$

The French physicist Edme Mariotte independently repeated Boyle's experiments about 15 years later; and Mariotte added the important qualification that the inverse relationship between P and V holds only if the temperature of the gas is constant, a fact Boyle neglected to mention. In Great Britain and in America, equation (2-1) is called Boyle's law, whereas in France it is called Mariotte's law (Asimov, 1964).

A quantitative study of the effect of temperature on the volume of a gas was first published in 1802 by Joseph Gay-Lussac, who reported that the volume of a gas expands by 1/267 of its volume at 0°C for each 1°C rise in temperature. These same observations were apparently made by Jacques Charles about 15 years earlier, but Charles did not publish his results. One complication is that the coefficient of expansion is not exactly the same for every gas, but the differences between gases become much less when measurements are made at pressures well below one atmosphere (1 atm). The expansion coefficient was subsequently measured more accurately at low pressures by Henri Regnault in 1847 and was found to be 0.003663 deg^{-1} ($=1/273$), a value that provided a thermodynamic temperature scale based on gas expansion. In algebraic terms,

$$V = V_0(1 + \alpha T) \tag{2-2}$$

where V_0 is the volume of the gas at 0°C, and α is the coefficient of thermal expansion. Equation (2-2) shows that the volume of a gas varies directly with temperature, or

$$\frac{V}{T} = \text{constant} \tag{2-3}$$

The opposing effects of pressure and temperature on the volume of a gas can be easily combined as

$$\frac{PV}{T} = \text{constant} \tag{2-4}$$

We need only determine the constant in equation (2-4). This can be done by measuring the PV product of a number of gases at very low pressure and fixed temperature:

$$\lim_{P \to 0} \frac{PV}{n} = \theta$$

where n is the number of moles. Experiments have shown that θ is a function of temperature only and does not depend on which gas species is measured. Furthermore, it is found that $\theta \to 0$ at low temperatures, which implies that there exists an absolute zero of temperature (see Chapter 3). Because absolute zero cannot be measured, the establishment of an absolute temperature scale must depend upon a reliable property of matter that is easy to measure.

For many years, temperature scales based on gas expansion (i.e., on equation 2-2) were used in many laboratories. They had the significant disadvantage that every gas has a slightly different coefficient of thermal expansion at 1 atm—thus, the nitrogen temperature scale is slightly different from the hydrogen temperature scale. As often happens in matters of dispute, these ambiguities regarding temperature scales had to be settled by international agreement. The first International Temperature Scale was adopted in 1927 and was subsequently revised in 1948 and in 1968. The latter meeting in Paris resulted in the International Practical Temperature Scale of 1968 (IPTS-68), for which it was decided that the unit of temperature—the kelvin (K)—would be defined as the fraction 1/273.16 of the thermodynamic temperature uniquely fixed by the equilibrium coexistence of ice, liquid water, and water vapor. This so-called triple point for H_2O is 0.01 K greater than the freezing point of pure water at 1 atm, defined as zero degrees on the Celsius scale ($0°C$). Accordingly, the thermodynamic temperature in kelvins, computed as

$$K = °C + 273.15$$

must be used in all calculations involving temperature. The concept of a temperature scale based on a single fixed point (the triple point) had actually been agreed upon about 15 years previously. The importance of the IPTS-68 meeting was that it formalized the earlier agreement and chose eleven other fixed points ranging between 13.81 K and 1337.58 K. Virtually all temperature measurements used in science at present are based on the IPTS-68 scale. For a brief discussion, see McGlashan (1979, pp. 46–47).

We may now write $\theta = RT$, where R is the *gas constant* per mole. The recommended value of R is $8.31441 \, J \cdot K^{-1} \, mol^{-1}$ (Cohen and Taylor, 1973) and was based on measurements of θ/T for many gases at many temperatures. We can now write

$$\boxed{PV = nRT} \qquad (2\text{-}5)$$

which is the *equation of state* for the perfect or *ideal gas*. The ideal-gas concept arises time and again in thermodynamics and has far-reaching applications.

Equations of state describe a quantitative relationship that exists between intensive (mass-independent) system parameters—here, pressure and temperature—and the extensive (mass-dependent) system parameters—in this case volume and number of moles. Such equations of state are the bricks from which the edifice of thermodynamics is constructed, and they are models that represent some aspects of the behavior of physicochemical systems. As with all models, they are not universally applicable. In the particular case of the ideal gas law, the gas constant R was derived from measurements made on a few specific gases at very low pressures; accordingly, one should not expect *a priori* either that every conceivable gas should follow the ideal gas law at low pressures or that any particular gas should behave ideally at pressures outside the very limited range used to measure the gas constant.

In subsequent chapters, we show that those thermodynamic properties of gases essential for calculation of mineral–gas phase equilibria (such as the fugacity or chemical potential) depend ultimately on knowledge of the molar volume of the gas phase. These essential data are obtained either by direct measurement over a range of temperatures and pressures or, less satisfactorily, by prediction based on some appropriate algorithm. The search for such an algorithm may proceed in one of two ways that differ significantly on philosophical grounds. The first approach is a "brute-force" method whereby measured gas volumes are fit by computer to an empirically derived polynomial function in P and T. The function chosen must be capable of calculating volumes that are acceptably close to the measured volumes. This method has the obvious advantage of providing "automatic" interpolation between data points, but it has one important weakness: the algorithm cannot be used with confidence at pressures and temperatures that lie outside the database.

An alternative approach is to discover an equation of state that adequately describes the observed gas behavior. The benefits of this approach may be considerable. For one thing, if it can be shown that the equation of state chosen has some theoretical basis, then extrapolation beyond the measured database may be warranted. Also, the same equation of state may be valid for many different gas molecules and may be adapted to gas mixtures. Finally, the equation of state will commonly be much simpler in form than an empirically derived high-order polynomial. In any case, the ultimate test of whether the specific equation of state chosen is an appropriate predictor of molar volume requires comparison of the calculated volume with the measured volume, so the measurement of the molar volume of a gas remains among the most fundamental of thermodynamic data. Thus, we begin our investigation of real gases by testing the most simplistic model—the ideal gas equation of state—against measurement.

2-2   REAL GASES AND THE SEARCH FOR AN EQUATION OF STATE

How closely do real gases follow the ideal gas equation of state? The answer depends largely on your perspective. If the geochemical problems that interest you most are found on the earth's surface where pressure is essentially fixed at 1 atm, then the answer is "fairly well, but not exactly." However, if you feel more at home deep in the earth's crust in the realm of regional metamorphism, then the answer is "just terribly!" How useful can an equation of state be if, when tested in the real world, the results range from "almost" to "awful"? In fact, the ideal gas law is extremely useful because it provides a reference state for gas behavior from which the deviations of real gases can be measured. Reference states such as this are used in virtually every area of thermodynamics. As we shall see, thermodynamic calculations depend fundamentally on comparisons between physical parameters in real systems and the same parameters in some clearly defined reference state.

One of the most useful measurements of the deviations of a gas from ideal behavior is the compressibility factor Z, defined as

$$\boxed{Z = \frac{P\bar{V}}{RT}} \tag{2-6}$$

where $\bar{V}$ is the volume of one mole of real gas. Note that the compressibility fac-

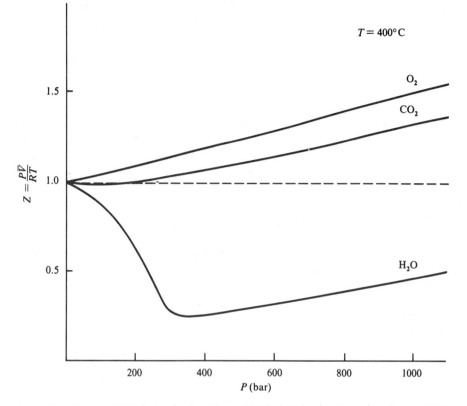

Figure 2-1. Compressibility factors for O_2, CO_2, and H_2O plotted as functions of pressure at $400°C$. The dashed line ($Z = 1.0$) represents ideal gas behavior.

tor of an ideal gas is equal to unity under any conditions.

Figure 2-1 shows compressibility factors for O_2, CO_2, and H_2O plotted as a function of pressure at $400°C$. Clearly, all three gases deviate markedly from $Z = 1.0$ at relatively low pressures: for instance, the molar volume of steam at 400 bar is one-fourth of the volume predicted from the ideal gas law.

In a qualitative way, this general behavior is easy to understand. The ideal gas constant was calculated from experiments at pressures well below 1 bar, where the density of the gas phase is much less than that at higher pressures. As pressure increases, the opportunity for molecular collisions and for repulsive or attractive interactions between molecules increases. These intermolecular effects are different for every gas and are reflected in the differences in compressibility factors shown in Figure 2-1. Conversely, as a gas is heated, the distance between molecules increases. The opportunity for intermolecular interactions decreases simply because the molecules are farther apart, and gas behavior becomes more nearly ideal. Thus, real gases have the best chance of approaching ideal behavior at very low pressures and very high temperatures; conversely, most gases deviate strongly from ideality at high pressures and low temperatures.

What are the most obvious deficiencies of the ideal gas law? When written in the form

$$\bar{V} = \frac{RT}{P} \tag{2-7}$$

it is evident that the equation predicts that, at any finite pressure, a gas at 0 K has zero volume. This is clearly nonsense. We know from experience that, at low-enough pressure, any gas will condense and eventually solidify; the hypothetical volume of a gas at absolute zero must be roughly similar to that of the solidified gas. The most straightforward solution is to write

$$\bar{V} = \frac{RT}{P} + b \tag{2-8}$$

where b represents the hypothetical volume of the gas at 0 K and thus serves as a first-order correction for the real volume of gas molecules.

The fact that individual gas molecules have finite volumes has two effects: for a given container, it limits the "free volume" of the system (i.e., the actual space that molecules are free to move about in), and it also limits how closely two molecules can approach one another before repulsive forces resulting from overlapping electron orbitals become significant. Thus, we may rearrange equation (2-8) as

$$P = \frac{RT}{\bar{V} - b} \tag{2-9}$$

which can be interpreted in molecular terms as a gas pressure that is corrected for weak repulsive forces resulting from closest approach.

Of course, attractive forces between molecules must also be significant and must predominate when conditions are favorable for condensation to a liquid phase to occur. To evaluate this effect, consider two volume elements V_1 and V_2 in a container of gas. Whatever attractive forces exist between molecules in these volume elements must be proportional to the number of molecules (or the concentration c_i) in each element. Because the concentration must be the same in each element, the attractive forces will be proportional to $c_1 \times c_2 = c^2$. Because $c = n/V = 1/\bar{V}$, the attractive forces are also proportional to $1/\bar{V}^2$. These positive forces must reduce the total pressure of the gas, so we may now write

$$P = \frac{RT}{\bar{V} - b} - \frac{a}{\bar{V}^2} \tag{2-10}$$

where a is the proportionality constant for attraction. This constant will generally depend on temperature and, to a lesser extent, on pressure.

Equation (2-10) is the *van der Waals equation*, named in honor of the first person to quantify the dual effects of repulsive and attractive intermolecular forces in gases. Subsequently, many alternative equations of state have been proposed that also are based on the sum of repulsive and attractive terms. One that has found special favor among geochemists was introduced by Redlich and Kwong in 1949:

$$P = \frac{RT}{\bar{V} - b} - \frac{a}{T^{1/2}\bar{V}(\bar{V} + b)} \tag{2-11}$$

Note that the *Redlich–Kwong equation* differs from the van der Waals equation in expressing the attractive potential as a more complicated function of temperature and molar volume.

A very different form of equation of state expresses the compressibility factor of a gas as a power series written in terms of density (or volume). The *virial equation* is

$$\frac{P\bar{V}}{RT} = 1 + \frac{B}{\bar{V}} + \frac{C}{\bar{V}^2} + \frac{D}{\bar{V}^3} + \cdots \tag{2-12}$$

where B, C, and D are the second, third, and fourth virial coefficients and must be determined empirically. This equation does not converge well at high gas densities, and other simpler equations of state may be equally adequate. Nonetheless, the virial equation has special significance because it is the only equation of state for gases that has a sound *theoretical* foundation based in statistical mechanics (for details, see Mason and Spurling, 1969). Moreover, each virial coefficient is related to molecular interactions: the second virial coefficient represents deviations from ideality resulting from two-molecule interactions, the third virial coefficient represents deviations from ideality resulting from three-molecule interactions, and so on.

Note that the van der Waals, Redlich–Kwong, and virial equations of state differ from the ideal gas equation in that each includes adjustable parameters that must be determined before the equations can be used. Before discussing how these parameters might be obtained, we must study the behavior of real gases in the process of condensation to liquid or the inverse process of boiling. The fact that the boiling curve for any liquid does not extend infinitely is a complication that has an important impact on the thermodynamics of gases.

2-3 *P–V* ISOTHERMS OF REAL GASES, AND CRITICAL PHENOMENA

Anyone familiar with automobile radiators knows that the boiling point of water increases with pressure—a powerful argument for not removing the radiator cap from an overheated engine. Of course, automobile radiators are relatively low-pressure systems; if the pressure could be increased in these systems to several hundred bars, a point would be reached beyond which boiling could not occur at any temperature. This is a fascinating phenomenon, and it could not have been predicted from classical thermodynamics.

In order to investigate the boiling process more closely, consider what happens when 1 mol of low-pressure steam is compressed along an isotherm. Figure 2-2 shows the $P–V–T$ relations for the H_2O system projected onto both

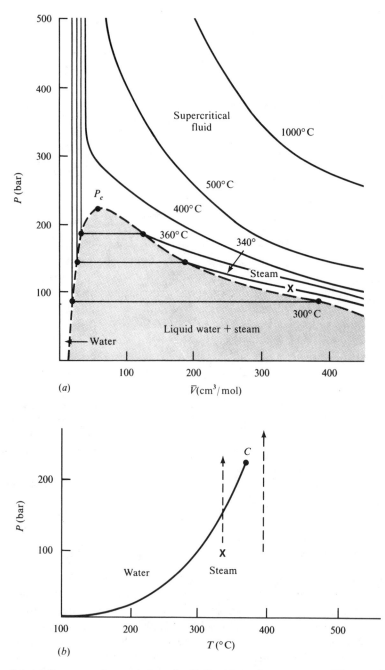

Figure 2-2. (*a*) Pressure–volume relations for H_2O with isotherm contours. Water and steam are defined only within the shaded area; supercritical fluids exist above P_c. (*b*) Boiling curve for H_2O, showing location C of the critical point.

the $P-V$ and $P-T$ planes. Let compression begin at 340°C at a pressure of 100 bar, marked by the X in Figures 2-2(*a*) and 2-2(*b*). As pressure is increased from 100 to 145 bar, the molar volume of the steam decreases from 320 cm³ to 190 cm³, as can be seen by following the isotherm. At 145 bar, the boiling curve is intersected, and steam begins to condense; the volume of the condensed liquid phase is about 25 cm³ mol⁻¹. Because the boiling curve is single-valued for every combination of pressure and temperature, steam will condense to liquid with no change in pressure: thus, the 340°C isotherm is horizontal at 145 bar and becomes a *tie line* that connects the volumes of the coexisting water and steam phases. When all the steam has condensed to liquid, pressure can increase once again; the isotherms are nearly vertical because liquid water is much less compressible than steam. If the compression process is repeated at somewhat higher temperature, the densities of boiling liquid water and steam are seen to approach one another; at about 374°C, they become identical. For all temperatures above 374°C, condensation can never occur, and the H_2O phase is everywhere homogeneous. The difference is seen dramatically in Figure 2-2(*a*) by comparing the 360°C isotherm with the 400°C isotherm (both are indicated also by dashed lines in Figure 2-2(*b*)); the lower isotherm shows two distinct breaks in slope marking the densities of the liquid and steam phases that coexist during boiling. In contrast, the upper isotherm is continuous at all pressures. The last occurrence of boiling in a pure system is called the *critical point* for the system. It is characterized by a critical pressure P_c, a critical temperature T_c, and a critical volume V_c. Note that the boiling curve terminates abruptly at the critical point.

Because of the homogeneous nature of matter beyond the critical point, it is inappropriate to refer to these supercritical phases as either liquids or gases; the

TABLE 2-1
CRITICAL CONSTANTS FOR SELECTED GASES

	T_c (K)	P_c (bar)	$\bar{V}_c$ (cm³/mol)	Z_c
H_2O	647.3	220.4	56.0	0.229
CO_2	304.2	73.7	94.0	0.274
CH_4	190.6	46.0	99.0	0.288
H_2S	373.2	89.3	98.5	0.284
SO_2	430.8	78.8	122	0.268
HF	461	65	69	0.12
HCl	324.6	83.1	81.0	0.249
NH_3	405.6	112.7	72.5	0.242
H_2	33.2	13.0	65.0	0.305
O_2	154.6	50.4	73.4	0.288
N_2	150.8	48.7	74.9	0.291

Data adapted from Reid et al. (1977), pp. 630–635.

density differences implied by these terms no longer exist. The more neutral term "fluid" is preferred. Fortunately, all thermodynamic equations that apply to gases can be applied equally well to supercritical fluids. The critical constants for geologically important gases are given in Table 2-1. The critical compressibility factor Z_c is equal to $P_c V_c / R T_c$.

2-4 CORRESPONDING-STATE THEORY

As first pointed out by van der Waals in 1873, any equation of state for a real gas that is written in the form $V = f(P,T)$ is subject to mathematical constraints at the critical point. Figure 2-3 is a schematic enlargement of PV isotherms in the critical area. Note that the volumes of the coexisting gas and liquid phases on the boiling curve merge to a common horizontal tangent at the critical point. Because the liquid and gas phases are continuous at the critical point, this critical volume must also correspond to an inflection point in the critical isotherm T_c. Thus we

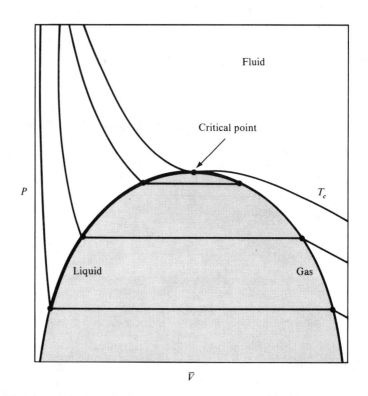

Figure 2-3. Schematic isotherms in the two-phase region for a pure boiling fluid. Note the inflection point in the critical isotherm (T_c) at the critical point.

must have

$$\left(\frac{\partial P}{\partial V}\right)_{T_c} = 0 \tag{2-13}$$

and

$$\left(\frac{\partial^2 P}{\partial V^2}\right)_{T_c} = 0 \tag{2-14}$$

When these constraints are applied to the van der Waals equation, it is possible to solve for the a and b parameters in equation (2-10) in terms of the critical constants of the gas. The results are

$$a_{vdW} = 0.4219 \frac{R^2 T_c^2}{P_c} \tag{2-15}$$

$$b_{vdW} = 0.1250 \frac{R T_c}{P_c} \tag{2-16}$$

The same constraints can be applied to the Redlich–Kwong equation to give

$$a_{R\text{-}K} = \frac{0.4275 R^2 T_c^{2.5}}{P_c} \tag{2-17}$$

$$b_{R\text{-}K} = \frac{0.0866 R T_c}{P_c} \tag{2-18}$$

Note that the a and b parameters calculated in this manner are constants.

At first glance, the preceding derivations may seem like a mysterious use of the critical constants of a gas, but these calculations are based on the *theory of corresponding states*, which asserts that the PVT properties of all gases are related to their critical properties in some consistent manner. In order to make use of the theory, it is necessary to normalize pressure, temperature, and volume with respect to their appropriate critical values. The normalized parameters are called *reduced variables*—i.e., reduced pressure P_r, reduced temperature T_r, and reduced volume V_r:

$$P_r = \frac{P}{P_c} \qquad T_r = \frac{T}{T_c} \qquad V_r = \frac{V}{V_c}$$

To understand the significance of this theory, divide the compressibility factor Z by the critical compressibility factor Z_c:

$$\frac{Z}{Z_c} = \frac{(P/P_c)(V/V_c)}{T/T_c} \tag{2-19}$$

Solving this expression for reduced volume, we find

$$V_r = \frac{V}{V_c} = \frac{(Z/Z_c)(T/T_c)}{P/P_c} = f(Z_r, P_r, T_r) \tag{2-20}$$

Calculation of the critical compressibility factor of a great many gases has shown that Z_c is nearly constant (ranging between 0.27 and 0.29), although there are notorious exceptions. Assuming for a moment the constancy of Z_c, we would have $Z = f(P_r, T_r)$. Note that V_r does not appear in this function because only two of the three parameters (P, V, T) are independent variables, and it is much easier to measure the pressure and temperature of a gas than to measure its volume. This equation may be seen as an algebraic expression of the theory of corresponding states, because it implies that a universal equation of state exists by which the compressibility of any gas could be calculated in terms of reduced pressure and reduced temperature. Of course, such an equation would apply only for those gases which have exactly the same values of Z_c; the last column of Table 2-1 shows how well our geochemically significant gases stack up in this regard. We see that Z_c values for Ar, N_2, O_2, and CH_4 are all very close to 0.29: H_2 is a little larger (0.305), and H_2S, SO_2, and CO_2 are a little smaller (0.27–0.28). Molecules that show more extreme deviation are NH_3 (0.242), H_2O (0.229), and the halogen acids HCl (0.249) and HF (0.12). The deviant molecules have strongly polar tendencies and, when such strong intermolecular forces (e.g., hydrogen bonding) are present, corresponding-state theory does not work very well. The theory works best for nearly spherical molecules and for those nonspherical molecules that do not assume preferred orientations. Nonetheless, corresponding-state theory has been the major tool used by chemists and engineers to predict the compressibility factors of gases, and it works remarkably well if its limitations are always kept in mind. The challenge is to find the function of P_r and T_r that best describes the largest number of gases. The most common approach is empirical and is based on analyzing $P–V–T$ data for a very large number of gases. The results are presented in graphical format. Figure 2-4 is an example of such a diagram that can be used to predict compressibility factors as a function of reduced temperature and reduced pressure. For most nonpolar molecules, the Z read from this figure is within 5% of the measured Z.

Although this kind of accuracy may be anticipated for more than one-half of the gases in Table 2-1, we must emphasize that this method is inappropriate for the most important of the geochemical gases, H_2O.

To illustrate the point, Figure 2-5 shows an enlarged section of Figure 2-4 drawn for reduced isotherms 1.00, 1.20, and 1.50. The data points are *measured* compressibility factors for H_2O taken at each of the reduced isotherms. The agreement between observation and prediction based on corresponding-state theory is very poor. The implication is that *any* equation of state for H_2O that relies on corresponding-state theory to calculate adjustable parameters will be subject to unacceptable errors. The magnitude of these errors will be explored in Problem 3. In contrast, Figure 2-6 shows the same graph with CO_2 data. The fit between observation and prediction is much more satisfactory in this case, but more significant deviations occur at higher reduced pressures. Because the

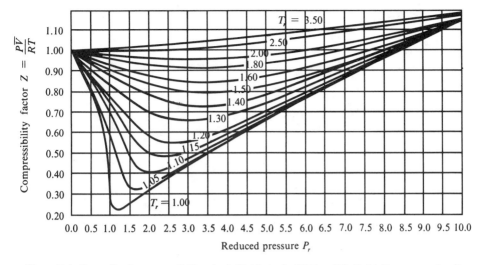

Figure 2-4. Generalized compressibility chart (Reid et al., 1977, p. 28). Solid lines are reduced isotherms; dashed lines are lines of constant reduced volume. Based on data from Nelson and Obert (1954).

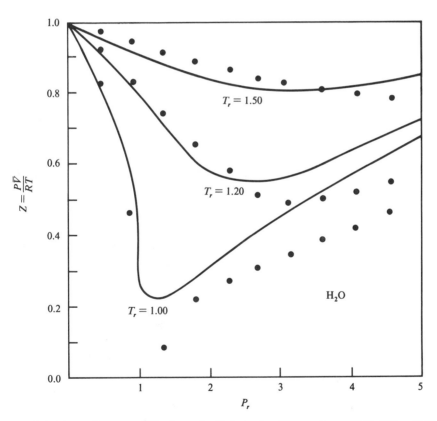

Figure 2-5. Measured compressibility factors for H_2O at reduced temperatures of 1.00, 1.25, and 1.50 compared to predicted compressibility factors based on corresponding-state theory and Figure 2-4.

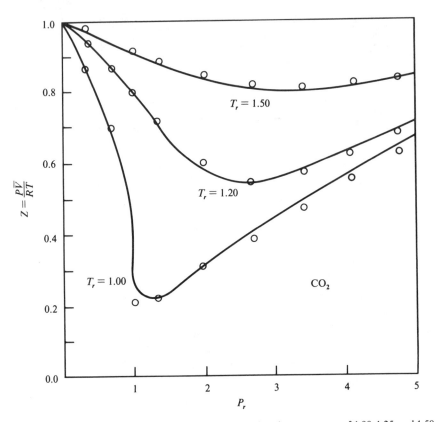

Figure 2-6. Measured compressibility factors for CO_2 at reduced temperatures of 1.00, 1.25, and 1.50 compared to predicted compressibility factors based on corresponding-state theory and Figure 2-4.

critical pressure of CO_2 is much lower than that of H_2O, the maximum *experimental* pressure shown in Figure 2-6 is 350 bar, as compared to 1200 bar in Figure 2-5.

2-5 MODIFIED REDLICH–KWONG EQUATIONS

It is an unfortunate fact of life that accurate molar-volume measurements of gases are difficult to obtain, especially at very high pressures. For that reason, it ultimately becomes necessary to predict Z (or V) for some conditions where no data exist. We have just seen that corresponding-state theory must be used with great caution. Are there better alternatives for determining the adjustable a and b parameters? Which equation of state is best?

Over the past 30 years, literally hundreds of papers have been published in the chemical and engineering literature in an attempt to find equations of state that can accurately predict molar volumes of gases over a wide range of conditions. Reid et al. (1977, pp. 11–56) present a cogent summary of this search. Of all the so-called two-parameter equations (i.e., $V = f(P,T)$ only), there is general agreement that the Redlich–Kwong equation works best, or at least as well as any other. However, virtually no one uses this equation of state as originally proposed (i.e., equation 2-10); most resort instead to one of several modified Redlich–Kwong (MRK) equations. There is no general agreement regarding the best way to modify the original equation, but all modifications are significant improvements over the corresponding-state version discussed in the preceding section. All MRK equations are based on revised expressions for either the attractive term or the repulsive term or both. This modification can be done either by writing new expressions for these terms, or by finding different ways to calculate the a and b terms while preserving the form of equation (2-10). The most noteworthy examples are the following.

1. Here a is a function of temperature only, and b is constant. For example, de Santis et al. (1974) defined $a(T) = a^\circ + a_1(T)$ for both H_2O and CO_2. Both a° and a_1 were obtained by empirical fits to available volume data. Holloway (1977) took the same form for $a(T)$ but derived different values for a° and a_1 based on the volume data for H_2O of Burnham et al. (1969a,b).

2. Here b is constant, but the repulsive term is rewritten as

$$\frac{RT}{\bar{V}} (1 + y + y^2 - y^3)/(1 - y)^3$$

where $y = b/4\bar{V}$ (Carnahan and Starling, 1969). This expression is an example of a "hard-sphere" model of nonattracting spheres, and it can be justified on statistical mechanical grounds. The derivation of this expression has nothing whatever to do with the Redlich–Kwong equation (it is much more closely allied to the virial equation), but many workers have merged the Carnahan–Starling repulsive term with Redlich–Kwong. The result of this marriage is a "hard-sphere modified Redlich–Kwong" (HSMRK) equation.

3. Here a is a function of *both* T and P, the factor b is constant, and the Carnahan–Starling repulsive term applies. This approach has been championed by Kerrick and Jacobs (1981), who claim that this rather complex HSMRK equation gives a better fit to the existing H_2O and CO_2 data than do any of its predecessors. The revised equation is

$$P = \frac{RT(1 + y + y^2 - y^3)}{\bar{V}(1 - y)^3} - \frac{a(P,T)}{T^{1/2}\bar{V}(\bar{V} + b)} \qquad (2\text{-}21)$$

and is virtually impossible to solve manually. The b coefficient and the $a(P,T)$ polynomial were once again empirically derived. For H_2O, this equation reproduces the observed data very well. In the ranges of $P = 1$–8.4 kbar and $T = 300$–925°C, the deviation of calculated volume from measured volume is less than $\pm 1\%$ in more than 90% of the observations. This degree of accuracy is satisfactory for most geo-chemical calculations.

Another significant advantage of the Redlich–Kwong and modified Redlich–Kwong equations is that they can easily be adapted to handle mixtures of gases. This is significant because gas mixtures (especially CO_2–H_2O) are very important in high-temperature geochemical application. We defer until Chapter 6 treatment of gas mixtures by means of MRK equations of state.

SUMMARY

- An equation of state describes a quantitative relationship that exists between intensive (mass-independent) and extensive (mass-dependent) variables in a thermodynamic system.

- The ideal gas law, $P\bar{V} = RT$, is an equation of state that approximately describes the behavior of real gases only in the limiting case of very low pressure and high temperature.

- The compressibility factor Z ($= P\bar{V}/RT$) is a convenient measure of the deviation of real gases from ideal behavior. Such deviations occur both because of repulsive forces related to the physical space occupied by gas molecules and because of attractive interactions that occur between adjacent molecules.

- The van der Waals equation of state is the simplest example of how to account for both repulsive and attractive intermolecular potentials. The virial equation of state is more satisfying theoretically, because it is based on fundamental statistical mechanics. Neither equation is useful at the high pressures (and high gas densities) commonly encountered in petrology.

- Every gas has a unique critical point (critical pressure and critical temperature) beyond which no physical distinction is possible between liquid and gaseous states.

- The theory of corresponding states says that all gases tend to follow the same equation of state when pressure and temperature are normalized with respect to their critical values. The theory has been used to derive adjustable parameters in the repulsive and attractive terms of many equations of state, but polar molecules such as water, ammonia, and the halogen acids HF and HCl do not follow it exactly.

- Various modifications of the Redlich–Kwong equation of state,

$$P = \frac{RT}{\bar{V} - b} - \frac{a}{T^{1/2}\bar{V}(\bar{V} + b)}$$

can successfully predict the properties of gases to fairly high pressures, and these are currently the most popular equations of state for gases in the geochemical literature. The best fits are obtained when the a and b parameters are empirically derived from PVT measurements of the gas in question.

PROBLEMS

1. Calculate the molar volume of N_2 at $0°C$ and 1 bar, assuming that the ideal gas law is obeyed. Compare your value with the measured density of 1.25055 g/1 at one atmosphere. Conclusions?

2. Use Figure 2-4 to estimate the compressibility factor of CO_2 at 250 bar for 400, 500, 600, 700, and 800 K.   (*Ans:* at 400 K, 250 bar, $Z = 0.68$)

3. Calculate the a and b parameters for H_2O in the van der Waals equation, assuming the validity of the theory of corresponding states.   (*Ans:* $a = 5.545 \times 10^6$, $b = 30.52$)

4. Repeat Problem 3 for the Redlich–Kwong equation.   (*Ans:* $a = 1.429 \times 10^8$, $b = 21.15$)

5. Problems like the following can be solved exactly with a microcomputer, by numerical methods (e.g. Newton's method), or by trial and error, which is sometimes faster.

   Both the van der Waals and Redlich–Kwong equations of state are cubic equations in either V or Z. When written in the customary cubic form

$$Z^3 + X_1 Z^2 + X_2 Z + X_3 = 0$$

where Z is compressibility factor, the coefficients X_1, X_2, and X_3 have the following values for the van der Waals equation:

$$X_1 = -\left(\frac{Pb}{RT} + 1\right)$$

$$X_2 = \frac{aP}{(RT)^2}$$

$$X_3 = -\frac{abP^2}{(RT)^3}$$

For the Redlich–Kwong equation, they have the following values:

$$X_1 = -1$$

$$X_2 = BP\left(\frac{A^2}{B} - BP - 1\right)$$

$$X_3 = -\frac{A^2}{B}(BP)^2$$

where $B = b/RT$ and $A^2 = a/R^2 T^{2.5}$ (cf Edmister, 1968). Write a program that will calculate the coefficients at any pressure and temperature given the a and b parameters. The cubic equation can be solved for Z in any of the standard ways—the largest real root is the answer you want. Take 100-bar increments from 1000 to 10,000 bar at two different isotherms, $400°C$ (673.15 K) and $800°C$ (1073.15 K).

The problem is to compare compressibility factors for H_2O calculated according to three different equations of state:

(1) van der Waals using a and b parameters calculated in Problem 3;
(2) Redlich–Kwong using a and b parameters calculated in Problem 4; and
(3) Redlich–Kwong using a and b parameters of Holloway (1977), which were obtained from empirical fits to measured molar volumes. They are

$$a(T) = 1.668 \times 10^8 - 1.9308 \times 10^5 T + 186.4 T^2 - 0.071288 T^3 \ (T \text{ in } °C)$$

$$b = 14.6$$

Plot the results (Z vs P) for both isotherms and compare with the following *measured* compressibility factors for H_2O (calculated from molar volumes in Burham et al., 1969). Conclusions?

P (bar)	400°C	800°C
1000	0.466	0.876
2000	0.814	0.950
5000	1.728	1.456
10000	3.062	2.281

(*Ans:* at 400°C, 1000 bar (1) $Z = 0.820$, (2) $Z = 0.656$, and (3) $Z = 0.441$).

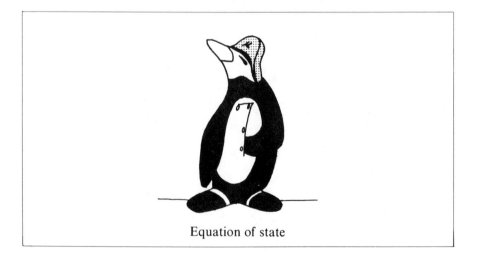

Equation of state

3

THE FUNDAMENTAL
RELATIONSHIPS

*The science of thermodynamics represents an attempt to summarize analyt-
ically the sum total of man's experiences with respect to processes involving
energy changes attending the interplay of heat and work.*

ARNOLD REISMAN (1970)

Thermodynamics as a unified science grew from the search for the fundamental
independent variables that determine the state of a system. The zeroth law of
thermodynamics introduces the state function known as temperature. The first
law proclaims the existence of a state function called the internal energy. The
second law defines a state function known as the entropy. Pressure and volume
also are state functions, and we could say that their existences as independent
variables of state constitute two more laws of thermodynamics. Although this
suggestion is consistent with the formulation of the laws of thermodynamics, it is
unnecessary because pressure and volume have already been defined in classical
mechanics, whose historical development preceded that of thermodynamics.
Furthermore, not only pressure and volume are intuitively and empirically more
familiar but, for a closed system of fixed mass, two state functions are sufficient to
describe completely the thermodynamic state of the system. In Chapter 2, we saw
that the molar volume of a gas is uniquely defined by the two state functions
pressure and temperature: $\bar{V} = f(P,T)$. However, all the other thermodynamic
parameters of the gas—such as internal energy, entropy, and free energy—also
are fixed at constant P and T. The choice of fixed P and T is not unique; we could
have chosen entropy and volume to be constant, and this choice would then have
defined all the remaining state functions.

Temperature is a less familiar concept than pressure or volume, and a *zeroth
law* of thermodynamics has been used to define it empirically. Consider a
substance 1 for which the function $f(P_1,V_1)$ defines a temperature T. If this

substance is in thermal equilibrium with another substance 2, then

$$f(P_1,V_1) = f(P_2,V_2) \tag{3-1}$$

If substance 1 is also in thermal equilibrium with substance 3, such that

$$f(P_1,V_1) = f(P_3,V_3) \tag{3-2}$$

then substance 2 will be in thermal equilibrium with substance 3:

$$f(P_2,V_2) = f(P_3,V_3) \tag{3-3}$$

This zeroth law may be stated as follows: two bodies (2 and 3) each in thermal equilibrium with a third body (1) must also be in thermal equilibrium with each other. Statements (3-1) through (3-3) are intuitively and empirically obvious, but they must be stated explicitly in order to construct a scale for measuring temperatures and to identify temperature as a state function.

All state functions behave the same way. For a state function Y and its differential dY, the following statements are valid.

1. The differential dY is mathematically exact (see Appendix B).
2. For any process $Y_1 \rightarrow Y_2$, the change ΔY is *independent of the path* chosen to effect that change. Thus, ΔY can be calculated from the initial and final states of the system only; i.e.,

$$\int_1^2 dY = Y_2 - Y_1 = \Delta Y \tag{3-4}$$

3. In any closed cycle, the net change in Y is always zero:

$$\oint dY = 0 \tag{3-5}$$

These three statements are equivalent, and they form the operating basis for most thermodynamic calculations. Classical thermodynamics can compare only differences between initial and final states. The details of actual *processes* that describe the passage between these two states depend on many unpredictable factors such as nucleation rates, kinetics, and so on. Thus, we wish to find as many relevant state functions as we can, because these functions will provide the framework for all subsequent calculations.

3-1   THE FIRST LAW OF THERMODYNAMICS

The first law relates two important properties of a system: heat and work. The work done by applying a force to a body is simply defined from classical mechanics:

$$W = \int \mathbf{F} \cdot d\mathbf{x} \tag{3-6}$$

where $\mathbf{F}$ is the force vector and $d\mathbf{x}$ is the differential displacement. One special kind of work is compression, or "PV work," which is especially important in thermodynamics. Consider a gas that is held under pressure by a piston in a vertical cylinder. If the piston is released, the gas will expand a distance dx against the external pressure, which is equivalent to the atmospheric force on the piston divided by the cross-sectional area A of the cylinder. For a differential amount of work, we have

$$dW = -F\,dx = -P_{ex}\,dA\,dx = -P_{ex}\,dV \qquad (3\text{-}7)$$

If the change is defined to be reversible, then the external pressure P_{ex} is equal to the pressure P on the system, and

$$dW_{rev} = -P\,dV \qquad (3\text{-}8)$$

If any energy leaves the system, as in the performance of work by an expanding gas, then the sign of the energy term is negative. Although this sign convention is arbitrary, it is consistent with the idea that a decrease in energy is negative and an increase in energy is positive. By this convention, $dW < 0$ for an expanding gas. Because $dV > 0$ for the expansion, the pressure must have a negative sign, and $dW = -P\,dV$. This relationship may seem a little strange because, when we measure P, we almost always obtain a positive quantity. Keep in mind, however, that the pressure is measured as a force applied by the surroundings on the system. The system's pressure on the surroundings must be equal and opposite in sign to the surroundings' pressure on the system. Other important kinds of geochemical work, including chemical and electrochemical work, are discussed in later chapters. When we speak of work, PV work will be assumed unless stated otherwise. The SI unit for work is the joule (J): 1 J equals 1 kg^2 s^{-2}, and 1 kilojoule = 1 kJ = 10^3 J.

We can now determine whether work is a state function by applying the usual mathematical *test for exactness* to 1 mol of ideal gas. Expressing $\bar{V}$ as a function of P and T, we have

$$\bar{V} = \frac{RT}{P} \qquad (3\text{-}9)$$

Because V is a state variable, it has an exact differential:

$$\bar{V} = \bar{V}(P,T) \qquad (3\text{-}10)$$

$$d\bar{V} = \left(\frac{\partial \bar{V}}{\partial P}\right)_T dP + \left(\frac{\partial \bar{V}}{\partial T}\right)_P dT \qquad (3\text{-}11)$$

such that

$$\left(\frac{\partial^2 \bar{V}}{\partial T\,\partial P}\right)_{T,P} = \left(\frac{\partial^2 \bar{V}}{\partial P\,\partial T}\right)_{P,T} \qquad (3\text{-}12)$$

To test that $\bar{V}$ is an exact differential, we first take the partial derivatives of equation (3-9),

$$\left(\frac{\partial \bar{V}}{\partial P}\right)_T = -\frac{RT}{P^2} \qquad \left(\frac{\partial \bar{V}}{\partial T}\right)_P = \frac{R}{P} \tag{3-13}$$

and then test for cross-differentiation:

$$\left[\frac{\partial(-RT/P^2)}{\partial T}\right]_P = -\frac{R}{P^2} = \left(\frac{\partial(R/P)}{\partial P}\right)_T \tag{3-14}$$

Thus, $d\bar{V}$ is an exact differential. Now consider an amount of work dW. Is the work differential exact? Because $dW = -P\,d\bar{V}$, we have

$$dW = -P\left[-\frac{RT}{P^2}\,dP + \frac{R}{P}\,dT\right]$$

so that

$$dW = \frac{RT}{P}\,dP - R\,dT \tag{3-15}$$

The test for exactness shows

$$\left[\frac{\partial(RT/P)}{\partial T}\right]_P = \frac{R}{P} \tag{3-16}$$

$$\left(\frac{\partial R}{\partial P}\right)_T = 0 \tag{3-17}$$

Thus, dW is not an exact differential, and W is not a state function.

Heat is a concept much more difficult to understand. It obviously is related to temperature, and these two properties set thermodynamics apart from classical mechanics. The problem arises because heat, like work, is not a state function. A fixed body at rest cannot contain a finite amount of heat, and heat can be defined only in terms of some *process*. Exactly the same statement applies to work—work also is defined only with respect to a specific process. Just as there is no "absolute-zero" scale for work, there is no "absolute-zero" scale for heat within the framework of thermodynamics, and any process that causes changes in heat must be defined relative to arbitrary limits. This limitation introduces the necessity to set so-called *standard states* for all thermodynamic processes. Perhaps the most useful definition of heat is the following: heat is a form of energy that flows in a system as the result of a difference in temperature. This statement explicitly describes the link between heat and temperature that we must explore quantitatively.

As mentioned previously, energy lost by the system is a negative quantity, so that "heat given off" is negative by convention. Chemical reactions that liberate heat are known as *exothermic* reactions. The reverse, a chemical reaction that absorbs heat from the surroundings, is *endothermic* and carries a positive sign.

Unfortunately, no such uniformity in convention exists for work. We have chosen the same convention for work as that which exists for heat: a system that does work on the surroundings "loses work," so "work done by the system" carries a negative sign. If the surroundings perform work on the system, the sign is positive (see Figure 3-1).

The fundamental unit of heat/work was named in honor of J. P. Joule, who investigated the relationship between heat and work in a series of experiments in the mid-1800s. His observations can be summarized as follows: in a system that is thermally isolated from its surroundings, a fixed amount of work introduced into the system produces the same temperature rise in the system, regardless of how the work is performed. Joule's system was essentially a jacketed water bath that was heated in various experiments by mechanical means (a paddle wheel), by an electrical current, by gas compression, and by friction of two iron blocks. The observation of a path-independent temperature rise suggested the existence of some state function defined in terms of *both* heat and work. This state function has come to be defined as the internal energy U. The change in internal energy $U(2) - U(1)$ may be written as

$$\Delta U = Q + W \qquad\qquad (3\text{-}18)$$

which is a mathematical expression of the first law. Alternatively, the differential

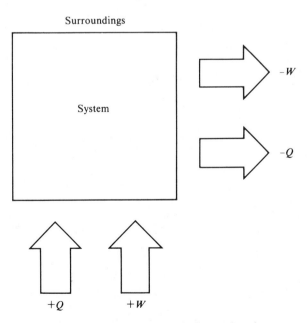

Figure 3-1. Sign conventions for heat and work.

expression is often used:

$$dU = dQ + dW$$ (3-19)

where dU is an exact differential, and dQ and dW are inexact differentials. Because Joule's system was thermally insulated, $dQ = 0$, and the change in internal energy was equal to the work performed; this explains why the changes in the system were independent of the way in which the work was done. Such thermally insulated systems are called *adiabatic*, and the work performed in an adiabatic system becomes a state function.

Joule's experiments, sometimes described as a measurement of "the mechanical equivalence of heat and work," can be generalized to universal dimensions. Over many years of experimentation and theoretical effort, a fundamental postulate has never been disproved: *energy is conserved in any transformation*. This statement of the first law is one of the most common; it is simply a restatement of the law of the conservation of energy. In thermodynamics, we speak of the total energy as that energy internal or intrinsic to the system. The absolute amount of this *internal energy* is not known. It is safe to say, of course, that the internal energy is the sum of the kinetic and potential energies, but these quantities cannot be measured on an absolute scale. What can be measured is the difference in internal energy after some change in state has taken place. The transformation of 1 mol of carbon from graphite to diamond represents a significant change in the internal energy, and this value can be calculated. But the internal energy of 1 mol of graphite at 25°C and 1 bar pressure is not known without reference to some arbitrary energy scale.

Because U is a state function, the conservation of energy principle can be stated mathematically as

$$\oint dU = 0$$ (3-20)

Energy is always conserved in any cyclic process. This empirical observation places the first of two restrictions on the fruitless search for a "perpetual-motion machine" and will be discussed further.

Geothermal reservoirs provide an example of an energy transformation that includes both heat and work energy. Consider a deep subsurface geothermal reservoir in which the temperature may be around 200°C and the pressure may be 100 bar. If the fluid ($CO_2 + H_2O$) finds a weak fracture, it will tend to migrate up the fracture and eventually will reach areas in the shallow subsurface or the surface itself where throttling can occur. Throttling will cause fluid pressures and temperatures to decrease by expansion. Very often, the conduit for a geothermal fluid has self-sealed by the precipitation of dissolved silica, and the expansion process may be largely adiabatic. The fluid will separate into steam (perhaps 10%) and water. If no work has been performed (e.g., in turning turbines) and no heat has been lost by convection, then the internal energy has remained unchanged. If the migrating fluid is tapped by a pipeline and run into a turbine generator for the production of energy, then the expansion process has been harnessed to perform work. During the production of energy, a considerable amount of heat is irreversibly lost through thermal conduction in the pipelines and in the turbine generating plant. If all of these heat losses and the energy produced (and the energy lost during chemical reactions such as scaling) were tallied up, they would equal the change in the internal energy of the fluid as it went from the geothermal reservoir to the waste products of the effluent water and the gases escaping to the air.

3-2   THE SECOND LAW OF THERMODYNAMICS: SEMANTICS

The second law of thermodynamics is one of the most fascinating empirical statements in all of science. Few laws have aroused more curiosity, provoked more controversy, motivated more books and articles, and created more misunderstanding. It applies directly to most physico-chemical processes, and some have tried to apply it (perhaps overzealously) to such subjects as biological evolution and economic theory. Historically, the second law grew from the search for a state function that describes the tendency for all processes to change in a certain direction. For example, water always flows from high pressure to low pressure, rocks always slide from a high gravitational potential to a low potential, dissolved salts always diffuse from more concentrated solutions to less concentrated ones, and heat always flows from higher temperatures to lower temperatures. All natural processes are unidirectional, and the second law expresses this tendency.

The central character in the second law is a new state function that Clausius named "entropy," from the Greek word meaning "transformation." Entropy actually measures the tendency for spontaneous heat flow and, when combined with the first law, leads to many powerful quantitative expressions that can describe the spontaneous direction for chemical reactions. This unidirectional tendency has led to several qualitative statements of the second law. One of the most popular classical statements was given by Clausius:

> *It is impossible to construct a machine that is able to convey heat by a cyclical process from one reservoir (at a lower temperature) to another at a higher temperature unless work is done by some outside agency.*

This statement emphasizes the obvious fact that heat always flows from a hot body to a cold body, although there is nothing in the first law to prevent the reverse. Refrigerators are possible only because a machine is present in the system to pump heat in the reverse, or unnatural, direction.

All the early work on the second law in the nineteenth century related to the way in which machines, both hypothetical and real, function. No doubt, this had much to do with prevailing fascination about the industrial revolution. The strong interest in increasing the efficiency of engines led to another statement of the second law:

Heat cannot be extracted from a body and turned entirely into work.

Every machine loses energy during the conversion of heat to work, and this inevitable loss prevents the construction of a 100%-efficient engine, although there is nothing in the first law to prevent such a possibility. Some energy is always dissipated and therefore unavailable for useful work. This result is clearly not desirable if your objective is to make better engines, and the second law was irrefutably shown to be a fundamental fact of nature through the continued efforts to disprove it. An additional problem confronts those who want to extract "useful" work from heat energy. The greatest amount of work that can be achieved in any system is *reversible* work, but reversibility cannot be fully attained in any real system performing work. For this reason, real work is always less than reversible work. As shown by Clausius, the same applies for the heat exchanged in any cyclic process: reversible heat is maximum heat. You can prove this yourself by writing the first law for a cyclic process.

An entirely different view of the second law arose early in the twentieth century, when statistical mechanics was first developed for thermodynamic expressions. By using a molecular rather than a macroscopic view of systems, Boltzmann and others showed that there is a relationship between the spontaneous direction of natural processes and probability theory:

Every system left to itself will, on the average, change toward a condition of maximum probability.

This statement makes the connection between (1) energy distributed on a molecular and electronic scale and (2) energy distributed on a macroscopic scale, and it has had great impact on many areas of science. It can be shown to be equivalent to the two statements quoted earlier.

Perhaps the most vivid statement of the second law, and certainly a no-less-accurate one, is this:

You can't shovel manure into the rear end of a horse and expect to get hay out of its mouth.

This barnyard analogy of unknown origin clearly demonstrates the uni-directional nature of natural processes—something we all take for granted.

Other statements of the second law embodied in the Carnot cycle or the principle of Caratheodoroy are more satisfying for those requiring the mathematical rigor. Numerous texts may be consulted for either or both of these treatments (e.g., Margenau and Murphy, 1956; Klotz and Rosenberg, 1972; Denbigh, 1981).

3-3 THE SECOND LAW OF THERMODYNAMICS: MATHEMATICS

Our next step is to express these statements of the second law in mathematical forms. Remember that our goal is to find a state function that expresses the spontaneous direction of heat flow in a system. From the first law, we have

$$dU = dQ - P\,dV - \sum dW' \tag{3-21}$$

where $\sum dW'$ represents the sum of all non-PV work such as electrical work, gravitational work, magnetic work, and chemical work. For closed systems with only PV work,

$$dU = dQ - P\,dV \tag{3-22}$$

We recall that, for a system in which both heat and work energy are exchanged with its surroundings, dW is an inexact differential. However, by dividing dW by $-P$ so that

$$\frac{-dW_{\text{rev}}}{P} = dV \tag{3-23}$$

we have transformed it into an exact differential. In so doing, we discover another interesting property of differential equations: the integrating factor. This factor has the property of converting an inexact differential into an exact one. It can be proven mathematically that any function of only two independent variables has an integrating factor. Therefore, an integrating factor should exist for dQ. Because dW_{rev} is a function of an intensive variable and an extensive variable—i.e., the driving force P and the measure of displacement dV—we conclude that dQ should be a function of the driving force for heat flow, T, and the measure of thermal displacement dS, the differential entropy change. By analogy with dW_{rev}, the integrating factor for dQ_{rev} should be $1/T$, which yields a new state function, the entropy S:

$$\boxed{\frac{dQ_{\text{rev}}}{T} = dS} \tag{3-24a}$$

where the subscript rev reminds us that the definition holds only for reversible processes, just as it did for PV work. Equation (3-24a) is the mathematical statement of the second law and is valid only for a closed system. If the system is open, thus allowing simultaneous reversible transfer of both energy and mass between system and surroundings, a modification of (3-24a) is required. For dn_j moles of phase j exchanged between system and surroundings, and assuming an entropy of S_j for that amount of phase, then the differential change in the entropy of the open system is

$$\frac{dQ_{\text{rev}}}{T} + \sum_j S_j\,dn_j = dS \tag{3-24b}$$

This important modification of (3-24a) was brought to the attention of geochemists by Tunnell (1977), although the equation was first derived by Gillespie and Coe (1933). All subsequent discussions in this chapter, however, continue to assume closed-system conditions.

For any cyclic process *in which irreversibility enters at any step*, we must have

$$\oint \frac{dQ}{T} < 0 \quad \text{or} \quad dS > \frac{dQ_{\text{irrev}}}{T} \tag{3-25}$$

because $Q_{\text{rev}}/T > Q_{\text{irrev}}/T$. This relationship is the *inequality of Clausius*, which we shall used later to identify the direction of spontaneous reactions. Another way of stating this inequality is that in any process the thermal energy $T\,dS$ is always greater than or equal to the heat flow Q:

$$T\,dS - dQ \geq 0 \tag{3-26}$$

Thus, a more complete expression of the second law is

$$\boxed{dS \geq \frac{dQ}{T}} \tag{3-27}$$

Let us test the exactness of dS to make sure it qualifies as a state function. We can rearrange the first law to obtain

$$dQ = dU + P\,dV \tag{3-28}$$

Writing the internal energy in terms of T and V, we have

$$U = U(T,V)$$

Therefore,

$$dU = \left(\frac{\partial U}{\partial T}\right)_V dT + \left(\frac{\partial U}{\partial V}\right)_T dV \tag{3-29}$$

Substituting into equation (3-28), we have

$$dQ = \left(\frac{\partial U}{\partial T}\right)_V dT + \left(P + \left(\frac{\partial U}{\partial V}\right)_T\right) dV \tag{3-30}$$

The property of exactness can be demonstrated by application to an ideal gas. Remember, the first definition of an ideal gas is that it follows the equation of state $P\bar{V} = RT$. A second (and equivalent) definition is that the internal energy of an ideal gas is a function of temperature only, so

$$\left(\frac{\partial U}{\partial V}\right)_T = \left(\frac{\partial U}{\partial P}\right)_T = 0 \tag{3-31}$$

Thus, we may write (3-30) for the ideal gas as

$$dQ = \left(\frac{\partial U}{\partial T}\right)_V dT + \frac{RT}{V} dV \qquad (3\text{-}32)$$

We can see that dQ is inexact because

$$\left(\frac{\partial^2 U}{\partial V \partial T}\right)_{V,T} = 0 \qquad (3\text{-}33)$$

$$\frac{\partial(RT/V)}{\partial T} = R/V \qquad (3\text{-}34)$$

Using the integrating factor, $1/T$, we obtain

$$dS = \frac{dQ}{T} = \frac{1}{T}\left(\frac{\partial U}{\partial T}\right)_V dT + \frac{R}{V} dV \qquad (3\text{-}35)$$

The rule of cross-differentiation shows that

$$\left[\frac{\partial\left(\frac{1}{T}\frac{\partial U}{\partial T}\right)}{\partial V}\right]_T = 0 \qquad (3\text{-}36)$$

$$\left(\frac{\partial(R/V)}{\partial T}\right)_V = 0 \qquad (3\text{-}37)$$

Therefore, dS is an exact differential, and S is a state function.

We will now continue our discussion using the ideal gas as a model. Consider these three different ways to expand an ideal gas.

1. *Reversible isothermal expansion.*   What is ΔQ_{rev}? Because $\Delta T = 0$, we know that ΔU also must equal 0. Thus, $Q_{rev} = -W_{rev}$, so we need calculate only the reversible work done in the isothermal expansion of an ideal gas. We have

$$W_{rev} = \int dW_{rev} = -\int P\,dV = -RT\int_{V_1}^{V_2}\frac{dV}{V} = -RT\ln\frac{V_2}{V_1} \qquad (3\text{-}38)$$

where V_1 is the initial volume and V_2 is the final volume. The entropy change for the expansion must be

$$\Delta S = \frac{Q_{rev}}{T} = R\ln\frac{V_2}{V_1} = R\ln\frac{P_1}{P_2} \qquad (3\text{-}39)$$

2. *Free expansion.*   A free expansion is just as imaginary as a reversible expansion. In the free expansion, the gas instantaneously expands from V_1 to V_2, so it has no opportunity to work against the opposing piston. Because no work is done, $W_{free} = 0 = Q_{free}$. What is the entropy change

TABLE 3-1

ISOTHERMAL EXPANSION OF AN IDEAL GAS BETWEEN V_1 AND V_2

	Reversible	Free	Real
Q	$RT \ln \dfrac{V_2}{V_1}$	0	$RT \ln \dfrac{V_2}{V_1} > Q > 0$
W	$-RT \ln \dfrac{V_2}{V_1}$	0	$-RT \ln \dfrac{V_2}{V_1} < W < 0$
ΔU	0	0	0
ΔS	$R \ln \dfrac{V_2}{V_1}$	$R \ln \dfrac{V_2}{V_1}$	$R \ln \dfrac{V_2}{V_1}$

Note that the reversible heat is the maximum possible heat, and the reversible work is the minimum possible work. These extreme limits for both reversible heat and reversible work, when contrasted with irreversible processes, form the basis of the second law of thermodynamics.

for the free expansion? If you said zero, you are wrong! Because S is a state function, ΔS must have the same value as that calculated for the reversible process; state functions are independent of path.

3. *Real expansion.*   Every real expansion of an ideal gas must be intermediate between the reversible expansion and the free expansion. There will be some work done against the piston, but it will not be the maximum work possible.

Table 3-1 summarizes the three cases. The most important point is that the two state functions U and S have the same values, regardless of how the expansion was carried out. This is why the imaginary reversible process is so important in thermodynamics—if we can imagine a reversible process that will allow us to calculate some change in a state function, then that change will be equally valid for any irreversible process, regardless of how it is carried out.

3-4 COMBINED FIRST AND SECOND LAWS

Substituting the second-law equation (3-27) into equation (3-22), we obtain the fundamental equation relating the first and second laws:

$$\boxed{dU \leq T\,dS - P\,dV} \tag{3-40}$$

for either reversible or irreversible changes in a closed system. We now have all the primary variables needed to describe a system: P, V, T, U, and S. All other thermodynamic quantities can be derived from these five variables. Equation (3-40) tells us that $U = U(S,V)$ is the most convenient form for the internal energy; S and V are the "natural" variables for U because they provide the simplest form of the exact differential. We can also determine how the internal energy changes with a change in entropy or volume:

$$U = U(S,V) \tag{3-41}$$

$$dU = \left(\frac{\partial U}{\partial S}\right)_V dS + \left(\frac{\partial U}{\partial V}\right)_S dV \tag{3-42}$$

By comparison with equation (3-40), and at equilibrium

$$\left(\frac{\partial U}{\partial S}\right)_V = T \tag{3-43}$$

$$\left(\frac{\partial U}{\partial V}\right)_S = -P \tag{3-44}$$

Planck (1926) described state variables in the following manner: if S and V are the chosen independent variables, then U is the *characteristic function*. In geochemistry, we are interested in the independent variables P and T, and now we seek the characteristic function, $f(P,T)$.

3-5 ENTHALPY

Consider any process carried out at constant pressure. If the only work is PV work, then the change in internal energy is

$$\Delta U = U_2 - U_1 = Q + W = Q - P(V_2 - V_1) \tag{3-45}$$

or

$$(U_2 + PV_2) - (U_1 + PV_1) = Q_P \tag{3-46}$$

where subscript P serves as a reminder that pressure is constant. Q_P must be a state function (because it is defined in terms of state functions U, P, and V) and it is called the *enthalpy* (H):

$$H = U + PV \tag{3-47}$$

Taking the total derivative of (3-47),

$$dH = d(U + PV) = dU + P\,dV + V\,dP \tag{3-48}$$

and substituting equation (3-40), we obtain at equilibrium

$$\boxed{dH = T\,dS + V\,dP} \tag{3-49}$$

Thus the enthalpy is the characteristic function for the independent variables S and P. The quantity dQ is an exact differential under this specific set of conditions: constant pressure, and PV work only. This set of conditions is the basis for calorimetric measurements from which a great many enthalpy values are obtained. The practical application of equation (3-46) is in measuring the heat of a chemical reaction with a constant-pressure calorimeter. This heat is the enthalpy change for that reaction. If the heat of a chemical reaction is measured at constant volume, then, by application of the first law,

$$\Delta U_V = Q_V \tag{3-50}$$

Equation (3-50) tells us that the internal energy of a chemical reaction can be determined by measuring the heat in a constant-volume calorimeter. The actual measurement of constant-volume and constant-pressure heats of reaction will be discussed in Chapter 11.

Enthalpy changes for chemical reactions are very important quantities that can be used in many ways. The enthalpy change for a process determines whether a chemical reaction will be exothermic or endothermic. It usually contributes the major portion of the energy change for any transformation. Extensive tabulations of enthalpies are available in many different publications (see Chapter 12) but, before these can be used, it is necessary to understand standard-state conventions.

Enthalpy, like internal energy, is not an absolute quantity. We cannot measure the intrinsic enthalpy of a substance, but only the enthalpy resulting from a change in state. In order to assign enthalpy values to substances, we define the *standard-state enthalpy of formation from the elements*, ΔH_f°. The standard state refers to a specific form, the most stable form, of a substance at a specified pressure and temperature. The standard-state pressure is usually fixed at 1 bar, and the reference temperature of 298.15 K is often chosen as the standard-state temperature. The standard-state enthalpies of the elements are arbitrarily assigned a value of zero. For the well-known, highly exothermic reaction of liquid water forming from the combustion of hydrogen and oxygen,

$$H_{2(g)} + \tfrac{1}{2}O_{2(g)} \rightarrow H_2O_{(l)} \qquad \Delta H_{f,298}^\circ = -285.83\,\text{kJ} \cdot \text{mol}^{-1} \tag{3-51}$$

Heat is given off during this combustion (thus energy is leaving the system), and there is a net decrease in enthalpy (sign is negative). A closely related reaction is the formation of water vapor from the elements:

$$H_{2(g)} + \tfrac{1}{2}O_{2(g)} \rightarrow H_2O_{(g)} \qquad \Delta H_{r,298}^\circ = -241.81\,\text{kJ} \cdot \text{mol}^{-1} \tag{3-52}$$

Note that the subscript on enthalpy is r (indicating reaction) rather than f (for formation), because water vapor is not the most stable state for H_2O at 25°C. We can make use of the fact that enthalpy is a state function to calculate the enthalpy change for $H_2O_{(g)} \rightarrow H_2O_{(l)}$ (i.e., for the condensation of one mole of H_2O)

from the data in (3-51) and (3-52). By setting up the cycle

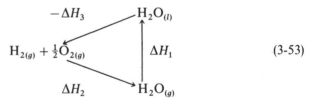

$$(3\text{-}53)$$

we see that $\Delta H_1 + \Delta H_2 - \Delta H_3 = 0$ because enthalpy is a state function and the process is cyclic. Note that ΔH_3 is negative because it is written in the reverse direction of (3-51), so the sign of the heat exchanged must be opposite. We now can calculate ΔH_1:

$$\Delta H_1 = -\Delta H_2 + \Delta H_3 = -44.02 \text{ kJ} \cdot \text{mol}^{-1} \qquad (3\text{-}54)$$

This equation may also be interpreted as the enthalpy of the products minus the enthalpy of the reactants, and is sometimes known as Hess's law. The same procedure will work for any state function. For a reaction involving i species, we may generalize (3-54) to

$$\Delta H_r = \sum_i \nu_i \Delta H_{f,i}^{\circ} \qquad (3\text{-}55)$$

where the ν_i values are the *stoichiometric coefficients* for all species. The sign of the stoichiometric coefficient is positive for products and negative for reactants. Thus ν_{H_2O} in (3-52) is $+1$, whereas ν_{O_2} is $-1/2$.

This principle of addition of known enthalpies to calculate an unknown enthalpy is so important that we will give another example. Suppose we want to calculate ΔH_r for the hydration of corundum (Al_2O_3) forming one mole of gibbsite ($Al(OH)_3$) from the following standard data:

$$\Delta H_{f,298}^{\circ}(Al_2O_3) = -1675.70 \text{ kJ} \cdot \text{mol}^{-1}$$
$$\Delta H_{f,298}^{\circ}(Al(OH)_3) = -1293.13 \text{ kJ} \cdot \text{mol}^{-1}$$
$$\Delta H_{f,298}^{\circ}(H_2O_{(l)}) = -285.83 \text{ kJ} \cdot \text{mol}^{-1}$$

We could set up the following table of enthalpies of formation:

$$\frac{3}{2} \times (H_2O_{(l)}) \rightarrow H_{2(g)} + \tfrac{1}{2}O_{2(g)} \qquad +428.75 \text{ kJ} \qquad (3\text{-}56)$$
$$\tfrac{1}{2} \times (Al_2O_{3(c)}) \rightarrow 2Al_{(c)} + \tfrac{3}{2}O_{2(g)} \qquad +837.85 \text{ kJ} \qquad (3\text{-}57)$$
$$Al_{(c)} + \tfrac{3}{2}O_{2(g)} + \tfrac{3}{2}H_{2(g)} \rightarrow Al(OH)_{3(c)} \qquad -1293.13 \text{ kJ} \qquad (3\text{-}58)$$
$$\overline{\tfrac{1}{2}Al_2O_{3(c)} + \tfrac{3}{2}H_2O_{(l)} \rightarrow Al(OH)_{3(c)}} \qquad -26.53 \text{ kJ} \qquad (3\text{-}59)$$

The sum of reactions (3-56) through (3-58) is the desired reaction (3-59). ΔH_r is -26.53 kJ per mole of gibbsite formed, and the reaction is exothermic. Note that all the elements in the summation exactly cancel, a necessary result of conservation of mass in the cycle. Because of this, the answer could have been more easily found by resorting to equation (3-55):

$$\Delta H_r = \sum_i \nu_i \, \Delta H_{f,i}^\circ = -\tfrac{1}{2}(\Delta H_{f,Al_2O_3}^\circ) - \tfrac{3}{2}(\Delta H_{f,H_2O(l)}^\circ) + 1(\Delta H_{f,Al(OH)_3}^\circ) \quad (3\text{-}60)$$

which is the usual procedure for calculating the changes in any state function due to a chemical reaction.

One final comment regards correct usage of units. It is clear that enthalpies are extensive properties, dependent on the amount of each substance present. The standard enthalpies of formation are always written in such a way as to define 1 mol (or, more generally, 1 gram formula weight) of the substance in question. However, when more than one product is formed in a chemical reaction, ΔH_r cannot be defined in terms of moles. For instance, the reaction shown in Problem 3-2

$$2\,FeS_{2(c)} + 7\,O_{2(g)} + 2\,H_2O_{(l)} \rightarrow 2\,Fe_{(aq)}^{2+} + 4\,SO_{4(aq)}^{2-} + 4\,H_{(aq)}^{+} \quad (3\text{-}61)$$

has an enthalpy change that can be expressed in terms of kJ relative to the number of moles of all species present, but the unit $kJ \cdot mol^{-1}$ is not appropriate. We will express the enthalpies of such reactions in terms of kJ only, but you need always to keep in mind the implied mass dependence.

3-6   FREE ENERGY FUNCTIONS

To derive another thermodynamic function, let us consider the rearrangement of the fundamental equation for reversible processes only

$$dU - T\,dS = -P\,dV \quad (3\text{-}62)$$

Here $-P\,dV$ is a work function and $T\,dS$ is a heat function subtracted from the total energy function dU. Because the heat is unavailable for work, the energy term TS is called unavailable energy, or bound energy, and the term

$$dU - T\,dS \quad (3\text{-}63)$$

is available energy, or *free energy*—energy available for work. The general function

$$\boxed{U - TS = A} \quad (3\text{-}64)$$

is called the *work content*, or *Helmholtz free energy*. The total differential is

$$dA = dU - T\,dS - S\,dT \quad (3\text{-}65)$$

and substituting equation (3-40) for dU leads to

$$\boxed{dA = -S\,dT - P\,dV} \quad (3\text{-}66)$$

Thus, for the characteristic function A, the independent variables are T and V.

The most frequently encountered condition in the laboratory and in real systems is constant P and T. To derive the characteristic function, $f(P,T)$, we will go back to the equation for enthalpy and subtract the bound energy, $d(TS)$, from both sides:

$$dH - d(TS) = T\,dS + V\,dP - d(TS) \tag{3-67}$$

$$d(H - TS) = T\,dS + V\,dP - S\,dT - T\,dS \tag{3-68}$$

$$d(H - TS) = V\,dP - S\,dT \tag{3-69}$$

By analogy with the Helmholtz free energy, this function should be called the available enthalpy, or "free enthalpy." It is more commonly called the *Gibbs free energy*

$$\boxed{G = H - TS = U + PV - TS} \tag{3-70}$$

From (3-69), we see that

$$\boxed{dG = V\,dP - S\,dT} \tag{3-71}$$

This thermodynamic function was named in honor of Josiah Willard Gibbs, a brilliant physical chemist who lectured at Yale University from 1871 to 1903. Gibbs' publications on the thermodynamics of heterogeneous phase equilibria were revolutionary and laid the foundations for virtually all subsequent work on the subject.

The Gibbs free energy is one of the most useful functions in geochemical thermodynamics, and it is a measure of the available non-PV work. If we represent all non-PV work (electrical, gravitational, chemical, etc.) by W' and take the total differential for G,

$$dG = dU + P\,dV + V\,dP - T\,dS - S\,dT \tag{3-72}$$

we find, by substitution with the combined first-and-second-law equation

$$dG = V\,dP - S\,dT + \sum dW' \tag{3-73}$$

At constant pressure and temperature, this equation reduces to

$$dG = \sum dW' \tag{3-74}$$

Therefore, non-PV work (e.g., work energy available from geochemical reactions) can be determined by the Gibbs free energy. Like H and A, G must also be a state function, because it is defined in terms of other state functions.

Free energies of substances can be obtained from enthalpies and entropies at constant temperature and pressure. By differentiation of the defining equation

$G = H - TS$ at constant temperature, we obtain

$$dG = dH - T\,dS \qquad (3\text{-}75)$$

which becomes, by integration,

$$\boxed{\Delta G = \Delta H - T\,\Delta S} \qquad (3\text{-}76)$$

This is one of the most useful relations in thermodynamics. When applied to the formation of 1 mol of substance from the elements in their standard states, it yields the standard Gibbs free energy of formation:

$$\Delta G_f^\circ = \Delta H_f^\circ - T\,\Delta S_f^\circ \qquad (3\text{-}77)$$

Values of ΔG_f° may be obtained also from compilations in standard tables. The free energy change for any reaction may be determined from the standard values as

$$\Delta G_r = \sum_i v_i \Delta G_{f,i}^\circ \qquad (3\text{-}78)$$

3-7 CRITERIA FOR SPONTANEITY AND EQUILIBRIUM

We are now in a position to answer two important questions about chemical reactions:

1. Under what conditions are products and reactants at equilibrium?
2. If not at equilibrium, in which direction will a reaction proceed?

The second law of thermodynamics provides the answer to both questions. The algebraic link is the inequality of Clausius (cf. 3-26), which can be rewritten as

$$dQ - T\,dS \le 0 \qquad (3\text{-}79)$$

The inequality applies to irreversible process, and the equality applies to reversible processes. When the equality applies, we have the equilibrium condition. Both free energy functions A and G are easily related to the Clausius inequality. At constant temperature, (3-65) is written

$$dA = dU - T\,dS \qquad (3\text{-}80)$$

which is equivalent to

$$dA = dQ + dW - T\,dS \qquad (3\text{-}81)$$

Because of (3-79), we have

$$dA - dW \le 0 \qquad (3\text{-}82)$$

which reduces (for PV work only) to

$$(dA)_{T,V} \le 0 \qquad (3\text{-}83)$$

Because of (3-70), we see that G and A are related:

$$G = U - TS + PV = A + PV \tag{3-84}$$

At constant pressure, we may write

$$dG = dA + P\,dV \tag{3-85}$$

For PV work only, we substitute (3-82) into (3-85) to obtain

$$\boxed{(dG)_{P,T} \le 0} \tag{3-86}$$

The constant-temperature subscript applies because $dT = 0$ was assumed in the derivation of (3-82).

Equation (3-86) is the relation we seek. It is a quantitative expression for the direction of a spontaneous reaction. Recalling the significance of the inequality sign, we may summarize:

A chemical reaction will proceed in the direction of lower Gibbs free energy (i.e., $\Delta G_r < 0$),

and

products and reactants are in equilibrium when their Gibbs free energies are equal (i.e., $\Delta G_r = 0$).

Both statements apply for conditions of constant temperature and pressure. Thus we need only calculate the sign of the Gibbs free energy (1) to see whether a reaction may proceed as written and (2) to test for equilibrium between products and reactants.

The inequality of Clausius can also be applied to master variables U, S, and H, but these functions are not used in geochemistry to test for reaction directions because the fixed conditions applied to the differentials cannot be realized in nature. Nonetheless, the equations analogous to (3-83) and (3-86) are

$$(dU)_{S,V} \le 0 \tag{3-87}$$

$$(dH)_{S,P} \le 0 \tag{3-88}$$

$$(dS)_{U,V} \ge 0 \tag{3-89}$$

3-8 AN APPLICATION OF THE COMBINED FIRST AND SECOND LAWS: THE DANIELL CELL

The equations derived in this chapter will have much more meaning if applied to a specific example with real numbers. A good example is afforded by the dissolution reaction of zinc in a copper sulfate solution:

$$Zn_{(c)} + CuSO_{4(aq)} \rightarrow Cu_{(c)} + ZnSO_{4(aq)} \tag{3-90}$$

Observation demonstrates that this reaction proceeds spontaneously with the plating of copper on a metallic zinc surface. We can measure the heat of the reaction in an irreversible manner, with reference to 298.15 K and 1 bar, in a calorimeter. We find that $Q_{irrev} = -218.16\ kJ \cdot mol^{-1}$. In the calorimeter, the reaction can do no work, so

$$\Delta U_r = Q_{irrev} \qquad (3\text{-}91)$$

Furthermore, because

$$\Delta H_r = \Delta U_r + P\,\Delta V \qquad (3\text{-}92)$$

we have

$$\Delta H_r = \Delta U_r = -218.16\ kJ \cdot mol^{-1} \qquad (3\text{-}93)$$

Thus, if the calorimeter is a constant-volume system, we see that both the internal energy and the enthalpy are equal to the heat of reaction. We cannot proceed any farther until we have measured the same reaction by a reversible path. We can set up a reversible path by using a zinc electrode in a zinc sulfate solution, separated by a porous partition from a copper sulfate solution with a copper electrode in it (Figure 3-2). If those electrodes are then connected through a potentiometer with an opposing external electromotive force (EMF) so that negligible current is drawn, the system will be as close to reversibility as can be approached in the real world. The EMF of the cell is used as a measure of reversible work (see Chapter 10):

$$W_{rev} = nFE \qquad (3\text{-}94)$$

where n is the number of electrons involved in the oxidation–reduction couple ($n = 2$ in this case, because both Zn and Cu ions are divalent), F is the Faraday constant, and E is the EMF of the cell. For the Daniell cell, $E = 1.103\ V \cdot mol^{-1}$, and $W_{rev} = 212.78\ kJ$. This work is being performed on the system by the surroundings and is positive. From the system's point of view, the work is negative and

$$W_{rev} = -212.78\ kJ \cdot mol^{-1} \qquad (3\text{-}95)$$

Now we can calculate the heat of reaction for the reversible path:

$$Q_{rev} = \Delta U - W_{rev}$$
$$= -218.16 + 212.78 = -5.38\ kJ \cdot mol^{-1} \qquad (3\text{-}96)$$

The enthalpy is greater than the heat of the reversible reaction. Where did the extra heat go? It became work energy.

From equation (3-24a), we can calculate the entropy change for this reaction:

$$\Delta S_{rev} = \frac{Q_{rev}}{T} = \frac{-5.38}{298.15} = -18.0\ J \cdot mol^{-1} \cdot K^{-1} \qquad (3\text{-}97)$$

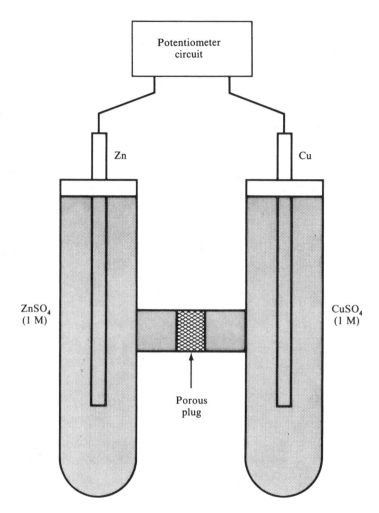

Figure 3-2. The Daniell cell.

Because state functions are independent of path,

$$\Delta S_{irrev} = \Delta S_{rev} = -18.0 \, \text{J} \cdot \text{mol}^{-1} \cdot \text{K}^{-1} \qquad (3\text{-}98)$$

What then is Q_{irrev}/T? Here, Q_{irrev} represents the total amount of available heat that can be transferred to the surroundings; Q_{rev} is the exchangeable heat which is unavailable for work; and Q_{irrev}/T represents the entropy change seen in the surroundings. The total change in entropy is the sum of the entropy changes for the system and the surroundings:

$$\Delta S_{total} = \Delta S_{sys} + \Delta S_{surr} \qquad (3\text{-}99)$$

For the irreversible process,

$$\Delta S_{total} = \frac{Q_{rev}}{T} + \frac{Q_{irrev}}{T} \tag{3-100}$$

$$= -18.0 + 731.7$$

$$= 713.7 \text{ J}\cdot\text{mol}^{-1}\cdot\text{K}^{-1}$$

For the reversible process,

$$\Delta S_{total} = -18.0 + 18.0 = 0 \tag{3-101}$$

In the irreversible process, the surroundings saw $731.7 \text{ J}\cdot\text{mol}^{-1}\cdot\text{K}^{-1}$ even though the system changed only by $-18.0 \text{ J}\cdot\text{mol}^{-1}\cdot\text{K}^{-1}$. The amount seen by the surroundings is the "exchanged," or "external," entropy ΔS_e, and ΔS_i is the "internal" entropy, which is the entropy change within the system due only to irreversible processes. For the irreversible transformation the *system's* change in entropy is the sum of the external and internal entropy:

$$\Delta S_{irrev, sys} = \Delta S_e + \Delta S_i = -18.0 \text{ J}\cdot\text{mol}^{-1}\cdot\text{K}^{-1} \tag{3-102}$$

Because the external entropy is just opposite in sign from that seen by the surroundings, we can calculate the internal entropy produced:

$$\Delta S_i = \Delta S_{irrev, sys} - \Delta S_e$$

$$= -18.0 - (-731.7) = 713.7 \text{ J}\cdot\text{mol}^{-1}\cdot\text{K}^{-1} \tag{3-103}$$

Therefore, we discover that an extra $713.7 \text{ J}\cdot\text{mol}^{-1}\cdot\text{K}^{-1}$ of entropy was produced or created *internally* for a system undergoing an irreversible change. For the reversible change the amount of entropy exchanged to the surroundings was the same as the amount by which the system changed, so that no internal entropy was produced. This conclusion leads to the strengthened form of the second law:

$$\boxed{dS_i \geq 0} \tag{3-104}$$

We say "strengthened" because no restrictions whatever are applied to it. However, it is difficult or impossible to calculate ΔS_i for many reactions of geochemical interest because it is not a state function, and the Gibbs free energy (carrying restrictions of constant temperature and pressure) is used instead.

What other thermodynamic quantities do we know or can we calculate from the Daniell cell measurements? The available work obtained from the system was non-PV work; therefore, it is the same as the Gibbs free energy, and

$$\Delta G = W_{rev} = -212.78 \text{ kJ}\cdot\text{mol}^{-1} \tag{3-105}$$

This quantity could also be calculated from the enthalpy and entropy if the

TABLE 3-2
THERMODYNAMIC VALUES FOR THE DANIELL CELL
REACTION AT 25°C

Property	Reversible path	Irreversible path
Q	-5.38 kJ	-218.16 kJ
W	-212.78 kJ	0
ΔU	-218.16 kJ	-218.16 kJ
ΔH	-218.16 kJ	-218.16 kJ
ΔS	-18.0 J$\cdot$K^{-1}	-18.0 J$\cdot$K^{-1}
ΔS_i	0	713.7 J$\cdot$K^{-1}
ΔS_e	-18.0 J$\cdot$K^{-1}	-731.7 J$\cdot$K^{-1}
ΔG	-212.79 kJ	-212.79 kJ
$T\Delta S_i$	0	212.79 kJ

available non-PV work was not measured or known:

$$\Delta G = \Delta H - T\Delta S = -218.16 - (298.15 \times -0.018) = -212.79 \text{ kJ}$$

The small difference is due to rounding errors in calculation of ΔS.

Table 3-2 summarizes all the thermodynamic values that can be derived from only two careful measurements of a Daniell cell: one measurement under irreversible conditions, the other under reversible conditions. Note that the internal and external entropies are not state functions and depend on the chosen path. These functions are, in fact, a direct measure of the degree of reversibility of a reaction.

3-9   A FIRST LOOK AT TEMPERATURE DEPENDENCE

Most tables of standard thermodynamic properties refer to 25°C. Because many important geochemical reactions occur at higher temperatures, it is important to be able to calculate the effect of temperature on enthalpy, entropy, and free energy. A direct result of applying heat to a given mass of a substance is that the temperature rises. If the heat supplied and the temperature rise are measured, it is found that these quantities are proportional:

$$Q \propto \Delta T \quad \text{or} \quad Q = C\Delta T \tag{3-106}$$

where C is the proportionality constant. This constant is known as the *heat capacity*. The heat capacity itself is also found to be a function of temperature. Therefore, it is more rigorous to define heat capacity as the differential change in heat with respect to temperature:

$$C = \frac{dQ}{dT} \tag{3-107}$$

The value of the heat capacity depends on the conditions under which the heat is transferred, so equation (3-107) is inexact. If the pressure is held constant, then

$$dQ_P = dH_P \tag{3-108}$$

and

$$C_P = \left(\frac{\partial H}{\partial T}\right)_P \tag{3-109}$$

If the volume is held constant, then from equation (3-18)

$$dQ_V = dU_V \tag{3-110}$$

and

$$C_V = \left(\frac{\partial U}{\partial T}\right)_V \tag{3-111}$$

For relatively incompressible substances such as minerals, the difference between C_P and C_V is generally small.

For a limited number of simple substances, heat capacities can be calculated from more fundamental data. More commonly, heat capacities are obtained by direct measurement and fitted to an empirical equation such as a simple power series for minerals:

$$C_P = a + bT + cT^2 + \cdots \tag{3-112}$$

Maier and Kelly (1932) found that the three-term expansion

$$C_P = a + bT - cT^{-2} \tag{3-113}$$

gives a better fit than a regular power series. Many other equations have been proposed for the fitting of empirical heat-capacity data. The implications and consequences of using different heat-capacity expressions will be discussed in Chapter 12.

From equation (3-109), we can derive the effect of temperature on enthalpy at constant pressure. By integration, we obtain

$$\int_{T_r}^{T} dH = \int_{T_r}^{T} C_P \, dT = \int_{T_r}^{T} (a + bT - cT^{-2}) \, dT \tag{3-114}$$

$$H_T - H_{T_r} = a(T - T_r) + \frac{b}{2}(T^2 - T_r^2) + c\left(\frac{1}{T} - \frac{1}{T_r}\right) \tag{3-115}$$

where T_r is some reference temperature. For instance, if the heat capacity of α-quartz is given by $C_{P(\alpha\text{-}qz)} = (44.603 + 3.7754 \times 10^{-2}T - 1.0018 \times 10^6 T^{-2}) \, J \cdot K^{-1}$, then we may substitute the coefficients of this expression into (3-115) to calculate the change in enthalpy that results from heating α-quartz

TABLE 3-3

HEAT-CAPACITY DATA FOR FORMATION OF α-QUARTZ FROM THE ELEMENTS

$$C_P \text{ expression } (J \cdot K^{-1})$$

$Si_{(c)}$	$31.778 + 5.3878 \times 10^{-4}T - 1.7864 \times 10^2 T^{-0.5} - 1.4654 \times 10^5 T^{-2}$
$O_{2(g)}$	$48.318 - 6.9132 \times 10^{-4}T - 4.2066 \times 10^2 T^{-0.5} + 4.9923 \times 10^5 T^{-2}$
$SiO_{2(\alpha\text{-}qz)}$	$44.603 + 3.7754 \times 10^{-2}T \qquad\qquad\qquad\qquad - 1.0018 \times 10^6 T^{-2}$

$$\Delta C_{P_r} = -35.493 + 3.7907 \times 10^{-2}T + 5.993 \times 10^2 T^{-0.5} - 1.3545 \times 10^6 T^{-2}$$

$$\Delta(H_T - H_{T_r}) = -35.493(T - T_r) + 1.8954 \times 10^{-2}(T^2 - T_r^2) + 1.1986$$

$$\times 10^3 (T^{0.5} - T_r^{0.5}) + 1.3545 \times 10^6 \left(\frac{1}{T} - \frac{1}{T_r}\right)$$

Source: Robie et al. (1978)

from 298.15 K to 844 K. We find $H_{844} - H_{298} = 33.94 \text{ kJ} \cdot \text{mol}^{-1}$. If, on the other hand, we wish to know ΔH_f° for α-quartz at 844 K, then we must calculate the change in heat capacity for the *reaction*

$$Si_{(c)} + O_{2(g)} = SiO_{2(\alpha\text{-}qz)} \tag{3-116}$$

which may be evaluated from

$$\boxed{\Delta C_{P,r} = \sum_i v_i C_{Pi}} \tag{3-117}$$

Be advised that, although the elements in (3-116) have zero enthalpies of formation, they have finite heat capacities. Thus, if we evaluate

$$\int_{T_r}^{T} d\,\Delta H = \int_{T_r}^{T} \Delta C_P \, dT = \int_{T_r}^{T} (\Delta a + \Delta b\,T + \Delta c\,T^{-2})\,dT \tag{3-118}$$

we will obtain the change in ΔH as a result of heating products and reactants from T_r to T.

Table 3-3 gives heat-capacity data for reaction (3-116) and shows the integrated form of the $\Delta C_P\,dT$ equation. The C_P expressions for Si and O_2 contain four terms instead of the three terms of the Maier–Kelley expansion. Thus, the integrated equation shown in the table contains one more term than does equation (3-115). By substituting $T = 844$ and $T_r = 298.15$, we find $(H_{844} - H_{298}) = 3.629 \text{ kJ}$. The enthalpy of the reaction at 844 K is found by adding this value to the enthalpy at 298 K:

$$\Delta H_{r(T)} = \Delta H_{r(T_r)} + \int_{T_r}^{T} \Delta C_{P_r}\,dT \tag{3-119}$$

For the formation of α-quartz from the elements, we have

$$\Delta H^{\circ}_{f(844)} = -910.700 + 3.629 = -907.07 \text{ kJ} \cdot \text{mol}^{-1}$$

Figure 3-3 shows an example of heat-capacity data for two important rock-forming minerals: muscovite and pyrophyllite (Robie et al., 1976). The temperature dependence of the heat capacity is typically a well-behaved function up to a phase transition (polymorphic transition or melting point). As the temperature approaches absolute zero, the heat capacity approaches zero. Notice, also, that the heat capacities of these chemically different minerals are rather close over a large range of temperature. If two minerals are chosen that have exactly the same composition and structure but differ only in the amount of ordering—such as analbite (high albite) and low albite—then their heat capacities will be nearly

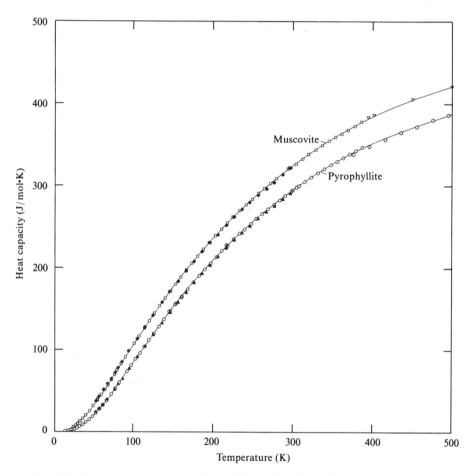

Figure 3-3. Low-temperature heat capacities of muscovite and pyrophyllite (Robie et al., 1976).

identical. The difference between their heat capacities will also decrease with decreasing temperature (see Figure 2 of Openshaw et al., 1976). In fact, all heat capacities of solid substances tend to converge as absolute zero is approached; i.e.,

$$\lim_{T \to 0} C_P = 0 \tag{3-120}$$

which means that

$$\lim_{T \to 0} \int \frac{C_P}{T} \, dT = 0 \tag{3-121}$$

Therefore,

$$\lim_{T \to 0} S = 0 \tag{3-122}$$

Equation (3-122) implies that, unlike enthalpy, entropy may be measured on an absolute scale. This discovery led to the third law of thermodynamics.

3-10   THE THIRD LAW OF THERMODYNAMICS AND FINITE ENTROPIES

The implication of equation (3-122), first stated by Nernst, is that the entropies of substances tend toward zero as zero temperature is approached. This general principle has proven valid for a great many solid substances, with relatively minor exceptions. In its most succinct form, stated by Lewis and Randall (1961, p. 130), this principle has become known as the third law of thermodynamics:

> If the entropy of each element in some crystalline state be taken as zero at the absolute zero of temperature, every substance has a finite positive entropy, but at the absolute zero of temperature the entropy may become zero, and does so become in the case of perfectly crystalline substances.

A perfectly crystalline substance is one that is characterized by a single quantum state at absolute zero, which means that there is only one possible arrangement (and only one possible energetic configuration) of its constituent atoms. The entropy of such substances can be measured on an absolute scale from heat-capacity data alone. From the second law, we have

$$dS = \frac{dQ_{\text{rev}}}{T} = \frac{C_P \, dT}{T} \tag{3-123}$$

Integration of (3-123) between 0 K and the reference temperature of 298.15 K yields

$$\int_0^{298} dS = S_{298} - S_0 = \int_0^{298} \frac{C_P}{T} \, dT \tag{3-124}$$

Because of (3-123), equation (3-124) states that the entropy of a perfectly crystalline substance has an absolute value at 298.15 K, and that this value can be determined by measuring the heat capacity of the substance from temperatures as near absolute zero as possible to ambient conditions.

Equation (3-124) does not completely define the third-law entropy for all substances in absolute terms, however, but rather reflects the so-called calorimetric contribution to the third-law entropy. Such heat-capacity measurements record the contribution due to lattice vibrations, which are not always the sole energetic contribution to the third-law entropy. All internal changes in the substance that are rapid enough to be detected during the course of the heat-capacity measurements will be automatically added into the calorimetric entropy term. One of the most important of these is the magnetic spin ordering that occurs in crystalline substances containing paramagnetic ions (see Figure 4-11). This effect may occur at very low temperatures (< 15 K) and was overlooked in some of the early heat-capacity measurements of minerals, because measurements were terminated at much higher temperatures.

Another important class of contributions cannot be determined from low-temperature heat capacities but rather must be calculated from additional information. The most important of these *residual-entropy* effects relates to mixing of two or more kinds of atoms in crystallographically equivalent sites. This disordering results in a large number of possible quantum states present at absolute zero; such substances are not perfectly crystalline in the third-law sense. One special example is isotopic mixing, which adds to the absolute entropy of crystals but is almost always neglected because these effects cancel in opposite sides of a chemical reaction.

A final consideration is the possibility of isothermal phase changes that might occur on cooling. The entropy contribution accompanying such phase changes is equal simply to $\Delta H_P / T_P$, where ΔH_P is the enthalpy change, and T_P is the reaction temperature. The thermodynamics of phase changes will be considered in Chapter 4.

In summary, the true third-law entropy of crystals is written as

$$S_{298} = \int_0^{298.15} \frac{C_P}{T}\, dT + S_0 + \Delta S_P \qquad (3\text{-}125)$$

where S_0 is the residual entropy, and ΔS_P is the entropy associated with any phase changes. The subject of the third-law entropy of minerals is treated with great clarity by Ulbrich and Waldbaum (1976).

The residual entropy contribution due to site mixing, called the *configurational entropy*, is a very important part of the third-law entropy of many common rock-forming minerals such as feldspars, feldspathoids, amphiboles, and micas. This configurational entropy contribution can be calculated by means of the Planck–Boltzmann equation:

$$S = \frac{R}{N_A} \ln W \tag{3-126}$$

where N_A is Avogadro's number, R is the gas constant, and W is the number of available configurations in the crystal. This fundamental equation is derived by combining probability theory with the third law of thermodynamics (e.g., Lewis and Randall, 1961, chaps. 8 and 12). The number of configurations can be directly related to the mole fractions of atoms on crystallographic sites. The important result (for derivation, see Ulbrich and Waldbaum, 1976, p. 3) is

$$S_{conf} = -R\sum_j m_j \sum_i X_{i,j} \ln X_{i,j} \tag{3-127}$$

where m_j is the total number of atoms occupying the jth crystallographic site (expressed in terms of atoms per formula unit) and $X_{i,j}$ is the mole fraction of the ith atom in the jth site.

For example, consider more specifically the difference between analbite and low albite discussed in Section 3-9: both are feldspars with composition $NaAlSi_3O_8$, but the Al and Si atoms in low albite occupy crystallographically distinct tetrahedral sites, whereas the Al and Si atoms in analbite are randomly distributed in crystallographically equivalent tetrahedral sites. Thus, equation (3-127) applies for analbite. Substituting $m_j = 4(1 \text{ Al} + 3 \text{ Si})$, $X_{Al}^{IV} = 1/4 = 0.25$, and $X_{Si}^{IV} = 3/4 = 0.75$, we find

$$S_{conf} = -4R(0.25\ln 0.25 + 0.75\ln 0.75) = 18.70 \text{ J} \cdot \text{K}^{-1} \cdot \text{mol}^{-1}$$

The third-law entropies of analbite and low albite, respectively, are $226.40 \text{ J} \cdot \text{K}^{-1} \cdot \text{mol}^{-1}$ and $207.40 \text{ J} \cdot \text{K}^{-1} \cdot \text{mol}^{-1}$ (Openshaw et al., 1976). The difference between the two (equal to $19.00 \text{ J} \cdot \text{K}^{-1} \cdot \text{mol}^{-1}$) is slightly different from the configurational contribution and reflects small differences in the observed low-temperature heat capacities of the two minerals.

SUMMARY

- The first law of thermodynamics is a statement of conservation of energy. It is an empirical principle that has never been disproved. It can be succinctly stated as

$$\oint dU = 0$$

where U is the internal energy of a system. For any process, the internal-energy change is the sum of the work done on the system and the heat absorbed by the system:

$$dU = dQ + dW$$

where dU is an exact differential, and dQ and dW are inexact. However, dQ and dW can be made exact by multiplying by integrating factors ($1/T$ and $-1/P$, respectively) for reversible changes.

- The second law of thermodynamics defines a state function, entropy (S):

$$dS = \frac{dQ_{\text{rev}}}{T}$$

All real processes are irreversible, so

$$T\, dS - dQ_{\text{irrev}} < 0$$

which provides a link to the direction in which any irreversible reaction will proceed. Any irreversible process creates entropy, whereas a system at equilibrium does not. Thus,

$$dS_i \geq 0$$

where dS_i is the change in internal entropy in the system. The equality sign applies to equilibrium and the inequality sign to irreversibility.

- The state functions enthalpy (H), Helmholtz free energy (A), and Gibbs free energy (G) are defined as follows:

$$H \equiv U + PV$$

$$A \equiv U - TS$$

$$G \equiv A + PV$$

- The differentials of U, H, A, and G are

$$dU = T\, dS - P\, dV$$

$$dH = T\, dS + V\, dP$$

$$dA = -S\, dT - P\, dV$$

$$dG = V\, dP - S\, dT$$

- The inequality of Clausius applied to the definition of Gibbs free energy leads to the following condition for any reaction measured at constant temperature and pressure:

$$\Delta G_r(P,T) \leq 0$$

At equilibrium, the Gibbs free energies of products and reactants are equal. For spontaneous reactions, the Gibbs free energy change is negative. This change ΔG may be calculated from ΔH and ΔS as

$$\Delta G = \Delta H - T\, \Delta S$$

- The effect of temperature on both the enthalpy and entropy of a reaction is

$$\Delta(H_T - H_{T_r}) = \int_{T_r}^{T} \Delta C_P \, dT$$

$$\Delta(S_T - S_{T_r}) = \int_{T_r}^{T} \frac{\Delta C_P}{T} \, dT$$

where ΔC_P is the change in heat capacity measured at constant pressure for products minus reactants, and T_r is the reference temperature.

- Because heat capacities approach zero as temperature approaches zero, the entropies of substances can be defined on an absolute scale. This gives rise to the third law of thermodynamics,

$$\lim_{T \to 0} S = 0$$

which applies to perfectly crystalline substances. Any transformations that are not detectable by low-temperature heat capacity measurements (such as cation ordering, magnetic transitions, etc.) will give rise to a residual entropy at 0 K. These corrections are very important for many minerals.

PROBLEMS

1. Consider the change in state of 1 mol of ideal gas from P_1,V_1,T_1 to P_2,V_2,T_2. Using $V(P,T)$, calculate and compare the change in state for two different paths: (1) by a two-step operation involving an isothermal change in state to P_2 and then an isobaric change in state to T_2; and (2) a one-step operation going directly from state 1 to state 2. Draw a schematic $P-T$ diagram to show each path and each state. Before the integration is carried out for path 2, use the following relationship for the slope of the path P_1,T_1 to P_2,T_2:

$$\frac{T_2 - T_1}{P_2 - P_1} = \frac{T - T_1}{P - P_1}$$

so that

$$T = T_1 + \frac{T_2 - T_1}{P_2 - P_1}(P - P_1)$$

Show that the change in volume is independent of path, but that the work is not.

2. The weathering of pyrite in coal mines and in metal sulfide ore deposits can be represented by the following reaction:

$$2\,FeS_{2(c)} + 7\,O_{2(g)} + 2\,H_2O_{(l)} = 2\,Fe^{2+}_{(aq)} + 4\,SO^{2-}_{4(aq)} + 4\,H^+_{(aq)}$$

Given the following enthalpies, should miners be prepared to fight fires or to wear very warm clothing?

Species	ΔH_f° (kJ·mol^{-1})
$FeS_{2(c)}$	-171.54
$H_2O_{(l)}$	-285.83
$Fe^{2+}_{(aq)}$	-89.1
$SO_{4(aq)}^{2-}$	-909.27
$H_{(aq)}^+$	0.0

3. Given that the enthalpy for oxidation of diamond by oxygen to carbon dioxide is -395.4 kJ·mol^{-1} and that the same reaction for graphite has an enthalpy of -393.5 kJ·mol^{-1}, find the enthalpy of reaction for graphite going to diamond. Values given are in their standard states. Is this reaction exothermic or endothermic? What does this imply about the chemical bonding in diamond compared to graphite? What is the enthalpy of formation of diamond from the elements under standard-state conditions?

4. (a) Given that $\Delta S_r^\circ = -1666.9$ J·mol^{-1}·K^{-1} for the reaction shown in Problem 2, find the Gibbs free energy of reaction under standard-state conditions. Is this reaction spontaneous or not?

   (b) Given that $\Delta S_r^\circ = -3.36$ J·mol^{-1}·K^{-1} for graphite $\rightarrow$ diamond, determine whether diamonds should be spontaneously converted to graphite or vice versa. Does the reaction proceed at a measurable rate? Why (or why not)?

5. Consider the following standard enthalpies of formation from the elements:

$$\Delta H_{f(298)}^\circ \text{ (kJ·mol}^{-1})$$

$CaSiO_{3(c)}$	-1635.22
$CaO_{(c)}$	-635.09
$SiO_{2(\alpha\text{-}qz)}$	-910.70

Calculate the enthalpy of formation of $CaSiO_3$ from the *oxides* at 298.15 K; i.e.,

$$CaO + SiO_2 \rightarrow CaSiO_3$$

(*Ans:* -89.43 kJ·mol^{-1}). Why is this a much smaller heat than the enthalpy of formation from the elements?

6. Zinc melts at 692.66 K, with an enthalpy of fusion of 7.322 kJ·mol^{-1}. Using the following heat-capacity data, calculate the total change in enthalpy that accompanies the heating of 65.38 g of crystalline zinc from 298.15 K to molten zinc at 1000 K. (*Ans:* 27.75 kJ)

$$Zn(c): \quad C_P = (22.38 + 0.0100\,T)\text{ J·mol}^{-1}\text{·K}^{-1}$$

$$Zn(l): \quad C_P = 31.38 \text{ J·mol}^{-1}\text{·K}^{-1} \quad \text{(independent of temperature)}$$

Compare this value with the standard enthalpy of formation of Zn from the elements at 1000 K. (*Ans:* 0).

7. Evaluate the integral $\int_{T_r}^{T} (\Delta C_P / T)\, dT$, assuming that the heat capacities can be represented by the equation $C_P = a + 2bT + cT^{-2} + gT^{-1/2}$ (see Table 4-2). Then calculate the entropy of α-quartz at 844 K, if S_{298} for quartz is 41.46 J·mol^{-1}·K^{-1}. (*Ans:* 103.55 J·mol^{-1}·K^{-1}) Compare this value with the entropy of formation of quartz at 844 K, i.e.,

$$\Delta S_f = S_{SiO_2} - (S_{Si} + S_{O_2})$$

Then calculate ΔG_f° for α-quartz at 844 K.
Use the data given in Table 3-3, along with

$$S_{298(Si(c))} = 18.81 \text{ J·mol}^{-1}\text{·K}^{-1}$$

$$S_{298(O_2(g))} = 205.15 \text{ J·mol}^{-1}\text{·K}^{-1}$$

8. Prove for 1 mol of ideal gas that the internal energy is independent of pressure; i.e., show $(\partial U / \partial P)_T = 0$. Use the definition of Gibbs free energy in your derivation.

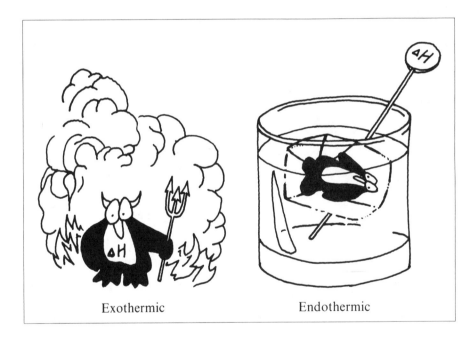

Exothermic   Endothermic

4

PHASE EQUILIBRIA
IN SIMPLE SYSTEMS

Most petrologic and geochemical processes involve transfers or exchanges of matter either between adjacent mineral grains, or between mineral grains and natural fluids with which they may come in contact. A convenient system of bookkeeping is thus essential to any adequate analysis of such natural processes: it must include both algebraic and geometric methods for the representation of the chemical composition of both homogeneous and heterogeneous systems.

J. B. THOMPSON, JR. (1982)

One of the major objectives in geochemistry is to predict equilibrium states of natural systems. These predictions can be formulated in three ways.

1. What substances can coexist at equilibrium under any given conditions of pressure, temperature, and composition? (That is, given a set of conditions, predict the minerals, magmas, or fluids that should be present.)
2. What equilibrium conditions can produce a given assemblage of minerals and fluids? (That is, given a set of minerals, magmas, or fluids, predict the temperature, pressure, or composition that must have been in effect to produce them.)
3. Perhaps most important—is a given mineral assemblage likely to have formed under equilibrium conditions or not?

The first two questions represent opposite ends of the same problem, and both can be answered through the methods of heterogeneous phase equilibria. The third question commonly requires a combination of geological observation and intuition, but an answer is required to tell us whether equilibrium thermodynamics is even applicable.

Assuming that we have answered the last question affirmatively, then the key to understanding this subject lies in the ability to predict how the free energy of a system varies in response to changes in pressure, temperature, and composition. Because this is both a complex and a critically important subject, we have chosen to divide it into two parts. This chapter considers only "simple" systems, defined herein as those systems in which the compositions of all substances present are fixed; thus, the only variables that can affect the free energy of simple systems are pressure, temperature, and the identity and number of moles of the various substances present. The second part of this subject, which considers how compositional variations can affect phase equilibria, appears in Chapters 8 through 10, following an introduction to elementary solution theory.

The tools required to make these predictions are obtained primarily from straightforward manipulation of a very few simple thermodynamic expressions—and from an implicit trust in the infallibility of the first and second laws. Additional restrictions on the state of a system are developed from several graphical and mathematical methods that depend only on the relative compositions of all substances present and that can be used in the absence of any thermodynamic data. Before we begin, it is essential to understand the precise meaning of some important terms that underlie all discussions of phase equilibria.

4-1 DEFINITIONS

Phase

A *phase* is a uniform, homogeneous, physically distinct, and mechanically separable portion of a system. These entities are real substances such as ice, water, pyroxene, halite, and carbon dioxide. Most substances can exist in at least three phases: gas, liquid, and solid form (e.g., ice, liquid water, and water vapor). In many cases, solid compounds of the same composition can exist in more than one crystalline phase, each of which has a characteristic crystal structure. Such compounds are called polymorphs. Examples are the $CaCO_3$ polymorphs calcite, aragonite, and vaterite and the SiO_2 polymorphs α-quartz, β-quartz, cristobalite, tridymite, coesite, and stishovite. Solids may also lack identifiable crystal structure (amorphous substances), but these amorphous forms are still considered phases if they have physically identifiable boundaries or surfaces and chemically identifiable compositions.

Phases may be quite distinct from *species*—both are real entities, but many different species could be present in a single phase. A pertinent example would be ionic species such as H^+, K^+, HCO_3^- or SO_4^{2-} dissolved in water: four species in a single, homogeneous phase. The same line of reasoning could be used to describe molecular species present in a single homogeneous silicate melt or supercritical fluid phase.

Component

A *component* is an algebraic operator. The operation is performed on the *compositions* of phases or species in a system in order to make chemical accounting convenient and internally consistent. Often, the best choice of component is simply the composition of some phase or species in the system. For instance, the single component SiO_2 describes the compositions of all six polymorphs of SiO_2 listed in the preceding paragraph and is an obvious choice for a component in that system. Most minerals, natural aqueous fluids, magmas, and gases have variable compositions and require more than one component to describe them. In nonreacting systems, the number of components is equal simply to the number of phases or species present. For instance, a closed system of hydrogen, oxygen, and water vapor molecules at 1 atm and 25°C is a three-component system so long as the molecules do not interact. Such nonreacting systems will not be considered further here.

If, on the other hand, we assume that chemical equilibrium exists for our system of interest, then we may define components either with respect to a single phase (homogeneous equilibrium) or with respect to more than one phase (heterogeneous equilibrium). Although the specific choice of components is arbitrary, *the minimum number of components in any equilibrium system is rigorously defined by thermodynamics.* A one-component system is called a unary system, a two-component system is a binary system, a three-component system is a ternary system, etc. The minimum number of components cannot be calculated unambiguously without reference to the actual chemical processes taking place in the system. For these reasons, the subject of components is one of the most critical and subtle problems in heterogeneous phase equilibria.

There are two primary reasons for these complexities. First, the definition of components depends ultimately on your scope of interest. For instance, a geochemist whose primary concern is the equilibrium distribution of isotopes between coexisting phases will choose components differently from a geochemist who wishes to study the behavior of trace elements in the same system. A petrologist interested only in major elements will view the identical system in an entirely different light and will arrive at yet another list of the components.

The second complication is that you must know what chemical reactions are possible within the framework of the system you have defined. The answer will not be a unique number because it depends on the level of complexity you wish to consider; the first and second parts of the problem are thus linked. Nonetheless, once you have decided on both the chemical scope of the system and the possible reactions that can occur in the system, the number c of components can best be determined by considering the number N of chemical species or substances (not necessarily the same as the number of phases) and subtracting the number R of *independent* chemical reactions that can occur between those phases or species: $c = N - R$.

For example, if a spark is introduced into the nonreacting gas mixture of H_2 and O_2 molecules described earlier, the large activation energy for the reaction

$$H_{2(g)} + \tfrac{1}{2}O_{2(g)} \rightarrow H_2O_{(g)}$$

is overcome, and equilibrium is attained rapidly (and sometimes explosively). For this reaction, $\Delta G°$ at 25°C is -228.6 kJ per mole of H_2O formed, emphasizing once again the important fact that kinetic barriers may prevent thermodynamically favorable reactions from occurring. Increasing the temperature of the system has the same effect as introducing a spark—at high temperatures, the equilibration of H_2, O_2, and H_2O vapor is virtually instantaneous. The three species are now related by one independent reaction, so we now have a two-component system ($3 - 1 = 2$). This system can also be treated graphically. If a line is drawn between the elemental H and O, then that line represents all possible molecular combinations of H_2 and O_2:

The composition of water vapor obviously falls on this line—two-thirds of the way between O and H, assuming that the line is constructed on an atomic or molal basis rather than a weight basis. This graphical approach leads to an alternative way to find the minimum number of components in a reactive system: it is *the smallest number of chemical entities that are required to describe completely the compositions of all phases or species present.* Unlike phases or species, components are abstract quantities, so they may be defined in any convenient manner. For the water vapor system, we could pick either H and O,

Components	Species
2H =	H_2
2O =	O_2
2H + O =	H_2O

or H_2 and $\tfrac{1}{2}O_2$,

$$H_2 = H_2$$
$$2 \times \tfrac{1}{2}O_2 = O_2$$
$$H_2 + \tfrac{1}{2}O_2 = H_2O$$

or, for that matter, H_2O and O, because

$$H_2O - O = H_2$$

$$2O = O_2$$

$$H_2O = H_2O$$

Note that the species H_2 is defined in terms of negative amounts of the component O. The important point is that the *number* of components in the system as defined is two, but the choice of components is arbitrary, so long as the compositions of all species can be described.

To emphasize the fact that chemical components need not represent compositions of real entities, we note that it is sometimes advantageous to define components containing negative masses of elements. This approach has been advocated by Thompson (1982), who refers to components with negative masses as exchange components. For example, we could write the exchange of Na^+ and K^+ in alkali feldspar as

$$NaAlSi_3O_8 + K^+ = KAlSi_3O_8 + Na^+$$

Alternatively, we define an *exchange operator* KNa_{-1} (where Na_{-1} represents -1 mol of Na ion) and write

$$NaAlSi_3O_8 + KNa_{-1} = KAlSi_3O_8$$

Such operators offer a very compact notation, which may be of considerable benefit in graphical applications. Other examples of exchange operators are HK_{-1} (hydrolysis), F_2O_{-1} (conversion of oxide to fluoride), $MgCa_{-1}$ (dolomitization), and $FeMg_{-1}$ (iron–magnesium exchange). The use of exchange operators has been extensively championed in the literature by D. M. Burt (e.g., 1975, 1981).

Here is an example of different ways to define components. Consider first a monoclinic pyroxene of variable composition $Ca(Mg,Fe^{+2})Si_2O_6$. Because Mg^{+2} and Fe^{+2} substitute for one another, we have two components. In molecular terms, they could be written as $CaMgSi_2O_6$ and $CaFeSi_2O_6$. Another system consisting of orthorhombic pyroxene $(Mg,Fe^{+2})_2Si_2O_6$ also has two components, and for the same reason. They could be defined as $Mg_2Si_2O_6$ and $Fe_2Si_2O_6$. However, if we want to consider how Mg^{+2} and Fe^{+2} distribute themselves in a system containing *both* orthorhombic and monoclinic pyroxene, then there are only three independent components in the two-pyroxene system because of the exchange reaction, which could be written as

$$CaMgSi_2O_6 + Fe_2Si_2O_6 = CaFeSi_2O_6 + Mg_2Si_2O_6$$

We could take the components to be any three of the four compositions shown in this expression, because the fourth composition can always be described in terms of the other three by simple algebraic rearrangement of the exchange equation.

Alternatively, exchange components could be defined to describe the two-pyroxene system. However, the minimum number of components must not change, no matter how their compositions are described. One possible choice in such a system would be $CaMgSi_2O_6$, $Ca_{-1}Fe$, and $FeMg_{-1}$. To prove that these less-conventional components truly describe the compositions of all molecular pyroxene compositions, consider the following equations:

$$CaFeSi_2O_6 = CaMgSi_2O_6 + FeMg_{-1}$$

$$Mg_2Si_2O_6 = CaMgSi_2O_6 + Ca_{-1}Fe - FeMg_{-1}$$

$$Fe_2Si_2O_6 = CaMgSi_2O_6 + Ca_{-1}Fe + FeMg_{-1}$$

It should be clear from this example that alternative choices for exchange components are equally acceptable (e.g., what would the exchange components be if $Mg_2Si_2O_6$ were chosen as the only molecular component?).

In simple petrologic systems, the number of components may often be simply determined by using the graphical approach. If the compositions of all phases plot on any straight line, the system is binary; if all compositions plot in a triangle, the system is ternary; and so on. Many examples are presented in the remainder of this chapter. In very complicated systems, graphical approaches are impractical.

Degrees of freedom (variance)

The number of degrees of freedom in a system is equal to the sum of the independent intensive variables (most commonly temperature and pressure) and the independent concentrations* of all components that must be fixed in order to define uniquely the state of a system. This number may be found by summing the total number of these T, P, and compositional variables in the system and then subtracting the total number of equations relating those variables. Using a mathematical analogy, this is equivalent to comparing the number of algebraic unknowns with the number of independent equations that contain the unknown terms. If the number of equations equals the number of unknowns, the system is uniquely defined because only one solution is possible, and the system is *invariant* (zero degrees of freedom). If there is one more variable than the number of equations, then the system is *univariant* (one degree of freedom). The significance of univariance can be seen by considering variables x and y related by the single equation $xy = 1$. The equation fixes only the product xy, not individual values of either x or y. However, once any single value is chosen for x (or for y)—this is the one degree of freedom—then the value of y (or of x) can be calculated, and the algebraic system is uniquely defined. Univariance is particularly significant because, as we shall see, many important reactions in mineral systems can be classed as univariant.

---

* More correctly, activities or chemical potentials. See Chapter 5.

These ideas can be related to phase equilibria by further consideration of the one-component system H_2O. At 1 atm and 100°C, liquid water and steam are in equilibrium (boiling). The system is invariant because any infinitesimal change in either pressure or temperature will alter the state of the system, resulting in either complete condensation of the steam or vaporization of the liquid water. However, as shown previously, boiling can extend to temperatures above (or below) 100°C, providing that the vapor pressure of water is increased (or decreased) by the exact amount needed to maintain equilibrium between liquid water and water vapor. These specific pairings of vapor pressure and temperature define the trace of the boiling curve for the H_2O system. Taken as a whole, the boiling curve is really univariant because only one variable (P or T) is required to define uniquely the position of the curve at that point. Boiling appears to be invariant at 1 atm only because pressure has been fixed by our environment, so one degree of freedom has already been sacrificed. The melting of ice defines a univariant curve that intersects 1 atm pressure at 0°C. The melting and boiling curves intersect at a unique point (an invariant point) corresponding to 273.16 K and 0.006 bar. The sublimation curve that describes the equilibrium between ice and water vapor also intersects this invariant point, extending downward to lower pressures (Figure 4-1). The figure shows one invariant point ($f = 0$), three

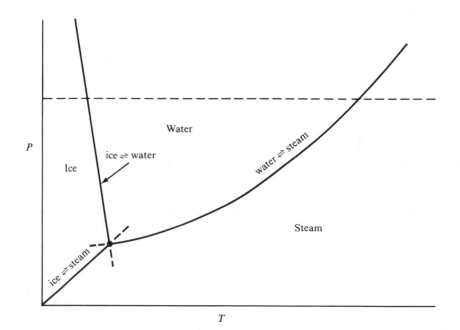

Figure 4-1. A $P-T$ projection for part of the unary system H_2O, showing three univariant lines, three divariant areas, and one invariant point. The 1-atm isobar is shown as a horizontal dashed line.

univariant ($f = 1$) lines, and three divariant ($f = 2$) areas. The areas are divariant because both pressure and temperature can be independently varied within those areas without changing the state of the system.

4-2   GIBBS PHASE RULE

The geometric elements of the H_2O system shown in Figure 4-1 are summarized in Table 4-1. These data can be obtained from purely empirical observations, without recourse to theoretical models.

Note that the sum of the number of phases that can coexist at equilibrium and the variance of the system is constant—if the variance increases, then the number of coexisting phases must decrease by the same amount. For a one-component system,

$$p + f = 3$$

where p is the number of phases, and f is the number of degrees of freedom, or the variance. Intuitively, one might expect that adding more components to the system should increase the constant in the above equation because increased compositional variation introduces more variables into the system. In fact, it can

TABLE 4-1
GEOMETRIC ELEMENTS OF SYSTEM SHOWN IN FIGURE 4-1

Number of coexisting phases	Variance	Number of geometric elements
1	2	3 divariant areas: (ice or water or steam)
2	1	3 univariant lines: ice + water (melting), or ice + steam (sublimation), or water + steam (boiling)
3	0	1 invariant point: (ice + water + steam)

be shown* that each additional component increases the constant by one, so that

$$p + f = c + 2 \quad \text{or} \quad \boxed{f = c - p + 2}$$

where c is the number of components.[†] This equation is the *Gibbs phase rule.*

The Gibbs phase rule may be seen as a logical consequence of evaluating the numbers of components and the thermodynamic variance of the system. For instance, let us return to the equilibrium system of H_2, O_2, and H_2O vapor. The state of the system is fully defined by the pressure and temperature (or any two state functions) and the number of moles of H_2, O_2, and H_2O present, making a total of five variables. In a closed system, the total number of moles is constant, or

$$n_{H_2} + n_{O_2} + n_{H_2O} = \text{Constant}$$

At high temperature, the existence of the equilibrium

$$H_2 + \tfrac{1}{2}O_2 = H_2O$$

adds a second restriction to the system. Thus, the number of degrees of freedom is three $(5 - 2 = 3)$. According to the phase rule, two components must be present:

$$f = c - p + 2$$
$$3 = 2 - 1 + 2$$

in agreement with our previous discussion.

Although the phase rule is elegant in concept, its application to systems with large numbers of components may be a difficult and unrewarding exercise. Fortunately, once the pertinent equilibrium reactions have been identified, calculations of geochemical reactions can continue without specific reliance on the phase rule.

4-3 CHEMOGRAPHY

The phase rule gives information about the maximum number of phases that may be present at invariant points, on univariant reaction boundaries, and in multivariant fields. Another vital aspect of phase equilibria not treated by the phase rule is the relationship between the bulk composition of a system (equivalent to the chemical analysis of the entire system) and the compositions of all possible phases present. Such compositional studies are known as *chemography.* The principles are simple, and the applications are powerful.

---

* A proper derivation of the phase rule requires considerably more thermodynamic functions than we have yet developed. See Section 5-3 for a discussion. For the time being, it is sufficient to accept the algebraic consequences of the phase rule, providing you are willing to wait for proof of its general applicability.

[†] Additional intensive variables such as gravitational force, electromagnetic force, etc. would each increase the degrees of freedom by one—thus for P, T, and E (electromagnetic force), $p + f = c + 3$.

One important aspect of chemographic analysis is that all possible univariant reactions in a system can be derived solely from knowledge of the compositions of the reacting phases and the number of components present. For example, let us look systematically at unary, binary, and ternary systems.

Unary systems

Because no compositional variation is possible in a unary system, chemographic analysis is irrelevant. The only equilibrium reactions involve melting, boiling, sublimation, or polymorphic transformations between two pure phases of identical composition.

Binary systems

The compositions of all phases in a binary system lie on a line connecting the compositions of the two components. Only one compositional variable is independent (i.e., the ratio of the two components), because the total number of moles of substances present in the system is constant. According to the phase rule, every time one component is added to a system, the number of phases present for a given variance must also increase by one. Thus, in a binary system, a univariant reaction involves three phases and a divariant field contains two phases. These conditions can be easily verified for a binary system by rearrangement of the phase rule:

$$p = c - f + 2 = 4 - f$$

The following sketch shows binary system a–b and the composition of three phases 1, 2, and 3, each characterized by a different a/b ratio. Any line connecting the compositions of phases is called a *join*, so this system is the binary join a–b.

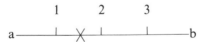

The X represents the bulk composition of the system. This composition could be a mixture of phases 1 and 2 or of 1 and 3, but not of 2 and 3. A mixture of phases is commonly called an *assemblage*. By mass balance, the only possible reaction that can be written between these three phases is $2 = 1 + 3$, because phase 2 can be made by combining phases 1 and 3, but neither phase 1 nor phase 3 can be made out of the other two phases. The reader should be alert to the fact that the equation written in the preceding sentence is a shorthand notation describing which reaction is possible; it has no mathematical significance. This reaction is an example of a *phase-elimination reaction*, so named because phase 2 is unstable relative to a mixture of phases 1 and 3 on one side of the reaction boundary.

In the next figure, the vertical axis shows the relative proportions of phases present as a function of bulk composition. For instance, the extreme ends of the

righthand diagram represent pure 1 at the left and pure 3 at the right. All intermediate compositions are composed of the divariant assemblage 1 + 3, with the proportion of each phase varying continuously as a function of the a/b ratio. For instance, the dashed line marked x is one quarter of the distance between pure 1 and pure 3, and it thus represents a system composed of 75% of phase 1 and 25% of phase 3. On the left side of the reaction, phase 2 becomes stable, and the possible divariant assemblages are 1 + 2 and 2 + 3. The presence of phase 2 has precluded the possibility of equilibrium between phases 1 and 3; the assemblage 1 + 3 cannot exist on this side of the reaction. The lefthand diagram shows that the same x composition is now represented by a 50/50 mixture of phases 1 and 2 under those conditions where phase 2 is stable. Phase 3 can appear only if the bulk composition of the system is richer in the b component than the composition of phase 2 (e.g., the horizontal line labelled y). This example underscores the point that the bulk composition of the system exerts a control on which phases will appear. The other important control is, of course, the state of the intensive variables such as pressure or temperature, which can cause a reaction to tip one way or another. In the present example, the latter control is signified by the fact that phase 2 is missing from the righthand assemblages.

As a specific example, consider the four phases corundum (c, Al_2O_3), diaspore (d, $AlOOH$), gibbsite (g, $Al(OH)_3$), and water (w):

$$\underset{Al_2O_3}{\underset{\rule{0pt}{0pt}}{c}} \qquad\qquad d \quad\quad g \qquad \underset{H_2O}{w}$$

From this chemography, you should see that four univariant reactions are possible:

$$3\,d = c + g$$

$$g = d + w$$

$$2\,g = c + 3\,w$$

$$2\,d = c + w$$

The chemographic analysis only identifies those phases that are distributed on opposite sides of a reaction. The stoichiometric coefficients must be added later to insure mass balance of all components. Also, the chemography reveals nothing about the reaction direction as a function of such variables as pressure and temperature. The righthand and lefthand sides of the reaction carry no significance regarding "product" or "reactant" assemblages.

Ternary systems

By adding a third component, we increase the number of independent compositional variables from one to two. Therefore, two dimensions are required to express reactions in a ternary system, and the appropriate compositional space is a triangle with each of the components occupying one corner.

The number of phases present for each variance will also increase by one relative to the binary system: univariant reactions in ternary systems will involve four phases, and each divariant assemblage will contain three phases. These divariant three-phase assemblages can be represented graphically as triangles lying within ternary compositional space, with the composition of each phase occupying one corner of a triangle. The orientations of the joins that cross compositional space, determining which phases combine to make up the divariant three-phase triangles, depend on which univariant reactions are possible. Thus, our immediate task is to determine how univariant reactions between four coplanar phases can be expressed chemographically. If we add the arbitrary restriction that no more than two phases can appear on the same join (equivalent to restricting all phase points to general positions* in the ternary system), then there are only the two possibilities shown here in Figure 4-2. The numbered points refer to the compositions of phases. Figure 4-2(a) shows a triangle with a single interior point, and Figure 4-2(b) shows a quadrilateral. Both figures are drawn within the confines of ternary system abc. Neither the interior triangle nor the quadrilateral need be regular, and either can assume any orientation within the ternary system.

These two distinct chemographies represent two different reaction types, and they are, in fact, the only generalized reactions possible in ternary systems.

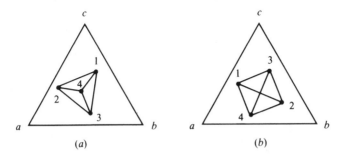

Figure 4-2. Diagrammatic representations of possible univariant reactions for the ternary system a–b–c. (a) The phase-elimination reaction $4 = 1 + 2 + 3$. (b) The join-crossing reaction $1 + 2 = 3 + 4$.

---

* A general position is a composition that can be expressed only by using all components present in the system.

Figure 4-2(a) represents a ternary example of a phase-elimination reaction:

$$4 = 1 + 2 + 3$$

In graphical terms, this reaction is shown by

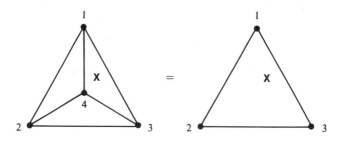

The righthand triangle shows that only the three-phase assemblage $1 + 2 + 3$ is stable over the complete range of bulk composition. The proportions of these phases will, once again, depend on the bulk composition: the point shown by X in the figure represents 60% phase 1, 10% phase 2, and 30% phase 3. However, when the reaction proceeds to the left, phase 4 appears in the center of the triangle. Now it is possible to divide the chemographic space into three triangles representing the three divariant assemblages that each contain phase 4; these are $1 + 2 + 4$, $1 + 3 + 4$, and $2 + 3 + 4$. Composition X now happens to fall within triangle 134, so the bulk composition would be represented under these conditions by a mixture of phases 1, 3, and 4. Once again, this is a phase-elimination reaction because one of the phases (4 in this case) is not stable for *any* bulk composition on one side of the equilibrium.

In contrast, all four phases shown in Figure 4-2(b) are stable on both sides of the reaction, but the opposite sides of the equilibrium are distinguished by different combinations of three-phase divariant triangles. These different combinations come about by rearranging existing joins between the phases, and the reaction type is known as a *join-crossing reaction*. It is written

$$1 + 2 = 3 + 4$$

and looks like this:

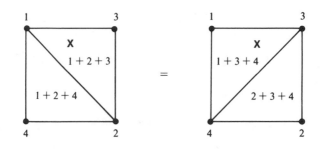

Join crossings are always characterized by a rearrangement of existing joins, never involving the loss or introduction of a phase. Bulk composition X is represented by assemblage 1 + 2 + 3 on one side and by 1 + 3 + 4 on the other. On crossing the reaction boundary from left to right, one might observe for bulk composition X the disappearance of phase 2 and growth of phase 4. Depending on the precise composition of X, phases 1 and 3 would probably change their relative proportions but would continue to coexist stably on both sides of the boundary.

Real systems invariably introduce complications. For instance, if we add SiO_2 to the simple $Al_2O_3-H_2O$ binary system considered earlier, we obtain the ternary system shown in Figure 4-3. In addition to corundum (c), diaspore (d), gibbsite (g), and water (w) that we considered previously, we find two additional polymorphs of $Al(OH)_3$ (bayerite, by, and nordstrandite, n) and one polymorph of $AlO(OH)$ (boehmite, bo) on the $Al_2O_3-H_2O$ join. Adding SiO_2 to the system gives the three binary compounds andalusite, kyanite, and sillimanite (a, k, and s, all polymorphs of Al_2SiO_5), and the two ternary compounds pyrophyllite (p, $Al_2Si_4O_{10}(OH)_2$) and kaolinite (ka, $Al_2Si_2O_5(OH)_4$) with polymorphs halloysite (ha), dickite (di), and nacrite (na), in addition to six polymorphs of SiO_2

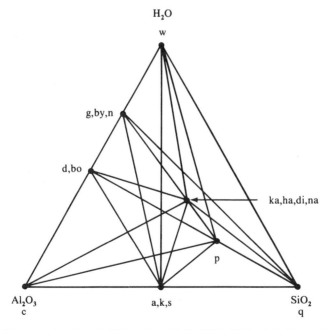

Figure 4-3. Chemography of part of the system $Al_2O_3-SiO_2-H_2O$, plotted in mole percentage. Compositions of phases are given in the text. (w = water; c = corundum; q = quartz; p = pyrophyllite; d,bo = diaspore and boehmite; g,by,n = gibbsite, bayerite, and nordstrandite; a,k,s = andalusite, kyanite, and sillimanite; ka,ha,di,na = kaolinite, halloysite, dickite, and nacrite).

(only quartz, q, is shown). Mullite is another phase that is found on the Al_2O_3–SiO_2 join, but it is not plotted in Figure 4-3 because it can vary widely in Al_2O_3/SiO_2 ratio. This ternary system contains a large number of phases but, according to the phase rule, only four phases at one time can participate in a univariant reaction, and only three phases can be present in a divariant assemblage. The figure also contains several examples of phase-elimination and join-crossing reactions. For instance, center your attention on the four phases kaolinite, quartz, diaspore, and water. Because the composition of kaolinite falls in the center of the triangle composed of quartz, diaspore, and water, we have the phase-elimination reaction

$$ka = d + q + w$$

Alternatively, we see that the phases diaspore, kaolinite, pyrophyllite, and andalusite are located at the corners of a quadrilateral, which implies the existence of the join-crossing reaction

$$ka + a = d + p$$

This brand of chemographic analysis serves only to identify the reactions that are possible from the point of view of mass conservation; it is not capable of providing information about whether the reaction is actually possible in a thermodynamic sense. The latter information requires either thermodynamic data or careful observations of mineral assemblages.

A further complication arises from the fact that most of the phases in Figure 4-3 do not occupy general positions in the ternary system because of the occurrence of the binary join Al_2O_3–H_2O (containing the phases c,d,bo,g,by,n,w), Al_2O_3–SiO_2 (phases c,a,k,s,q), diaspore–SiO_2 (with phases d,bo,p,q), and gibbsite–pyrophyllite (including phases g,by,n,k,ha,di,na,p). These special (i.e., nonternary) compositions result in univariant reactions that involve fewer than the maximum number of phases permitted by the phase rule for three-component systems. Such reactions are termed *degenerate*.

For instance, consider the four phases g, ka, p, and w as shown here.

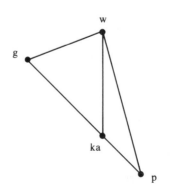

The only reaction possible is the binary phase elimination ka = g + p, because no univariant reaction can be written that involves water and all of the other three phases (you're welcome to try, but don't spend too long). The reaction ka = g + p is said to be degenerate in the ternary system Al_2O_3–SiO_2–H_2O (because the univariant reaction involves three instead of four phases) but is not considered degenerate in the context of the two-component gibbsite–pyrophyllite system. Degeneracy is always a function of how large a system you are willing to consider.

An even more advanced case of degeneracy is shown by the occurrence of the polymorphs of $Al_2Si_2O_5(OH)_4$, Al_2SiO_5, $Al(OH)_3$, and $AlO(OH)$ in Figure 4-3. Reactions including more than one polymorph cannot involve any other phase in the ternary system. For example, compositional points ka = ha = di = na, a = k = s, g = by = n, and d = bo are four unary systems embedded within two binary joins.

Higher-order systems

As the number of components becomes greater than three, the graphical approach to chemographic analysis becomes formidable because of the limitations of living in a three-dimensional world. The most complicated case that we can picture in its entirety is the quaternary ($c = 4$) system. The chemographic figure for a quaternary system is a tetrahedron with one component occupying each of the four corners. Because five phases are required for a quaternary univariant reaction, our two reaction types are now represented in three dimensions as sketched in Figure 4-4. The phase-elimination reaction 5 = 1 + 2 + 3 + 4 is shown as a tetrahedron with an interior point (Figure 4-4a). The join-cross reaction 1 + 2 = 3 + 4 + 5 is represented by two tetrahedra joined on a common face (shown in exploded form in Figure 4-4b).

For $c > 4$, visualization is no longer possible, and analytical methods must be used to discover possible reactions. Although exactly the same principles

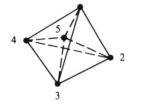

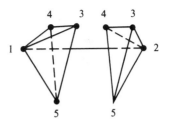

(a) Phase elimination 5 = 1 + 2 + 3 + 4   (b) Join cross (exploded) 1 + 2 = 3 + 4 + 5
(tetrahedron with interior point)   (two tetrahedra joined on a common face)

Figure 4-4. Univariant reactions in a quaternary system. (a) The phase-elimination reaction 5 = 1 + 2 + 3 + 4. (b) The join-cross reaction 1 + 2 = 3 + 4 + 5 (this diagram is shown in exploded form for clarity).

apply, the possible combinations for most systems are many and not always obvious. Linear algebra techniques adapted to computer programs may be needed to identify all possible reactions in systems with large numbers of components.

4-4 THE COMBINATORIAL FORMULA

Systems that contain more phases than can coexist at a single invariant point are called multisystems; the $Al_2O_3–SiO_2–H_2O$ system discussed previously is a good case in point. The phases in multisystems can be combined in various ways to make a number of invariant points, univariant lines, and divariant fields, although only a small number of these can be thermodynamically stable at one time in $P–T$ space. Nonetheless, it often is useful to be able to enumerate all possible phase elements in a system; in fact, this is commonly the first step in chemographic analysis. This can be done by using the *combinatorial formula* that states the number w of ways that k objects can be taken m at a time:

$$w = \frac{k!}{m!(k-m)!}$$

In chemographic analysis, we are interested in the number of ways that p phases can be taken m at a time, where m here stands for the number of phases that can coexist for invariant, univariant, and divariant equilibrium (e.g., $m = 3$ for a univariant line in a binary system, and $m = 5$ for an invariant point in a ternary system). Thus, we have

$$w = \frac{p!}{m!(p-m)!}$$

For example, if you want to find the number of univariant lines in a five-phase binary system, then substitute $p = 5$ and $m = 3$; you find that there are ten possible univariant reactions. At this point you should stop and convince yourself that there would also be five invariant points and ten divariant fields.

As you may well imagine, the combinatorial formula gives only an accounting of the number of different phase elements present; it says nothing about their orientations or relative stabilities. Questions of stability can be resolved only through consideration of free energy functions.

4-5 FREE ENERGY SURFACES

Because the Gibbs free energy $G(P,T)$ of a phase depends on temperature and pressure, the total differential of G written for a phase of fixed composition is

$$dG = \left(\frac{\partial G}{\partial P}\right)_T dP + \left(\frac{\partial G}{\partial T}\right) dT \tag{4-1}$$

By comparison with (3-71), the partial derivatives of G with respect to pressure and temperature are

$$\left(\frac{\partial G}{\partial P}\right)_T = V \tag{4-2}$$

$$\left(\frac{\partial G}{\partial T}\right)_P = -S \tag{4-3}$$

These equations allow the construction of surfaces in Gibbs free energy–pressure–temperature (G–P–T) space. Each surface may represent the free energy of either a single phase or (because free energies are additive) of an assemblage of phases. Because both the volume and entropy of phases or assemblages are finite positive quantities, equations (4-2) and (4-3) restrict the slopes of these surfaces to be always positive in isothermal section and always negative in isobaric section.

To get a feeling of how this works, consider the polymorphic inversion of kyanite (k) to sillimanite (s) in the one-component system Al_2SiO_5. Figure 4-5 shows the intersection of k and s free energy surfaces. In an attempt to maximize clarity, we have not drawn the figure to scale; it should be regarded as a cartoon. Also, you must imagine that both surfaces extend infinitely in all directions—they are shown with edges in the figure only for convenience. First, note the intersection of the k and s surfaces. Everywhere along this line of intersection, the free energies of kyanite and sillimanite are equal ($\Delta G = 0$), and kyanite and sillimanite must be in equilibrium. If this intersection is projected directly down onto the P–T plane, the resulting line defines the trace of the univariant curve k = s.

On the high-temperature side of the univariant curve, the s surface lies everywhere below the k surface—on the low-temperature side, the situation is reversed. Assuming that the cartoon is oriented such that G increases upward, then the difference in free energy calculated at any P,T point for the reaction k → s will be negative on the high-temperature side and positive on the low-temperature side. Because a positive sign for ΔG means that the reaction can proceed spontaneously only if written in reverse, (i.e., s → k), then the intersection of the two free energy surfaces also signifies a switching of two assemblages from relatively less stable to relatively more stable states. This switching of stabilities is the most critical feature of intersecting free energy surfaces. You should carefully study Figure 4-5 until you understand this principle.

We can make our model more interesting by adding a third surface representing the free energy of andalusite (a), the third Al_2SiO_5 polymorph. Figure 4-6 is a perspective drawing showing the intersection of the k and s surfaces with the a surface. Imagine that the plane of the page is parallel to the P–T plane, and that the G axis is normal to the page and increases away from you; you are looking directly into the surfaces having the lowest free energy. The intersections of the k, a, and s surfaces define three univariant lines, and their mutual intersection (where $G_k = G_s = G_a$) defines the system invariant point.

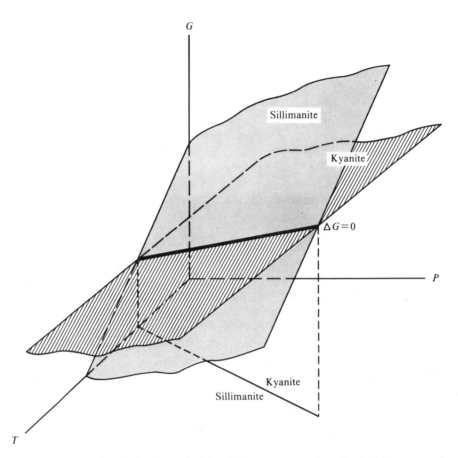

Figure 4-5. Intersection of kyanite and sillimanite free energy surfaces in $G-P-T$ space, and projection of the intersection onto the $P-T$ plane, creating the univariant line for kyanite = sillimanite equilibrium.

These intersections are projected onto the $P-T$ surface and, together with the relative positions of the k, a, and s surfaces, define the phase diagram for the Al_2SiO_5 system. Note, however, that the intersections between the surfaces continue indefinitely, extending behind the k, a, and s planes toward regions of higher free energy. These intersections are drawn with dashed lines in $P-T$ projection and are the *metastable extensions* of the univariant lines.

In geochemistry, metastable reactions and metastable equilibria are commonly as important as (or even more important than) stable equilibria. For this reason, it is important to investigate the phase equilibria that exist at relatively higher free energies. Consider schematic isobaric $G-T$ sections through the Al_2SiO_5 model. The intersections of each phase surface with isobaric sections corresponding to pressures both above and below the invariant point are shown in Figure 4-7(a,b). The arrow in Figure 4-7(a) corresponds to the arrow drawn in

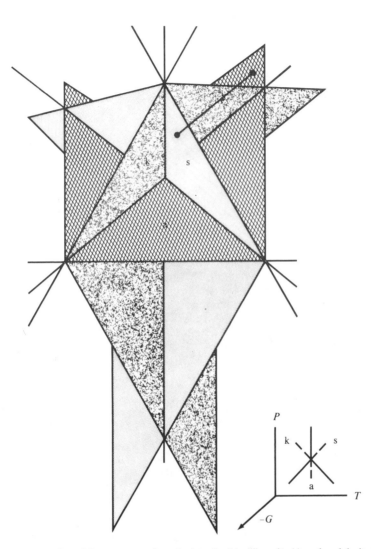

Figure 4-6. Intersection of free energy surfaces for kyanite (k), sillimanite (s), and andalusite (a). The $P-T$ plane is parallel to the page, and free energy increases away from the observer. The arrow shows a progression of surfaces that increase in G from s to k to a at constant temperature and pressure.

the perspective figure (Figure 4-6) and relates the sections to the three-dimensional model. Figure 4-7(c) shows how the $P-T$ projection for the Al_2SiO_5 system can be derived from the isobaric $G-T$ sections. Restricting our attention to section $P = P_1$, we note that T_2 is the temperature of intersection for the univariant line k = s, and that T_3 and T_1 are the temperatures of intersection for the metastable extensions of univariant lines a = k and a = s. Note that these metastable extensions lie at free energies greater than those for phases k and s, respectively. These three temperatures divide $G-T$ space into four sections

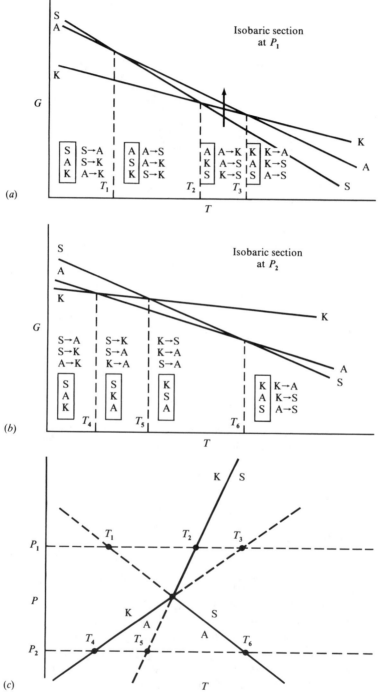

(a)

Isobaric section
at P_1

S S→A	A A→S	A A→K	K K→A
A S→K	S A→K	K A→S	A K→S
K A→K	K S→K	S K→S	S A→S
T_1	T_2	T_3	

(b)

Isobaric section
at P_2

S→A   S→K   K→S
S→K   S→A   K→A
A→K   K→A   S→A

S	S	K	K K→A
A	K	S	A K→S
K	A	A	S A→S
T_4	T_5	T_6	

(c)

characterized by four different sequences of relative stability. As labeled within rectangular boxes, the phases increase in relative stability from top to bottom, as can be read directly from the vertical sequence of the phase surfaces within each of the four sections. All possible spontaneous reactions (i.e., all those reactions having negative free energies) that can occur within the segment are also shown. Three features of these diagrams deserve special mention.

1. Within each section, it is possible to write reactions that result in the formation of phases different from the assemblage having the lowest free energy—e.g., for all temperatures below T_1, sillimanite can react to form andalusite within the kyanite field. As the number of components increases, the potential for spontaneous formation of metastable assemblages increases concomitantly. This possibility causes problems for experimentalists.

2. In terms of the relative stabilities of the two assemblages involved, metastable extensions are no different from stable univariant lines—i.e., metastable equilibrium is not "second-class" equilibrium. The only difference is that the equilibrium represented by a metastable extension always exists at a level of stability higher than that of some other phase or assemblage of phases.

3. Every segment of $G–T$ space can be arranged into different sets of three phases, each characterized by a different order of stability. Therefore, $P–T$ space can be broken down to three different stability levels. Based on the information in Figure 4-7(a,b), Figure 4-8 shows the relative stabilities of k, a, and s in each of six sectors around the invariant point. These can be redrawn as three different equilibrium phase diagrams differing only in their relative stabilities. In Figure 4-9, part(a) shows the most-stable and part(c) the least-stable arrangement of Al_2SiO_5 polymorphs in $P–T$ space.

Higher stability levels can be meaningfully studied both in the laboratory and in the field. For example, it is possible to study the reaction k = a on its metastable extension, well within the sillimanite field (shown as the shaded area in Figure 4-9(b); Richardson et al., 1969). This is possible because reaction rates in that system are slow. Metastable reactions between kyanite and sillimanite can be observed just so long as the rates of reaction between kyanite and andalusite are

---

Figure 4-7. Section through the $G–P–T$ model for the Al_2SiO_5 system of Figure 4-5 and 4-6. (A = andalusite; K = kyanite; S = sillimanite) (*a*) A $G–T$ section at $P = P_1$ (above the invariant point). The vertical arrow between T_2 and T_3 corresponds to the arrow shown in the three-dimensional model of Figure 4-6. The boxes between the isotherms show the relative free energy rankings of the three phases in that temperature range (the lowest is the most stable); the reactions shown in each range correspond to all possible spontaneous reactions for that combination of P and T. (*b*) A $G–T$ section at $P = P_2$ (below the invariant point). (*c*) Isothermal sections P_1 and P_2, showing intersections with univariant curves that generate the isotherms of parts (*a*) and (*b*).

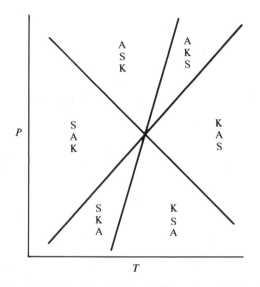

Figure 4-8. Intersections of univariant lines in the Al_2SiO_5 system, showing the relative stability rankings of three polymorphs (bottom is most stable) within each sector. These data were compiled from the two isobaric sections in Figure 4-7(a) and (b).

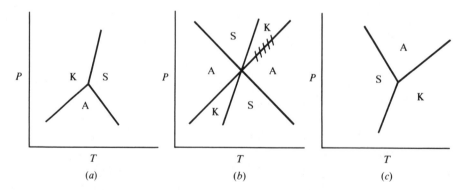

Figure 4-9. Three stability levels for the Al_2SiO_5 system. (a) The most stable level. (b) An intermediate level. The hatched line is discussed in the text. (c) The least stable level.

faster than the tendency for either phase to invert to the more stable phase sillimanite.

It is possible to add more surfaces (phases) to this unary model, but the phase rule requires that no more than three phases intersect at a point. The resulting geometries can be fairly complicated, and the number of possible stability sequences depends on the orientation of the model in $G-P-T$ space (Kujawa and

Eugster, 1966). Strictly geometric representations of stability sequences soon become impractical when $c > 1$ because the increasing number of independent compositional variables soon renders impossible three-dimensional mapping of any meaningful combinations of composition, pressure, temperature, and free energy.

4-6 MOREY–SCHREINEMAKERS RULES

In Section 4-5, we made considerable use of the one-component system Al_2SiO_5 to explain the concepts of metastable equilibrium and intersecting free energy surfaces. These same principles apply to multicomponent systems as well, because a free energy surface can represent the aggregate free energies of a number of phases as well as that of a single phase. Every univariant reaction can be expressed as the intersection of two free energy surfaces—one representing the product assemblage, and the other representing the reactant assemblage. This leads directly to the ability to make a number of powerful predictions regarding the placement of univariant curves around invariant points. These predictive tools fall under the general heading of Morey–Schreinemakers (M–S) rules, and they require no data other than knowledge of the compositions of the invariant phases and of the number of components present in the system.

The fundamental postulate on which all M–S techniques are based is already familiar to you:

A univariant line separates product and reactant divariant assemblages. On one side of the univariant line, one assemblage is relatively less metastable than the other assemblage; on the opposite side of the univariant line, the relative stabilities of the two assemblages are interchanged.

The proof of this rule falls directly from the geometry of intersecting free energy surfaces, as discussed in Section 4-5. To understand how this rule leads to prediction of the placement of univariant lines, we need to explore further some of the characteristics of invariant assemblages.

The phase rule requires that a c-component system always have $c + 2$ univariant lines radiating from each invariant point, similar to the spokes of a wheel radiating from a central hub. In phase-equilibrium jargon, this assemblage of radiating univariant lines is commonly called a *bundle*. Also, every nondegenerate univariant reaction involves exactly one less phase than the number present at the invariant point. This observation leads to an especially convenient way to identify univariant reactions in a specific bundle—such reactions are labeled according to the phase that is *missing* (i.e., *not* involved in the reaction). Thus, every univariant reaction can be unambiguously identified by a single phase label, regardless of the number of phases taking part. The same principle can be applied to the listing of phases stable within a divariant field—although, because one degree of freedom is gained, one additional phase must be used in the label. For instance, within the divariant field enclosed between univariant lines (P_1) and

(P_2), all invariant phases other than p_1 and p_2 can be in equilibrium. After a little practice, you will soon get accustomed to "missing-phase" nomenclature, although it may seem a little confusing at first.

Let us return once more to the Al_2SiO_5 system and list the univariant reactions according to this nomenclature. The missing phase in each case is customarily enclosed in parentheses; the reactions are

$$(A)\qquad k = s$$
$$(S)\qquad k = a$$
$$(K)\qquad a = s$$

The three univariant lines divide $P-T$ space into three angular sectors that represent all possible divariant assemblages that can exist at the lowest stability level. Using the missing-phase nomenclature, divariant field k (= kyanite present) must be located between the two univariant lines (A) and (S), and so on for the other two divariant fields. With this background, we can now state the most far-reaching of Morey–Schreinemakers rules:

No divariant assemblage can be stable within a sector that makes an angle of more than 180° measured between any two univariant lines in the same bundle.

When translated to the Al_2SiO_5 system, this rule means that we can have

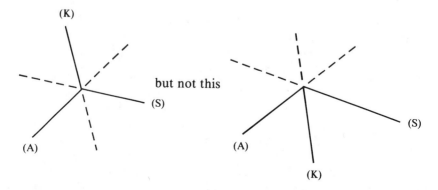

How does this rule relate to the fundamental postulate of intersecting free energy surfaces? Consider the (K) and (S) univariant lines in the Al_2SiO_5 system as shown here. We are confronted with two geometric possibilities for the placement of divariant assemblage (K,S) (i.e., andalusite stable). Based on the fundamental Morey–Schreinemakers rule, we can be rapidly convinced that the segment occupying the angular sector less than 180° is the only thermodynamic possibility. The alternative choice would require, starting counterclockwise from (K), that the metastable extension of (S) be crossed somewhere within the greater-than-180° segment. Because (S) represents the intersection of kyanite and andalusite free energy surfaces, such a crossing would require that the relative

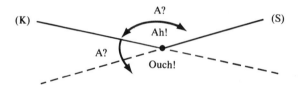

stabilities of kyanite and andalusite be interchanged—a logical impossibility if we require andalusite to be stable throughout the entire divariant segment. No assemblage can be both stable and metastable at the same time!

Note that the same impossible situation would have arisen if we considered a clockwise rotation from the stable end of univariant line (S): such a path would first cross the metastable extension of univariant line (K), which also would require a reversal in the relative stability of andalusite. Could any divariant field extend across a univariant line? Of course, just as long as such a crossing does not require any change in the stability of the divariant assemblage. In the present example, the divariant assemblage andalusite can extend only across the univariant line (A), because that univariant reaction (i.e., kyanite = sillimanite) does not affect the stability of andalusite. In like fashion, divariant assemblage kyanite extends only across univariant line (K), and divariant assemblage sillimanite extends only across univariant line (S). Thus, the only possible solutions are these. These two figures are enantiamorphous: Morey–Schreinemakers analysis cannot distinguish between mirror images.

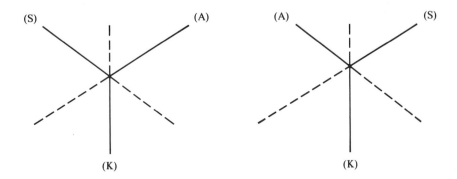

From the preceding arguments, the correct placement of divariant assemblages can be done by adherence to a simple corollary of M–S rules known as the overlap rule:

Any divariant assemblage that crosses univariant line (P) contains the phase p.

The divariant assemblage must, of course, adhere to the 180° rule, and the univariant line may be either stable or metastable. A further example of these rules will be shown for a binary system.

A binary system has four univariant curves radiating from the four-phase invariant point. Assuming that individual lines cannot be distinguished, the only geometric possibilities are those shown in part(*a*) of the following figure. The possibility shown in part(*b*)—a sequence of four stable ends and four metastable ends—is clearly incorrect because it violates the basic tenet of M–S rules: a single divariant area extends over 180°. Because the first possibility must be the right one, it remains only to correctly label all four univariant lines.

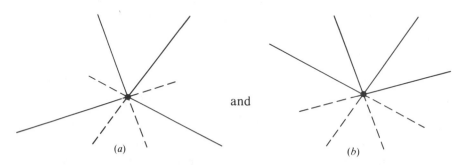

(*a*)   and   (*b*)

Consider a general case of a four-phase binary system a–b with chemography

and univariant lines

$$\text{(1)} \quad 3 = 2 + 4$$
$$\text{(2)} \quad 3 = 1 + 4$$
$$\text{(3)} \quad 2 = 1 + 4$$
$$\text{(4)} \quad 2 = 1 + 3$$

The only possibilities that satisfy M–S rules for the placement of divariant assemblages in binary systems are shown here; they differ only in that one is the

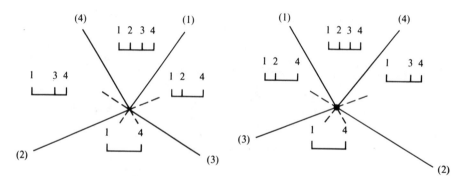

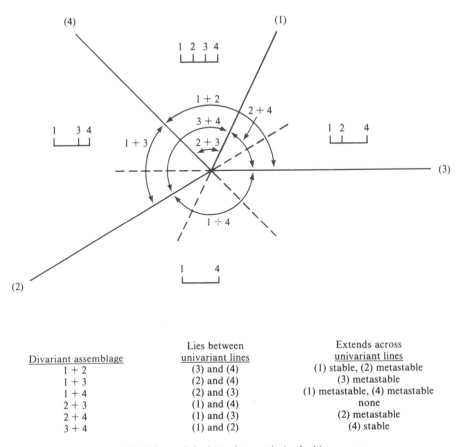

Divariant assemblage	Lies between univariant lines	Extends across univariant lines
1 + 2	(3) and (4)	(1) stable, (2) metastable
1 + 3	(2) and (4)	(3) metastable
1 + 4	(2) and (3)	(1) metastable, (4) metastable
2 + 3	(1) and (4)	none
2 + 4	(1) and (3)	(2) metastable
3 + 4	(1) and (2)	(4) stable

Figure 4-10. Morey–Schreinemakers analysis of a binary system.

mirror image of the other. Figure 4-10 shows the angular extent of each of the six divariant fields. Note that every divariant field is stable in a sector less than 180° and that the overlap rule is obeyed for every divariant assemblage. Because there is only one way to arrange the compositions of four nondegenerate phases on a binary join, *every* nondegenerate binary system will have the sequence of labeled univariant lines shown in Figure 4-10. The thermodynamic properties of the system will determine the slopes of the lines and will resolve the mirror-image ambiguity by distinguishing products from reactants. Sometimes it is possible to resolve the mirror-image problem by inspection. The best example of this possibility is a reaction that releases a single volatile phase (e.g., dehydration or decarbonation); the volatile (H_2O or CO_2) phase is almost always part of the high-temperature, low-pressure assemblage (see Problem 6).

One of the important applications of M–S rules is to the analysis of phase diagrams that are incomplete—in such cases, it is easy to predict from chemography alone the general location (within a segment) of one or more univariant lines that may be missing from a bundle because no data were available to calculate their locations. Conversely, "impossible" phase diagrams (i.e., those drawn in violation of M–S rules) may result from using either inconsistent thermodynamic data or results of incorrectly interpreted experiments. Such diagrams can be identified immediately by inspection and sent back to the laboratory for repair. In more complicated systems, a number of shortcuts—all variations on the same theme—are available to aid in a Morey–Schreinemakers analysis. Degeneracy is very common in geochemical systems and introduces some complications. E-an Zen's comprehensive study of the subject is required reading for any student who intends to use M–S techniques (Zen, 1966).

4-7 FIRST-ORDER AND SECOND-ORDER REACTIONS IN SOLIDS

Having been introduced to free energy surfaces and Morey–Schreinemakers rules, you may expect the actual calculation of univariant lines and invariant points for pure mineral assemblages to be rather straightforward, but this is not always true. Even apparently straightforward reactions between minerals of simple composition may be poorly understood in detail. This depressing state of affairs is partly due to complexities inherent in the crystal structures of many common rock-forming minerals—and to the effect those complexities have on the thermodynamic properties of the crystalline phases. Additional complicating factors are that reaction rates for many phase transitions involving minerals are agonizingly slow and that both the product and reactant phases may be incompletely characterized with respect to both chemical composition and crystal structure.

All preceding discussions relating to the intersection of free energy surfaces apply only to those reactions known as *first order*. First-order transitions always involve breaking and reforming of chemical bonds—all reactions involving transformation of a solid phase to a liquid or gas phase are first order. Reconstructive polymorphic transformations—such as the inversion of kyanite to sillimanite—are good examples of first-order reactions involving solids only. However, transformations in crystals may also occur in more subtle ways—for instance, by the stretching or rotation of chemical bonds as opposed to the breaking of them, or by the ordering of cations into crystallographically distinct sites. Certain of these processes are called *second order*.

The distinction between first-order and second-order transformations is traditionally based on the order of the derivative of the Gibbs free energy—for first-order transformations, we have

$$\left(\frac{\partial G}{\partial P}\right)_T = V \quad \text{and} \quad \left(\frac{\partial G}{\partial T}\right)_P = -S$$

whereas second-order transformations obey the equations

$$\left(\frac{\partial^2 G}{\partial P^2}\right)_T = -V\beta \quad \text{and} \quad \left(\frac{\partial^2 G}{\partial T^2}\right)_P = -C_P/T$$

The effects of these differences on the thermodynamic properties of the

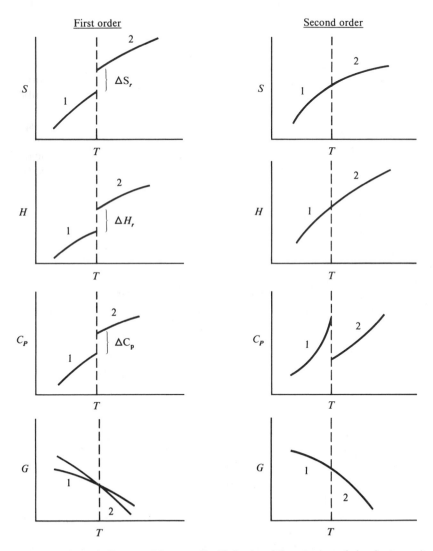

Figure 4-11. Schematic diagram of the generalized behavior of the entropy, enthalpy, heat capacity, and Gibbs free energy of first-order and second-order reactions. The dashed lines represent the temperatures of the reaction $1 \rightarrow 2$ for the respective first-order and second-order cases. (Adapted from Thompson and Perkins, 1981, p. 39.)

transformation are shown in Figure 4-11. At the transition temperature of first-order reactions, H, S, V, and C_P are discontinuous, and ΔH, ΔS, and ΔV are finite. In contrast, H, S, and V are continuous at the transition temperature of second-order reactions, but the slopes are discontinuous. This behavior causes the second order G–T and G–P curves to be continuous functions. Thus, second-order reactions that intersect invariant points do not have metastable extensions.

A special type of second-order behavior is known as the *lambda transition*. Lambda transitions are recognized by the shape of the heat capacity–temperature curve, which rises rapidly at temperatures approaching the inversion temperature, resembling the Greek letter λ. Most common lambda transitions in minerals are either order–disorder transformations (the temperature of the transition corresponds to complete disorder), or magnetic-spin alignments in minerals containing transition metals. Careful measurements of heat capacities are the best way to detect these transitions, providing that the rate of the transformation exceeds the rate of the measurement. Unfortunately, this assumption is rarely valid for order–disorder transformations in minerals, and heat-capacity measurements in such cases must be made on phases with varying metastable degrees of cation ordering. In contrast, reaction rates for magnetic transformations are rapid, and the characteristic lambda shape of the heat-capacity curve is evident. Figure 4-12 shows low-temperature heat-capacity measurements for fayalite, which has a lambda transition at 64.88 K. This transition is an important one to measure because it adds significantly to the third-law entropy of fayalite.

An especially complex situation arises when both first-order and second-order effects are present simultaneously. The very common polymorphic inversion between α-quartz and β-quartz is a case in point—this reaction has been extensively studied but is still incompletely understood. Much uncertainty still exists regarding values for the enthalpy and entropy of inversion and for the relative contributions of first-order and second-order energetics to the inversion. The brief review by Helgeson et al. (1978, pp. 81–85) is particularly informative.

A deeper consideration of these complexities is outside the scope of this text—the review article by Thompson and Perkins (1981) is recommended for further study. As we continue to discuss the techniques for the calculation of univariant lines, we urge you to keep in mind that, in real mineral systems, second-order effects may be significant contributors to the free energy of the reaction. Correct evaluation of these effects requires complete knowledge of all the structural and chemical properties of the phase you are dealing with. In the case of an alkali feldspar, for instance, this means knowing both the K/Na ratio and the exact degree of Si/Al disorder, combined with the insight to choose entropy and enthalpy values appropriate to these properties. Ignoring such effects may lead to significant errors.

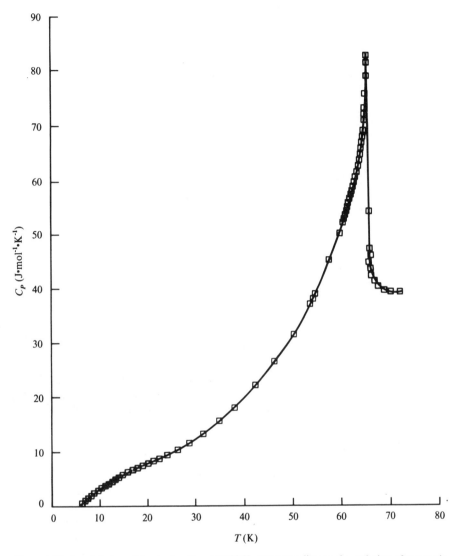

Figure 4-12. Lambda transition in fayalite at 64.88 K, corresponding to the ordering of magnetic moments of Fe^{2+} ions into a colinear antiferromagnetic arrangement. (Robie et al., 1982)

4-8 CALCULATION OF UNIVARIANT LINES IN SIMPLE SYSTEMS

In order to calculate univariant lines, we must find the free energy of a reaction as a function of both pressure and temperature and then set the free energy of the reaction equal to zero. With this in mind, let us recast (4-2) and (4-3) in terms of

ΔG for a reaction:

$$\left(\frac{\partial \Delta G_r}{\partial P}\right)_T = \Delta V_r \tag{4-4}$$

$$\left(\frac{\partial \Delta G_r}{\partial T}\right)_P = -\Delta S_r \tag{4-5}$$

The variation of ΔG_r with pressure and temperature can be found by solving the integrals

$$\int_{P_{ref}}^{P} \Delta G\, dP = \Delta G_{(P)} - \Delta G_{(P_{ref})} = \int_{P_{ref}}^{P} \Delta V_r\, dP \tag{4-6}$$

and

$$\int_{T_{ref}}^{T} \Delta G\, dT = \Delta G_{(T)} - \Delta G_{(T_{ref})} = -\int_{T_{ref}}^{T} \Delta S_r\, dT \tag{4-7}$$

or

$$\Delta G_{(T,P)} = \Delta G_{(T_{ref}, P_{ref})} + \int_{P_{ref}}^{P} \Delta V\, dP - \int_{T_{ref}}^{T} \Delta S_r\, dT \tag{4-8}$$

where $\Delta G_{(T,P)}$ is the free energy change for the reaction at the P and T of interest, and $\Delta G_{(T_{ref}, P_{ref})}$ is the free energy change for the reaction at some convenient reference temperature and pressure, commonly taken to be 298.15 K and 1 bar.

Solving the pressure integral

The pressure integral can be solved exactly only if the combined effects of temperature and pressure on the volume change of the reaction are known. If the reaction involves liquids or gases, this effect is very significant. However, as we shall see in Chapter 8, the volume terms for such easily compressible phases are handled separately from the volume terms for the relatively incompressible solid phases in cases like this. For now, we limit our consideration to cases that involve solids only.

The effects of pressure and temperature on volume are given by the compressibility β and thermal expansion α, respectively. For many minerals, the compressibility can be calculated from the polynomial

$$\frac{V_0 - V}{V_0} = aP - bP^2$$

where V is the volume at the pressure P of interest, V_0 is the volume at reference pressure (usually 1 bar), and a and b are constant. When this relation holds, we have

$$
\int_1^P \Delta V\, dP = \Delta V_0 (P - 1) - \Delta V_0 \left[\frac{\Delta a}{2} (P^2 - 1) - \frac{\Delta b}{3} (P^3 - 1) \right]
$$

$$
+ P \left(\int_{298}^T \alpha(T)\, dT\, \Delta V_{298,P} \right) \tag{4-9}
$$

The physical significance of this integral can be viewed as a two-step process: first the assemblage is compressed from 1 bar to P at 298.15 K, and then the assemblage is heated from 298.15 K to T at pressure P. Because crystalline phases decrease in volume when compressed and increase in volume when heated, these effects tend to cancel each other out—although the magnitude of the offset depends very much on the specific phases, P, and T chosen.

Although these corrections can be made easily, it is commonly impossible to make them because compressibility and thermal-expansion data are unavailable for many minerals. Fortunately, at modest temperatures and pressures, the volume change for a solid–solid reaction corrected for temperature and pressure differs from the volume change measured at 25°C and 1 bar by an amount that is comparable to the uncertainties in the molar volumes of the phases measured at room temperature and pressure. Even at more extreme temperatures and pressures (10 to 100 kbar, and temperatures near melting for silicates), these corrections are much smaller than the overall uncertainty in the Gibbs free energy of the reaction. Thus, if compressibility and thermal-expansion effects are ignored, then ΔV for the reaction is assumed to be independent of pressure and temperature, and

$$
\int_1^P \Delta V_s\, dP = \Delta V_s \int_1^P dP = \Delta V_s (P - 1) \tag{4-10}
$$

Remember that equation (4-10) holds only when ΔV refers to crystalline phases. To remind you of this fact, we have added a subscript s to ΔV; thus ΔV_s refers to the volume change for solids only.

Example. Calculate the change in the Gibbs free energy for the reaction

$$2 \text{ jadeite} \rightleftharpoons \text{analbite} + \text{nepheline}$$

if pressure is increased from 1 bar to 10 kbar. The molar volumes are

nepheline,	54.16 cm³
analbite,	100.43 cm³
jadeite,	60.40 cm³

The change in volume for the reaction is

$$\Delta V = 54.16 + 100.43 - (2)(60.40) = 33.79 \text{ cm}^3$$

Because $1 \text{ cm}^3 = 0.1 \text{ J} \cdot \text{bar}^{-1}$, we have

$$\Delta G_{(P,T)} - \Delta G_{(1,T)} = \Delta V_s(P - 1) = (3.379 \text{ J} \cdot \text{bar}^{-1}) \times (9999 \text{ bar}) = 33.79 \text{ kJ}$$

Note that this value is independent of temperature and contains no standard free energy information. In other words, we calculated only the change in free energy that would result from compressing products and reactants from 1 bar to 10 kbar, not the overall reaction energetics.

Solving the temperature integral

To expand equation (4-7), recall that the effect of temperature on the entropy change depends on the heat capacity:

$$\int_{T_{\text{ref}}}^{T} \Delta S = \int_{T_{\text{ref}}}^{T} \frac{dQ_{\text{rev}}}{T} = \int_{T_{\text{ref}}}^{T} \frac{d(\Delta H)}{T} = \int_{T_{\text{ref}}}^{T} \frac{\Delta C_P \, dT}{T} \qquad \text{(cf 3-123)},$$

We then must solve the double integral

$$-\int_{T_{\text{ref}}}^{T} \Delta S(T) \, dT = \int_{T_{\text{ref}}}^{T} \left(\Delta S_{(T_{\text{ref}})} + \int_{T_{\text{ref}}}^{T} \frac{\Delta C_P}{T} \, dT \right) dT \qquad (4\text{-}11)$$

The specific solution of the double integral depends on the particular power series that is chosen to describe the change in the heat capacities of the phases with temperature. Using the five-term polynomial proposed by Haas and Fisher (1976),

$$C_P = a + 2bT + cT^{-2} + fT^2 + gT^{-1/2}$$

we obtain

$$\Delta G_{(T)} - \Delta G_{(T_{\text{ref}})} = \Delta a(T - T \ln T) - \Delta b \, T^2 - \frac{\Delta c}{2T} - \frac{\Delta f \, T^3}{6} + 4 \Delta g \, T^{1/2}$$

$$- \Delta a \, T_{\text{ref}} - \Delta b \, T_{\text{ref}}^2 + \frac{\Delta c}{T_{\text{ref}}} - \frac{\Delta f}{3} T_{\text{ref}}^3 - 2 \Delta g \, T_{\text{ref}}^{1/2}$$

$$- T \Delta S_{(T_{\text{ref}})} + \Delta a \, T \ln T_{\text{ref}} + 2 \Delta b \, T T_{\text{ref}} - \frac{\Delta c \, T}{2T_{\text{ref}}^2}$$

$$+ \frac{\Delta f}{2} T T_{\text{ref}}^2 - \frac{2 \Delta g \, T}{T_{\text{ref}}^{1/2}} + T_{\text{ref}} \Delta S_{(T_{\text{ref}})} \qquad (4\text{-}12)$$

TABLE 4-2

CALCULATION OF FREE ENERGY, ENTHALPY, AND ENTROPY FOR ANY
REACTION AS A FUNCTION OF TEMPERATURE

If the heat capacities of all i phases involved can be expressed* as

$$C_{P,i} = a_i + 2b_i T + c_i T^{-2} + f_i T^2 + g_i T^{-1/2}$$

then

$$\Delta G_T = \Delta a\,(T - T \ln T) - \Delta b\,T^2 - \frac{\Delta c}{2T} - \frac{\Delta f\,T^3}{6} + 4\,\Delta g\,T^{1/2} + d - eT$$

$$\Delta H_T = \Delta a\,T + \Delta b\,T^2 - \frac{\Delta c}{T} + \frac{\Delta f\,T^3}{3} + 2\,\Delta g\,T^{1/2} + d$$

$$\Delta S_T = \Delta a \ln T + 2\,\Delta b\,T - \frac{\Delta c}{2T^2} + \frac{\Delta f}{2}\,T^2 - \frac{2\,\Delta g}{T^{1/2}} + e$$

where

$$d = \Delta H_{(T_{\text{ref}})} - \Delta a\,T_{\text{ref}} - \Delta b\,T_{\text{ref}}^2 + \frac{\Delta c}{T_{\text{ref}}} - \frac{\Delta f}{3}\,T_{\text{ref}}^3 - 2\,\Delta g\,T_{\text{ref}}^{1/2}$$

$$e = \Delta S_{(T_{\text{ref}})} - \Delta a \ln T_{\text{ref}} - 2\,\Delta b\,T_{\text{ref}} + \frac{\Delta c}{2T_{\text{ref}}^2} - \frac{\Delta f}{2}\,T_{\text{ref}}^2 + \frac{2\,\Delta g}{T_{\text{ref}}^{1/2}}$$

Source: After Haas and Fisher (1976, p. 532).

* This provision specifically rules out liquid water and aqueous species; these cases are discussed later in this text.

Adding to (4-12) the equation

$$\Delta G_{(T_{\text{ref}})} = \Delta H_{(T_{\text{ref}})} - T_{\text{ref}}\,\Delta S_{(T_{\text{ref}})}$$

we obtain the standard free energy for the reaction at T, which can be written in a somewhat more compact form by collection of the constant reference terms (Table 4-2). Table 4-2 also lists the algorithms for solving the standard enthalpy and entropy as a function of temperature, based on the five-term polynomial rather than the three-term Maier–Kelley equation given in Chapter 3.

    If heat-capacity data are unavailable for any of the phases in the reaction of interest, then it may be possible to estimate the unknown heat capacities (Robinson and Haas, 1983; see also Chapter 11). Lacking that possibility, it is then necessary to make the assumption that $\Delta C_P(\text{reaction}) = 0$, which means that the double integral in (4-11) collapses to

$$\Delta G_{(T)} - \Delta G_{(T_{\text{ref}})} = -\Delta S_{(T_{\text{ref}})}(T - T_{\text{ref}}) \tag{4-13}$$

This expression looks a lot easier to solve than equation (4-12), so it might save a lot of computation time to know how good an approximation it is. There is no easy answer. The fact is that the heat capacities for some minerals show fairly abrupt changes as a function of temperature, and the "nulling effect" that you might expect from the heat capacities of products tending to cancel out the heat capacities of reactants (the assumption that $\Delta C_P = 0$ would imply that they cancel exactly) cannot be tacitly assumed. For some reactions, the heat capacity terms in (4-12) can amount to tens of kilojoules over hundreds of kelvins and are significant contributors to the reaction free energy. In other cases, they might amount to much less than the overall uncertainty in the free energy of reaction. The best rule of thumb to follow is this: if reliable heat-capacity data are available for the reaction of interest, be sure to use them. On the other hand, lack of C_P data should not deter you from attempting geochemical calculations at high temperatures—just keep in mind that you might be introducing significant (and unknown) errors.

An equivalent expression for the temperature dependence of the Gibbs free energy can also be derived by the following route. At constant temperature and pressure, we have

$$\Delta G = \Delta H - T \Delta S \tag{3-76}$$

Recall that the temperature dependence of the enthalpy and the entropy of reaction are expressed by

$$\left(\frac{\partial (\Delta H)}{\partial T} \right)_P = \Delta C_P$$

$$\left(\frac{\partial (\Delta S)}{\partial T} \right)_P = \frac{\Delta C_P}{T}$$

Therefore, we can write the temperature dependence of the free energy as

$$\Delta G_{(T)} = \Delta H_{(T_{ref})} + \int_{T_{ref}}^{T} \Delta C_P \, dT - T \left(\Delta S_{(T_{ref})} + \int_{T_{ref}}^{T} \frac{\Delta C_P}{T} \, dT \right) \tag{4-14}$$

where the integrations are carried out at constant pressure.

What are the differences between equations (4-14) and (4-12)? To solve (4-12), you must know the free energy of the reaction at the reference temperature, whereas equation (4-14) requires that you know the enthalpy change for the reaction at the reference temperature. Each equation requires the entropy change at the reference temperature and, for a rigorous solution, the coefficients of the heat-capacity power series for all phases. The heat-capacity terms in equation (4-14) have not been expanded because the end result is identical to equation (4-12) (see Problem 5).

Combining the pressure and temperature integrals

Equation (4-8) can now be written in either of the following equivalent forms:

$$\Delta G_{(T,P)} = \Delta G_{(T_{ref},1)} - \int_{T_{ref}}^{T} \left(\Delta S_{(T_{ref})} + \int_{T_{ref}}^{T} \frac{\Delta C_P \, dT}{T} \right) dT + \Delta V_s(P - 1) \quad (4\text{-}15)$$

$$\Delta G_{(T,P)} = \Delta H_{(T_{ref},1)} - T \Delta S_{(T_{ref})} + \int_{T_{ref}}^{T} \Delta C_P \, dT - T \int_{T_{ref}}^{T} \frac{\Delta C_P \, dT}{T} + \Delta V_s(P - 1)$$

$$(4\text{-}16)$$

If the decision to neglect heat-capacity effects must be made, then both ΔH and ΔS become independent of temperature, so we have

$$\Delta G_{(T,P)} \approx \Delta G_{(T_{ref},1)} - \Delta S_{(T_{ref})}(T - T_{ref}) + \Delta V_s(P - 1) \quad (4\text{-}17)$$

$$\Delta G_{(T,P)} \approx \Delta H_{(T_{ref},1)} - T \Delta S_{(T_{ref})} + \Delta V_s(P - 1) \quad (4\text{-}18)$$

Example Problems

1. Calculate ΔG for the reaction

$$\text{jadeite} + \alpha\text{-quartz} \rightleftharpoons \text{analbite}$$

$$\text{NaAlSi}_2\text{O}_6 + \text{SiO}_2 \rightleftharpoons \text{NaAlSi}_3\text{O}_8$$

at 800 K and 20 kbar. The data at 298.15 K and 1 bar are

$$\Delta a = 325.67$$

$$2 \, \Delta b = -0.19460$$

$$\Delta c = 6.4151 \times 10^6$$

$$\Delta f = 3.6590 \times 10^{-5}$$

$$\Delta g = -5918.9$$

$$\Delta H = 15.86 \text{ kJ}$$

$$\Delta S = 51.47 \text{ J} \cdot \text{K}^{-1}$$

$$\Delta V_s = 1.7342 \text{ J} \cdot \text{bar}^{-1}$$

We have both data and an equation to use. From Table 4-2, we find

$$\Delta G_{(T,P)} = \Delta a(T - T \ln T) - \Delta b \, T^2 - \frac{\Delta c}{2T} - \frac{\Delta f \, T^3}{6} + 4 \Delta g \, T^{1/2}$$

$$+ \, d - eT + (P - 1) \Delta V_s$$

For $T = 800$ K and $P = 20,000$ bar, we have $\Delta G_{(800, 20kb)} = 9.86$ kJ. This answer can be contrasted with the free energy change for the same reaction at 800 K and 1 bar by subtracting the $P \Delta V_s$ term:

$$\Delta G_{(800, 1)} = 9.86 - 34.68 = -24.82 \text{ kJ}$$

Note the change in the sign of the free energy between 1 bar and 20 kbar. The calculations show that analbite should be stable relative to jadeite + α-quartz at 1 bar and 800 K. However, at 20 kbar and 800 K, analbite will break down to jadeite + α-quartz. The univariant line analbite = jadeite + α-quartz must be lurking somewhere between those two pressures.

2. Repeat the calculation for 1 bar assuming $\Delta C_P = 0$. Using equation (4-18), we find

$$\Delta G_{(800, 1)} \approx \Delta H_{(298)} - T \Delta S_{(298)} = 15.86 - (800 \times 0.05147) = -25.32 \text{ kJ}$$

Note that the approximation is within 500 J of the rigorous answer and thus that neglecting heat-capacity effects is not too serious at 800 K.

3. Bearing that in mind, calculate the univariant curve for the reaction analbite = jadeite + α-quartz assuming $\Delta C_P = 0$. At equilibrium, $\Delta G_{(P,T)} = 0$, so equation (4-18) can be rearranged to give

$$P_{eq} = T_{eq} \left(\frac{\Delta S_{298}}{\Delta V_s} \right) - \left(\frac{\Delta H_{298}}{\Delta V_s} \right) + 1 \qquad (4\text{-}19)$$

This is the equation for a straight line in P–T space—i.e., $P = mT + b$. *Univariant reactions involving only solids are linear functions of pressure and temperature if heat-capacity effects are neglected.* By substitution of the appropriate data, we obtain the approximate solution for equilibrium between jadeite, α-quartz, and albite:

$$P \text{ (kbar)} = 0.02968T - 9.146$$

Figure 4-13 plots the approximate univariant line for the reaction and contrasts it with the rigorous univariant line that was calculated using all the appropriate ΔC_P data. The rigorous line can be found by solving equation (4-19) at a number of temperatures, substituting $\Delta S(T)$ for ΔS_{298} and $\Delta H(T)$ for ΔH_{298}.

Cynics among you may well ask what significance this calculated curve has for the "real world." This is a pertinent question. Obvious complications that need further attention are (1) the effects of solid solution in both feldspar and pyroxene and (2) the degree of Al–Si disorder in albite. Because these questions have far-reaching significance for all calculated mineral equilibria, we shall return to this reaction in Chapter 8 to consider these effects. Finally, Appendix D discusses how these calculations can be reconciled with the experimentally measured P and T coordinates of the breakdown of albite at high pressure.

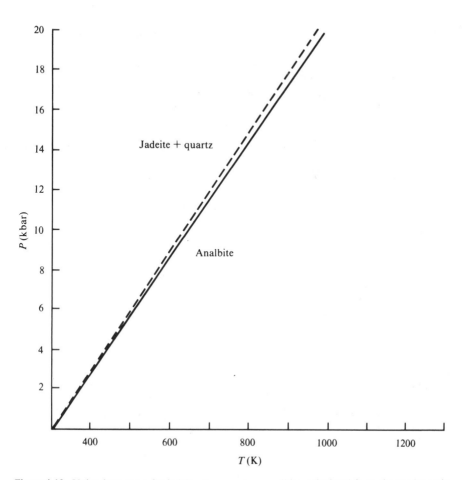

Figure 4-13. Univariant curve for jadeite + α-quartz = analbite, calculated from thermodynamic data given in text. The dashed line assumes $\Delta C_P = 0$ and was calculated from equation (4-19). The solid line incorporates C_P data and was calculated from the free energy equation in Table 4-2 with the $\Delta V_s(P - 1)$ term added.

4-9 THE CLAPEYRON EQUATION

A more general way to calculate the slope of a reaction line as a function of pressure and temperature follows directly from assuming equilibrium between products and reactants:

$$d(\Delta G) = \Delta V\, dP - \Delta S\, dT = 0 \tag{4-20}$$

which is equivalent to

$$\boxed{\frac{dP}{dT} = \frac{\Delta S}{\Delta V}} \tag{4-21}$$

This important relationship is known as the *Clapeyron equation* and is perfectly general because no approximations or simplifications were used in its derivation. Because no standard or reference values are implied for ΔH or ΔS, the equation records an instantaneous slope at any specific P and T of interest. Thus, it is possible to use the slopes of univariant lines that have been very accurately measured in the laboratory to set limits on the $\Delta S/\Delta V$ ratio for the reaction at that point on the curve. Alternatively, reaction slopes can be calculated from known entropy and volume data. It is even possible to use 1 bar, 298.15 K data if you choose to assume that both ΔV and ΔS are independent of temperature and pressure. For many reactions involving only solids of constant composition, this is a reasonable assumption.

SUMMARY

- Phases are real substances that are homogeneous, physically distinct, and ideally mechanically separable parts of a system. They may be solid (e.g., minerals), liquid (e.g., groundwater), or gas (e.g., steam).

- Components are imaginary chemical entities used to describe the compositions of phases or species that participate in a reaction; the minimum number c of components in a system at equilibrium may be found as $c = N - R$, where N is the number of phases or species present in the system, and R is the number of independent chemical reactions that can occur between all species.

- The Gibbs phase rule defines the number f of algebraic degrees of freedom that exists between the number p of phases and the number c of components in a thermodynamic system at equilibrium:

$$f = c - p + 2$$

- All possible heterogeneous reactions in a system can be derived if both the compositions of all phases present and the bulk composition of the system are known. Degenerate reactions occur when one or more phases plot in special (nongeneral) compositions; they are common in geochemistry.

- Phases or assemblages of phases exist as surfaces in $G-P-T$ space; the relative intersections of these surfaces define the positions of univariant lines and divariant fields, and they separate less-stable assemblages from more-stable assemblages.

- If the chemography of a system is known completely, it is possible to derive the sequence of univariant lines about an invariant point, which leads to a number of unimpeachable constraints on the construction of phase diagrams. The guiding principles of these Morey–Schreinemakers rules follow directly from an understanding of the intersection of free energy surfaces.

- The equilibrium $P-T$ coordinates of heterogeneous reactions involving solids only can be easily calculated from standard enthalpy, entropy, volume, and heat-capacity data. In many cases, heat-capacity effects can be neglected without introducing unacceptable errors. Second-order transitions in phases that participate in reactions may introduce significant corrections to the free energy of the reaction.

- The slope of a univariant reaction in $P-T$ space is defined by the Clapeyron equation:

$$\frac{dP}{dT} = \frac{\Delta S}{\Delta V}$$

Experimentally measured slopes may be used to place constraints on reaction entropies.

PROBLEMS

1. Consider a system of H_2, O_2, H_2O vapor, and liquid H_2O at 25°C and 1 atm. How many components are present? Explain your reasoning. (*Ans:* 3)

2. The triple point for the system H_2O is 273.16 K and 0.006 bar. The ice point (defined as equilibrium between ice and liquid water at 1 atm) is 273.15 K. If a mixture of ice and liquid water is open to the atmosphere, it *appears* as though we have a second combination of pressure and temperature for which liquid water, ice, and water vapor are in equilibrium, because some water vapor is always present in the atmosphere. The phase rule says that only *one* invariant point is possible in a three-phase, one-component system. Resolve the ambiguity, and explain the difference between the equilibrium at the ice point and that at the invariant point. (*Hint:* Consider the difference between the total pressure on the system and the partial pressure of H_2O).

3. Refer to Figure 4-3, and write balanced equations for all univariant reactions that are possible between phases kaolinite, andalusite, pyrophyllite, gibbsite, and water. Categorize each reaction as a phase elimination or a join crossing. Which reactions, if any, are degenerate?

4. Use $G-T$ and $G-P$ plots to prove the following statements:

   a) the polymorph stable at higher temperature has the larger finite entropy;

   b) the polymorph stable at higher pressures has the smaller molar volume.
   Can either of these statements be generalized to include product and reactant *assemblages* for reactions in multicomponent systems?

5. Prove that equations (4-12) and (4-14) are equivalent.

6. Consider the following minerals

$$\begin{array}{ll}
\text{gypsum} & CaSO_4 \cdot 2H_2O \\
\text{bassanite} & CaSO_4 \cdot \tfrac{1}{2}H_2O \\
\text{anhydrite} & CaSO_4
\end{array}$$

that differ in composition only by the amount of hydrated water. Both gypsum and anhydrite are common in sedimentary rocks, especially in marine evaporites. Bassanite is quite well known as the substance "plaster of Paris" but is very rare as a mineral. *Without consulting any thermodynamic data*, answer the following questions.

a) If water vapor is the only phase of pure water in the system, give the number of components and the total number of phases.

b) How many phases represent invariant equilibrium? How many phases represent univariant equilibrium? How many univariant reactions are possible? Write all univariant reactions, labeling each one according to the phase that does *not* participate in the reaction.

c) Draw a schematic $P-T$ diagram showing the relative positions of these equilibria and designating all divariant fields with their appropriate chemography. Note that there is no "mirror-image ambiguity" in this case because some of the reactions release H_2O vapor. From your knowledge of the occurrence of these three minerals, does the invariant point occur at pressures greater than or less than 1 atm?

d) Now suppose you want to know what the equilibrium phase diagram looks like for earth-surface conditions where the total pressure is 1 atm and the amount of H_2O vapor in the atmosphere can vary from 1 atm down to vanishingly small partial pressures. Draw another schematic diagram in terms of P_{H_2O} and T, and label each divariant field with the appropriate mineral name. Note that all mineral assemblages must be in equilibrium with H_2O vapor.

7. Consider the one-component system SiO_2 with phases α-quartz, β-quartz, tridymite, cristobalite, coesite, and stishovite. Calculate the numbers of possible invariant points, univariant lines, and divariant fields in the SiO_2 system. How many of the phase-diagram elements are metastable?

8. Consider the reaction

$$\text{grossular} + \alpha\text{-quartz} \rightleftharpoons \text{anorthite} + 2 \text{ wollastonite}$$
$$\text{Ca}_3\text{Al}_2\text{Si}_3\text{O}_{12} \quad \text{SiO}_2 \qquad \text{CaAl}_2\text{Si}_2\text{O}_8 \quad \text{CaSiO}_3$$

If the heat capacities of the four phases in the reaction are written according to the power series in Table 4-2, then the following data apply (Haas et al., 1981).

$$\Delta H_{(298,1)} = 49.67 \text{ kJ}$$
$$\Delta S_{(298,1)} = 63.89 \text{ J}\cdot\text{K}^{-1}$$
$$\Delta V_{(298,1)} = 32.66 \text{ cm}^3$$
$$\Delta a = 117.94$$
$$2\,\Delta b = -0.15806$$
$$\Delta c = 1.4151 \times 10^6$$
$$\Delta f = 8.0955 \times 10^{-5}$$
$$\Delta g = -1.4240 \times 10^3$$

a) Which assemblage is more stable at 298 K and 1 bar? Which assemblage is favored by increasing pressure? By increasing temperature?

b) Calculate the equilibrium pressure of the reaction at 600, 800, and 1000 K. (*Ans:* $-3.24, 0.83$, and 4.83 kbar).

c) Calculate both the equilibrium temperature at 1 bar and the $P-T$ slope of the univariant line, assuming $\Delta C_P = 0$. Write an equation for the univariant equilibrium that neglects heat-capacity effects in the form $P = aT + b$, where a and b are constants. Compare the approximate univariant line to the rigorous one you calculated in part b of this question. Is the $\Delta C_P = 0$ approximation justifiable?

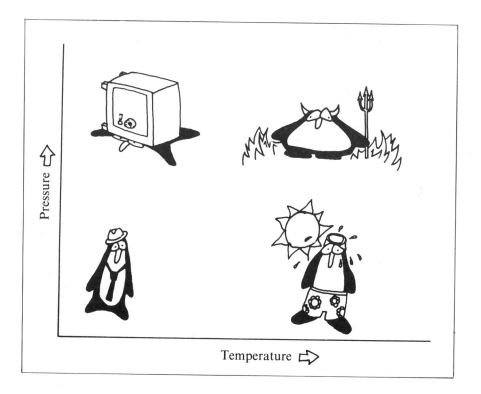

5

THERMODYNAMICS
OF SOLUTIONS

It is evident then that the study of systems of variable composition is geologically important...

R. KERN and A. WEISBROD (1967)

Phases of constant composition are rare in the natural world. We have only to look around us to realize that the air we breathe, the earth's oceans and surface waters, the magmas erupting from volcanoes, and most of the common rock-forming minerals are all solutions capable of wide-ranging compositional variations. Solutions are defined as homogeneous phases formed by dissolving one or more substances (solid, liquid, or gas) in another substance. The effect of solution composition on thermodynamic functions such as entropy, volume, and free energy is the subject of this chapter. This subject is a particularly cogent one for geochemists concerned with phase equilibria. Although we learned in Chapter 4 how to calculate Gibbs free energies for reactions involving minerals of constant composition, the widespread occurrence of mineral, liquid, and gaseous solutions in nature makes this a skill of fairly limited usefulness. Stated another way, Gibbs free energies depend on composition as well as pressure and temperature, so the position of any phase boundary that involves one or more solution phases depends on the compositions of the phases involved.

Compositional variation within a solution may be expressed either in terms of the number of moles of each species present or as the *mole fraction X*, defined for the *i*th species as the number of moles of *i* present divided by the total number of moles of all species in the solution:

$$X_i = \frac{\text{number of moles of } i}{\text{total moles}} \qquad (5\text{-}1)$$

Therefore, the sum of all mole fractions present in each phase must equal unity. For j species, we have

$$\boxed{\sum_{i=1}^{j} X_i = 1} \qquad (5\text{-}2)$$

Certain compositions in a solution have special significance. On the one hand, they may represent actual species which may be present—for example, O_2, H_2, H_2O, CO_2, CO, and CH_4 in a gaseous solution, or K^+, Na^+, and HCO_3^- in an aqueous solution. On the other hand, they may be idealized or "end-member" compositions that represent the limit of some possible solution phenomenon. Thus, $NaAlSi_3O_8$ in alkali feldspar represents the limit of $Na^+ = K^+$ substitution in feldspar solid solution, but $NaAlSi_3O_8$ does not exist as a discrete species in the solution. We refer to $NaAlSi_3O_8$ in the latter example as a *phase component*, because it is important to distinguish the components and/or species of a solution from the total number of components in the system, which must be the smallest number possible consistent with all homogeneous and heterogeneous equilibria in the system. For instance, the gas phase described earlier contains six species but, at high temperatures, all compositions may be described in terms of the three *system components* C–O–H. Mole fractions can be defined in terms of either species, phase components, or system components. It is important to understand the difference between these three entities.

Solutions can exist for all three common states of matter—gases, liquids, and solids. Of the three, gaseous solutions (and their supercritical counterparts) are perhaps the best understood from a thermodynamic viewpoint. Liquid solutions on the other hand are a more complicated matter—in fact, two very different classes of liquid solutions are important in geochemistry. Low-temperature aqueous solutions (liquid water containing dissolved species) are characterized by electrostatic and nonelectrostatic behavior involving solute–solute and solvent–solute interactions. The complications created by these interactions are unique and will be treated separately in Chapter 7. In contrast, silicate melts are liquid solutions of a very different sort. They exist only at very high temperatures and, for some compositions, are characterized by a remarkable degree of short-range ordering of discrete structural units. Understanding natural rock melts from a theoretical viewpoint has proved virtually impossible so far, but significant recent advances have been made using empirical approaches. Finally, crystalline mineral solutions such as feldspars, pyroxenes, and clay minerals are far more abundant than minerals such as quartz that do not significantly change composition. These "solid solutions" present unique problems for thermodynamic treatment because of the triperiodic ordered arrangement of ions that is characteristic of the crystalline state.

Building any solution out of its component parts is a two-step process. For example, suppose we want to make a homogeneous olivine of composition $(MgFe)SiO_4$, starting with pure forsterite (Mg_2SiO_4, fo) and pure fayalite (Fe_2SiO_4, fa). Our desired composition has an equimolar ratio of Mg to Fe, so a reasonable first step would be mechanically to mix together crystalline forsterite and crystalline fayalite in equimolar proportions. The resulting two-phase mixture has the desired bulk composition ($X_{fo} = X_{fa} = 0.5$), but it is not yet a solution. The molal free energy of this *mixture* must be simply halfway between the molal free energies of the end-member phases, because at this point no energetic interactions have occurred between forsterite and fayalite, and free energies of mechanical mixtures are strictly additive.

As a second step, we could heat the olivine mixture to as high a temperature as possible without producing any melt. Given enough time (a practical limitation that we don't have to worry about in this imaginary experiment), the forsterite and fayalite will homogenize to a single olivine phase, in response to the thermal energy introduced into the system. This transformation from a two-olivine mixture to a single homogeneous Mg–Fe olivine solution is a complex phenomenon involving breaking of some chemical bonds and ordering of Mg and Fe cations in the olivine structure. Clearly, this rearrangement of cations must involve changes in the entropy of the system. If the homogeneous olivine is more stable than the two-phase mixture, the free energy of the olivine solution must be less than that of the forsterite–fayalite mechanical mixture. Thus, the free energy of a solution can be represented in two parts—one term representing the free energy of mechanically bringing together the pure species to form a mixture with the appropriate bulk composition (equivalent to the molal free energies of each pure species multiplied by their respective mole fractions), and one term representing the homogenization processes, commonly called the free energy of mixing. These two steps are equally important in forming liquid and gaseous solutions, but they might not be physically recognizable as separate processes. We begin our investigation of solutions with some simple observations of the vapor pressures of gas mixtures.

5-1   VAPOR PRESSURES AND GAS MIXTURES: RAOULT'S AND HENRY'S LAWS

Some of the earliest studies of the thermodynamic properties of solutions centered on gases and the vapor pressures existing above liquid solutions. These studies came first because the instrumentation required for accurate vapor-pressure measurements is fairly simple, solution compositions can be easily measured, and much significant data can be gathered under essentially ambient conditions (laboratory temperature and atmospheric pressure).

One especially significant set of experiments was reported in 1886 by François Raoult, who observed that the vapor pressures of very similar components in solutions (such as ethylene bromide and propylene bromide) depend only on the mole fraction of each component present, providing that total pressure and temperature are held constant. Expressed for component A,

$$P_A = X_A P_A^\circ \qquad (5\text{-}3)$$

where P_A is the vapor pressure of A in the mixture and P_A° is the vapor pressure of *pure* A under the stated conditions. This formulation is known as *Raoult's law*. The pressure P_A is also called the *partial pressure* of A in the mixture of A and B; in general, P_i is the partial pressure of the ith gas in a mixture. It is assumed that the partial pressures are additive and that the sum of all the partial pressures is equal to the total gas pressure:

$$\sum_i P_i = P_A + P_B = P_{total} \qquad (5\text{-}4)$$

Using the equality

$$X_A + X_B = 1$$

and the observation that partial pressures are proportional to their mole fractions, we have

$$P_A + P_B = (X_A + X_B)P_{total} \qquad (5\text{-}5)$$

and

$$P_A = X_A P_{total}$$

or, for the general case,

$$P_i = X_i P_{total} \qquad (5\text{-}6)$$

which is the definition of the partial pressure for any ith gas in a mixture.

The molecular interpretation of Raoult's law is straightforward: a vapor pressure is a direct measurement of the escaping tendency of a gaseous component from a solution. Consider a hypothetical solution made by combining components A and B. If intermolecular forces A–A, A–B, and B–B are indistinguishable, then the tendency for any A molecules to escape into the vapor should depend for a given total pressure only on the amount of A present, and not on the relative proportions of A and B molecules surrounding it, nor on the chemical identity of the B molecules (CO_2 versus N_2, for instance). Noninteraction between species forms the molecular framework for the ideal-solution concept: an *ideal solution* is defined as one that obeys Raoult's law for all compositions. Experience has shown that such ideal behavior is seldom attained over a wide range of solution compositions. In practice, intermolecular forces are far from negligible, and deviations from Raoult's law are to be expected in most solutions. For example, Figure 5-1 plots vapor pressures for water–dioxane solutions; both species show large positive deviations from Raoult's law. These positive deviations are a macroscopic manifestation of sizable repulsive forces between water and dioxane molecules. Nonetheless, the figure also shows that Raoult's law is obeyed for the solvent when the solution is very dilute (X_B less than 0.08 when water is the solvent, and X_A less than 0.05 when dioxane is the solvent). On the other hand, the escaping tendency of the *solute* components (i.e., the dioxane in water-rich solutions, and the water in dioxane-rich solutions) is also proportional to its mole fraction for sufficiently dilute solutions, but the constant of proportionality is not equal to the vapor pressure of the pure solute component. Instead, it is dependent upon the particular solute and solvent being

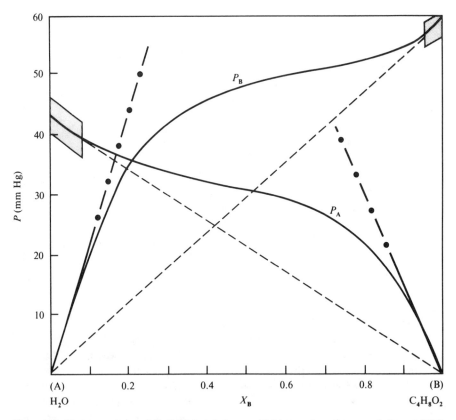

Figure 5-1. Vapor pressures of H_2O (P_A) and dioxane (P_B) in a water–dioxane solution at 35°C, showing positive deviations from Raoult's law. The shaded areas near pure solvent compositions emphasize compositions that follow Raoult's law. The dashed lines are the straight-line curves for an ideal solution. The dash–dot lines represent the Henry's law slopes for the solutes.

studied. Taking subscript B to represent all solute species, we may write

$$\boxed{P_B = hX_B}$$ (5-7)

where h is the Henry's-law constant, named in honor of William Henry, who established this relationship in 1803. Figure 5-1 shows Henry's-law slopes for both water and dioxane as dash–dot lines that originate at pure-solute (i.e., infinitely dilute) compositions.

We now have a working molecular model for the ideal mixing of ideal gases, have calculated the free-energy change associated with forming an ideal solution, and have seen experimental evidence of deviation of vapor pressures of solutions from ideal behavior. We cannot proceed much farther without introducing some new functions.

5-2 PARTIAL MOLAL PROPERTIES AND THE CHEMICAL POTENTIAL

To begin adapting our familiar thermodynamic functions for solution properties, it is convenient to isolate those variations in the extensive properties of a solution that depend only on composition. This can be done in two steps—first by holding pressure and temperature constant, and second by considering the effect of each phase component or species individually. Consider first the case of a binary system at constant temperature and pressure. Let Y represent any extensive thermodynamic property such as volume, entropy, or free energy. Because it is an extensive property, Y will depend on the number of moles of n_1 and n_2 present; i.e., $Y = f(n_1, n_2)$. If dY is an exact differential, we have

$$Y = n_1 \left(\frac{\partial Y}{\partial n_1} \right)_{P,T,n_2} + n_2 \left(\frac{\partial Y}{\partial n_2} \right)_{P,T,n_1}$$ (5-8)

The partial derivatives in (5-8) are known as *partial molal quantities* and are commonly written in abbreviated form for component i as

$$\bar{y}_i = \left(\frac{\partial Y}{\partial n_i} \right)_{P,T,n_j}$$ (5-9)

where the subscript n_j in a multicomponent solution means that the number of moles of *all* components other than i in solution are held constant. The significance of a partial molal quantity is that it describes the isothermal and isobaric variation of an extensive property with respect to one component only, regardless of the number of components present. Thus we can write (5-8) as

$$Y = n_1 \bar{y}_1 + n_2 \bar{y}_2$$ (5-10)

which can be written for any number of components as

$$Y = \sum_i n_i \bar{y}_i$$ (5-11)

To convert to mole fractions, we divide by $\sum_i n_i$ to obtain

$$\bar{Y} = \sum_i X_i \bar{y}_i \qquad (5\text{-}12)$$

For the case of a *pure* phase, the partial derivative in (5-9) is simply equal to Y divided by the total number of moles in the system. To demonstrate this, Figure 5-2 shows the volume of pure diopside ($CaMgSi_2O_6$) at 25°C and 1 bar plotted as a function of the number of moles present. The partial molal volume of pure diopside,

$$\bar{v}_{di} = \left(\frac{\partial V}{\partial n_{di}}\right)_{P,T}$$

is equal to the slope of the straight line in the figure. This slope at any point is simply the volume of diopside divided by the number of moles—i.e., the molar volume of diopside (66.09 cm^3).

One partial molal property of special interest is the partial molal Gibbs free energy—commonly called the *chemical potential*. Because of the significance of

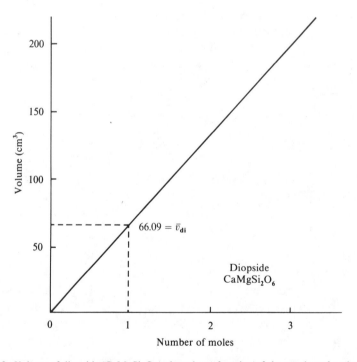

Figure 5-2. Volume of diopside ($CaMgSi_2O_6$) plotted as a function of the number of moles present. The partial molal volume of diopside ($\bar{v}_{di}$) is equal to the slope of the line at constant temperature and pressure (25°C and 1 atm in this case). The partial molal properties of any pure substance are constant and equal to the values per mole for each property—here, 66.09 cm$^3 \cdot$ mol^{-1}.

this function in phase equilibria, it has traditionally received the special symbol μ:

$$\mu_i = \left(\frac{\partial G}{\partial n_i}\right)_{P,T,n_j} \tag{5-13}$$

Because the state of a phase of fixed composition is totally specified by any two independent state functions such as U and V, S and V, or P and T, it is possible to make use of the chemical-potential function to demonstrate the effect of variable composition on the energy content of a solution phase. For a single phase composed of n_i moles of each of i species, we have $G(P,T,n_i)$ which leads to the following expression for the total differential of G in the solution:

$$dG = \left(\frac{\partial G}{\partial P}\right)_{T,n_i} dP + \left(\frac{\partial G}{\partial T}\right)_{P,n_i} dT + \sum_i \left(\frac{\partial G}{\partial n_i}\right)_{P,T,n_j} dn_i \tag{5-14}$$

By comparison with (3-71) and (5-13), we have

$$dG = V\,dP - S\,dT + \sum_i \mu_i\,dn_i \tag{5-15}$$

We can write similar expressions for dH, dA, and dU:

$$dH = T\,dS + V\,dP + \sum_i \mu_i\,dn_i \tag{5-16}$$

$$dA = -S\,dT - P\,dV + \sum_i \mu_i\,dn_i \tag{5-17}$$

$$dU = T\,dS - P\,dV + \sum_i \mu_i\,dn_i \tag{5-18}$$

These equations were first introduced by Gibbs (1906). Equation (5-18) is the most fundamental of the four, because the other three can be derived from it. For this reason, it is known as the *Gibbs equation*. All the μ_i terms in equations (5-16) through (5-18) are also known as chemical potentials, and they are defined individually as

$$\mu_i = \left(\frac{\partial H}{\partial n_i}\right)_{S,P,n_j}$$

$$\mu_i = \left(\frac{\partial A}{\partial n_i}\right)_{T,V,n_j}$$

$$\mu_i = \left(\frac{\partial U}{\partial n_i}\right)_{S,V,n_j}$$

This is a potentially confusing situation that requires some attention to definition. However, the chemical potential defined in (5-13) is the one almost exclusively used in geochemistry because it is the one defined in terms of constant P and T, which are the two most convenient intensive variables to keep constant in geochemical problems. In addition, it is the only chemical potential that is also a partial molal property, because partial molal properties are defined only in terms of constant P and T. It turns out that this is another major benefit, because it allows for some powerful analytical manipulations. Thus, when we refer to chemical potential for the remainder of this text, the definition of equation (5-13) will apply.

Chemical potentials impart very precise information regarding the state of a phase, and certain protocols regarding notation should be strictly observed. Each chemical potential must show a *subscript* indicating the composition of the component or species that defines it and, in heterogeneous systems, a *superscript* that identifies the phase in question. Examples of correct usage involving chemical potentials are $\mu_{H_2O}^{fluid}$, $\mu_{Mg_2SiO_4}^{olivine}$, $\mu_{KCl}^{seawater}$, and $\mu_{SiO_2}^{magma}$.

Although the correspondence between the sign of the free energy change for a given reaction and the equilibrium state of that reaction assemblage was well established in Chapter 4, it is important to see how these criteria apply to chemical potentials. To illustrate the point, consider two homogeneous phases a and b, in each of which the hypothetical component i is continuously variable. For the process of transferring dn_i moles of i from phase a to phase b at constant temperature and pressure, the free energy change must be the sum of the free energy changes for each of the phases:

$$dG = dG^a + dG^b = \mu_i^a dn_i^a + \mu_i^b dn_i^b \qquad (5\text{-}19)$$

If mass is transferred from phase a to phase b, then (by conservation of mass) $-dn_i^a = dn_i^b$. On substitution, we see that

$$dG = (\mu_i^b - \mu_i^a)\,dn_i^b \qquad (5\text{-}20)$$

What does this equation imply concerning the possibility of equilibrium between phases a and b? We know that at equilibrium $dG = 0$, which (because dn_i^b can have any value) requires that

$$\mu_i^a = \mu_i^b \qquad \begin{array}{l}\text{Equilibrium between a and b with respect}\\ \text{to component } i \text{ (Figure 5-3}a\text{)}\end{array}$$

In contrast, if the transference of dn_i moles from a to b is thermodynamically possible, then the net free energy change for that process must be negative $(dG < 0)$. By reference to (5-20), this requires that

$$\mu_i^a > \mu_i^b \qquad \begin{array}{l}\text{Spontaneous transfer of component } i \text{ from}\\ \text{a to b (Figure 5-3}b\text{)}\end{array}$$

This simple example underscores the most fundamental properties of chemical potentials:

1. *At equilibrium, the chemical potential of a species or phase component must be equal for every phase in which that species or component is present.*
2. *In any spontaneous process, components or species are distributed between phases in such a way as to decrease (or minimize) the chemical potentials of all components present.*

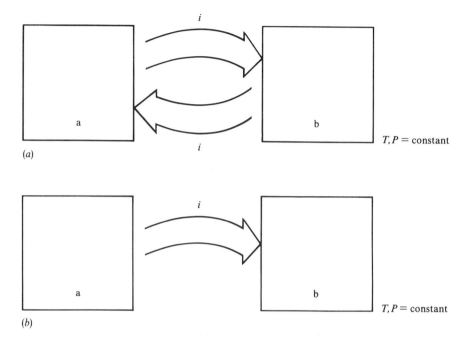

(a)

(b)

Figure 5-3. Transfer of phase component i between phases a and b.   (a) Equilibrium distribution of component i between phases a and b; $\mu_i^a = \mu_i^b$.   (b) Condition for spontaneous transfer of component i from phase a to phase b; $\mu_i^a > \mu_i^b$.

This is all you need to know in order to understand the principles of fairly complex phase relations. However, the calculation of chemical potentials and other functions that will relate chemical potentials to easily measured compositional variables can be a fairly difficult task. As this subject is encountered in future sections, it is important not to get bogged down in some of the details while losing sight of the simplicity and significance of the relations shown in Figure 5-3.

5-3 GIBBS–DUHEM EQUATIONS AND THE PHASE RULE

We continue our investigation of the chemical potential function by writing (5-10) in terms of Gibbs free energy:

$$G = n_1 \mu_1 + n_2 \mu_2 \tag{5-21}$$

On differentiation, we find

$$dG = n_1 \, d\mu_1 + \mu_1 \, dn_1 + n_2 \, d\mu_2 + \mu_2 \, dn_2 \tag{5-22}$$

Compare this expression with (5-15) written for two components:

$$dG = V \, dP - S \, dT + \mu_1 \, dn_1 + \mu_2 \, dn_2 \tag{5-23}$$

By equating (5-22) to (5-23), we obtain

$$V\,dP - S\,dT - n_1\,d\mu_1 - n_2\,d\mu_2 = 0 \tag{5-24}$$

It is clear that (5-24) can be written in general form to accommodate any number of components:

$$\boxed{V\,dP - S\,dT - \sum_i n_i\,d\mu_i = 0} \tag{5-25}$$

This very important equation is the *Gibbs–Duhem relation*. For a homogeneous phase at equilibrium, it governs the relation that must exist between the intensive variables pressure, temperature, and the chemical potentials of all components. If pressure and temperature are held constant, we see that the equilibrium state of a binary system is

$$n_1\,d\mu_1 + n_2\,d\mu_2 = 0 \tag{5-26}$$

Equation (5-26) shows that only one chemical potential in a closed binary system can vary independently. This is a very important constraint, and it is used frequently in phase-equilibrium calculations. It is the underlying principle on which activity diagrams are based, and we shall explore it in detail in Chapter 9.

The Gibbs–Duhem relation is also the only tool we need to derive the phase rule.* We have already shown that two intensive parameters are sufficient to describe completely the thermodynamic state of a pure phase, providing that the number of moles of substance is known. However, if the phase in question is allowed to vary its composition and contains c components, then the equilibrium state of that phase will be completely described by $c + 2$ variables (e.g., $P, T, \mu_1, \mu_2, \ldots, \mu_c$). However, only $c + 1$ of these variables are independent, because a Gibbs–Duhem equation similar to (5-26) must apply at equilibrium. Now, allow for the possibility of extending the equilibrium to p phases, all of which can be described by the same number c of components, although not every component need be present in every phase. The same $c + 2$ variables will apply to every phase, but they will now be reduced by p Gibbs–Duhem equations, one for each phase. Thus the number of independent variables in the system (which is the same as the number f of degrees of freedom) is $f = c - p + 2$, the familiar Gibbs phase rule.

5-4 WORKING WITH PARTIAL MOLAL QUANTITIES: A GRAPHICAL METHOD

Don't be discouraged if partial molal quantities seem like a hopelessly abstract concept at this point. One way to make them seem more real is to show how they could be calculated from easily obtained data. Because volumes of solutions are

---

* After Guggeheim (1967, p. 34).

easier to understand than free energies, we will use the measured volumes of ethanol–water solutions as an example. The principles involved will work just as well for any other extensive property.

Perform this simple thought experiment. Take 100 ml of distilled water and 100 ml of pure ethanol, and mix them together. The resulting volume of the solution is not 200 ml as you might think but is in fact closer to 193 ml if our imaginary laboratory is at 25°C. As you might suspect, this lower-than-anticipated volume is the result of attractive forces between H_2O and C_2H_5OH molecules in solution. Next, prepare about 20 more solutions ranging in composition from nearly pure water to nearly pure ethanol. By using measured densities of water and ethanol at 25°C, you can convert these compositions from volumes to numbers of moles of the end members, and finally to mole fractions.

Lastly, divide the measured volumes by the total number of moles present in each solution, and plot the results as a function of the mole fraction of ethanol (X_{EtOH}). Figure 5-4 is an exaggerated version of what this graph would look like, showing that the molal volumes of all solutions are less than those predicted if the volumes were strictly additive.

These data can be used to obtain the partial molal volumes of water and ethanol for any solution composition. Let water be component 1 and ethanol component 2. From (5-10), the total volume of the solution is related to the partial molal volumes by

$$V = n_1 \bar{v}_1 + n_2 \bar{v}_2 \tag{5-27}$$

Divide both sides by $n_1 + n_2$ to convert to mole fractions:

$$\bar{V} = X_1 \bar{v}_1 + X_2 \bar{v}_2 \tag{5-28}$$

Because $X_1 + X_2 = 1$, we may write

$$\bar{V} = (1 - X_2)\bar{v}_1 + X_2 \bar{v}_2 \tag{5-29}$$

Note the tangent to the molar volume curve drawn at composition X'_2 in Figure 5-4. Any such tangent will be given by the equation for a straight line with $\bar{V}$ as ordinate and X_2 as abscissa:

$$\bar{V} = \left(\frac{d\bar{V}}{dX_2}\right) X_2 + \text{constant} \tag{5-30}$$

Taking the derivative of (5-28) with respect to X_2, we obtain

$$\frac{d\bar{V}}{dX_2} = \bar{v}_2 - \bar{v}_1 \tag{5-31}$$

At every X_2, the molal volume given by the tangent line must be identical to the measured molal volume at that point. Thus, it is possible to equate the righthand sides of equations (5-29) and (5-30) to solve for the constant in (5-30). The result is

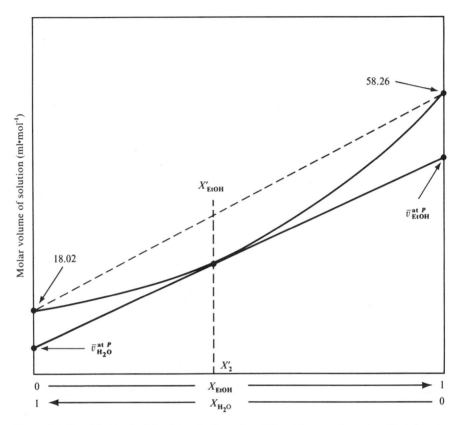

Figure 5-4. Graphical method for determination of partial molal properties, using ethanol–water solutions as an example. Curvature of the molal-volume function is significantly exaggerated for clarity. The molal volumes of pure H_2O and ethanol (EtOH) are 18.02 and 58.26 $ml \cdot mol^{-1}$ at 25°C, respectively. The partial molal volumes of H_2O and EtOH at X' are given by the intercepts of the tangent to the solution curve at X'.

that the constant equals $\bar{v}_1$, and (5-30) becomes

$$\bar{V} = (\bar{v}_2 - \bar{v}_1)X_2 + \bar{v}_1 \qquad (5\text{-}32)$$

The intercept of the tangent line at $X_2 = 0$ is equal to the partial molal volume of component 1 at X'_2. Conversely, if equation (5-29) were written in terms of X_1 instead of X_2, you would see that the intercept of the same tangent line at $X_1 = 0$ is equal to the partial molal volume of component 2 at X_2 (Figure 5-4). This graphical "method of intercepts" works for any partial molal property and will be used extensively later to solve for the chemical potentials of components in solution.

Figure 5-5 shows the partial molar volumes of water and ethanol plotted as a function of X_{EtOH}. By comparison with Figure 5-4, note that the partial molal

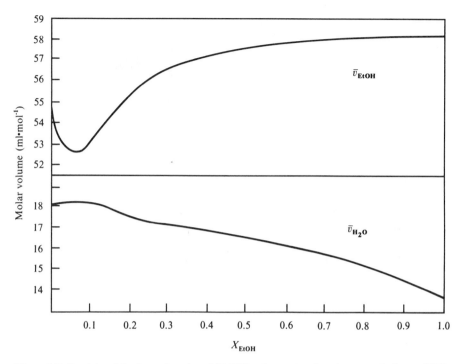

Figure 5-5. Partial molal volumes for ethanol (EtOH) and water in ethanol–water solutions at 20°C.

volumes of each of the *pure* phases are equal to the molal volumes of pure water and pure ethanol, respectively. Also, note that the slopes of the two curves have opposite signs: when the slope of $\bar{v}_{EtOH}$ vs X_{EtOH} is negative, the slope of $\bar{v}_{H_2O}$ vs X_{EtOH} is positive, and vice versa. This is a graphical example of a Gibbs–Duhem relation for partial molal volumes. If equation (5-25) is divided by $(n_1 + n_2)$ and is written in terms of partial molal volumes instead of chemical potentials, we obtain

$$X_1 \, d\bar{v}_1 + X_2 \, d\bar{v}_2 = 0 \qquad (5\text{-}33)$$

which states that the partial molal volumes of water and ethanol are not independent. We could have calculated one of the two curves in Figure 5-5 if we knew the partial molal volume of either water or ethanol as a function of composition.

5-5 THERMODYNAMIC PROPERTIES OF THE IDEAL SOLUTION

In the introduction to this chapter, we suggested that the mixing of two or more pure substances to form a solution can be conceptualized as a two-step process— the first step being the mechanical combination of those pure phases to yield the desired bulk composition, and the second step being the homogenization process. For any extensive property $\bar{Y}$ (per mole), we may write

$$\bar{Y} = \sum_i X_i \bar{Y}_i + \Delta Y_{\text{mixing}} \tag{5-34}$$

where $\bar{Y}_i$ is the molal value of the pure ith component for property $\bar{Y}$, and ΔY_{mixing} is the change in $\bar{Y}$ that occurs as a result of forming the solution. We will now calculate $\bar{Y}$ for the volume, enthalpy, entropy, and Gibbs free energy of the ideal solution, and we shall see that our qualitative molecular concepts of the ideal solution can be justified quantitatively.

Because the definition of the ideal solution is based on Raoult's law, it makes sense to return to a study of the vapor pressure of ideal gases. We need to relate these vapor pressures to the chemical potentials of the pure gaseous components. The fundamental relationship between the free energy and volume was given in Chapter 4 as

$$\left(\frac{\partial G}{\partial P}\right)_T = V \tag{4-2}$$

Replacing both G and V by their partial molal counterparts, we obtain

$$\left(\frac{\partial \mu}{\partial P}\right)_T = \bar{v} \tag{5-35}$$

If the gas behaves ideally, we may write

$$\left(\frac{\partial \mu}{\partial P}\right)_T = \frac{RT}{P} \tag{5-36}$$

Integrating both sides of (5-36) between limits P_0 and P, we obtain

$$\int_{P_0}^{P} \left(\frac{\partial \mu}{\partial P}\right)_T = \mu^P - \mu^{P_0} = RT \ln \frac{P}{P_0} \tag{5-37}$$

which can be rearranged to yield

$$\mu^P = \mu^{P_0} + RT \ln \frac{P}{P_0} \tag{5-38}$$

Note that, as a result of the lower limit of integration, μ^{P_0} is the chemical potential of the pure gas at the reference pressure P_0. This is known as a standard-state chemical potential, and it is commonly abbreviated simply as μ°. If we decide to let P_0 represent the vapor pressure of *pure i* while P is the vapor pressure of i in an ideal solution, then we may substitute Raoult's law (5-3) into (5-38) to obtain the relationship between the chemical potential of the ith gas and its mole fraction in the *ideal solution*:

$$\mu_{i,\,\text{ideal}} = \mu_i^\circ + RT \ln X_i \qquad (5\text{-}39)$$

Note that μ_i° is the chemical potential of pure i at the temperature of interest.

If we differentiate (5-39) with respect to pressure while keeping temperature constant, we will, according to (5-35), arrive at an expression for the partial molal volume of i. Because neither μ_i° nor $RT \ln X_i$ depend on pressure, $\bar{v}_i$ will be a constant, independent of X_i. Because $\bar{v}_i$ must equal the molar volume of pure i ($\bar{V}_i$) as X_i approaches unity, then $\bar{v}_i$ must also equal $\bar{V}_i$ at every other composition so long as $\bar{v}_i$ cannot change as a function of X_i.

If $\bar{v}_i = \bar{V}_i$, then the following summations must also be equal:

$$\sum_i X_i \bar{v}_i = \sum_i X_i \bar{V}_i \qquad (5\text{-}40)$$

By comparison with (5-34), we see that this implies

$$\Delta V_{\text{ideal mixing}} = 0 \qquad (5\text{-}41)$$

$$\bar{V}_{\text{ideal}} = \sum_i X_i \bar{v}_i = \sum_i X_i \bar{V}_i \qquad (5\text{-}42)$$

When two or more pure species mix ideally, there is no positive or negative volume change as a result of that mixing; the molal volume of the solutions is a linear function of the mole fractions of all species present.

Similar arguments can be made for the enthalpy of an ideal solution. The partial molal enthalpy of each component is independent of composition and equal to enthalpy of the pure component ($\bar{h}_i = \bar{H}_i$). Thus, no heat is evolved when an ideal solution is formed, so that

$$\Delta H_{\text{ideal mixing}} = 0 \qquad (5\text{-}43)$$

$$\bar{H}_{\text{ideal}} = \sum_i X_i \bar{H}_i = \sum_i X_i \bar{h}_i \qquad (5\text{-}44)$$

This result agrees with our molecular model of noninteraction; if molecules display no preferential energetic interactions as a result of mixing, then no heat should be exchanged as a result of that mixing.

Whenever a solution is made from pure components, there is a change in the state of ordering in the system. Thus ΔS of ideal mixing cannot equal zero. To find out how entropy changes on ideal mixing, we can use the ideal gas model once again. Recall from Chapter 3 that the entropy change accompanying the isothermal expansion of 1 mol of ideal gas from P_1 to P_2 is

$$\Delta S = R \ln \frac{P_1}{P_2} \qquad (3\text{-}39)$$

When ideal gases are mixed together to form a solution, the total entropy change per mole will be equal to the summation of the entropy change for each species multiplied by the mole fraction of that species. The entropy change for each species results from the expansion of that gas from its initial pressure (P_i°) to its partial pressure in the mixture (P_i). From (3-39),

$$\Delta S_{\text{ideal mixing}} = R \sum_i X_i \ln \frac{P_i^\circ}{P_i} \tag{5-45}$$

By substitution of Raoult's law, we obtain

$$\Delta S_{\text{ideal mixing}} = -R \sum_i X_i \ln X_i \tag{5-46}$$

which is the general expression for the entropy change that accompanies the formation of an ideal solution. Note the similarity of this equation to (3-127), which was derived for mixing on crystallographic sites. The expression for the molal entropy is then

$$S_{\text{ideal}} = \sum_i X_i \bar{S}_i - R \sum_i X_i \ln X_i \tag{5-47}$$

We are finally in a position to calculate the free energy of the ideal solution. We may write

$$\Delta G_{\text{mixing}} = \Delta H_{\text{mixing}} - T \Delta S_{\text{mixing}} \tag{5-48}$$

It is apparent from (5-43), (5-46), and (5-48) that

$$\Delta G_{\text{ideal mixing}} = RT \sum_i X_i \ln X_i \tag{5-49}$$

Finally, we have

$$\bar{G}_{\text{ideal}} = \sum_i X_i \mu_i^\circ + RT \sum_i X_i \ln X_i \tag{5-50}$$

where μ_i° is equal to the chemical potential of each pure (unmixed) component.

We now turn our attention to the free energies of ideal binary solutions. A logical starting point is to rewrite (5-50) in terms of components 1 and 2, substituting $X_1 = 1 - X_2$:

$$\bar{G}_{\text{ideal}} = (1 - X_2)\mu_1^\circ + X_2 \mu_2^\circ + RT[(1 - X_2)\ln(1 - X_2) + X_2 \ln X_2] \tag{5-51}$$

The last term in (5-51) is the free energy of ideal mixing, and it is a symmetrical parabola having a minimum at $X_1 = X_2 = 0.5$. As you can see, the size of the parabola is proportional to temperature (Figure 5-6). Thus, two factors govern the magnitude of the free energy of ideal mixing. For a given composition, ΔG_{mixing} becomes more negative as temperature increases. For a given temperature, ΔG_{mixing} equals zero for the pure (unmixed) phases and decreases symmetrically, reaching its most negative value at $X_1 = X_2 = 0.5$.

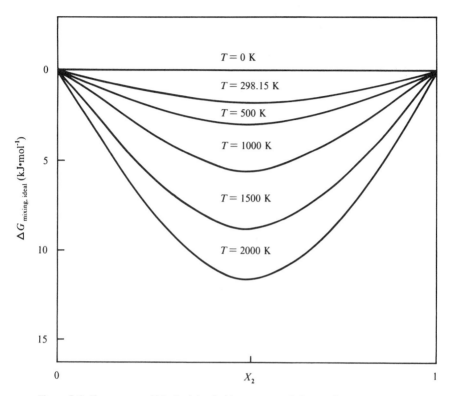

Figure 5-6. Free energy of ideal mixing in binary system 1–2 at various temperatures.

The first term in (5-51) can also be written as $X_2(\mu_2^\circ - \mu_1^\circ) + \mu_1^\circ$. If $\bar{G}$ is plotted against X_2, this is the equation for a straight line connecting μ_2° with μ_1°. If for every X_2 we add the free energy of the straight line (equivalent to mechanical mixing of pure 1 and pure 2) to the free energy of the parabola (equivalent to the homogenization of 1 and 2 to form a solution), we obtain the $G-X$ curve for the solution (Figure 5-7).

Because μ_1 and μ_2 are partial molal quantities, the chemical potentials of components 1 and 2 in the solution can be read from the intercepts of a tangent line to the $G-X$ curve at any point, as shown for composition X_2' in Figure 5-7. This graphical technique for the determination of chemical potentials is very handy in the construction of phase diagrams, as we shall see in Chapter 6. Because the method of intercepts makes no assumptions regarding the shape of the partial molal function, this technique will work regardless of the shape of the $G-X$ curve—i.e., it is equally effective for the determination of chemical potentials in nonideal solutions. The world of the nonideal or real solution is the subject of the remainder of this chapter.

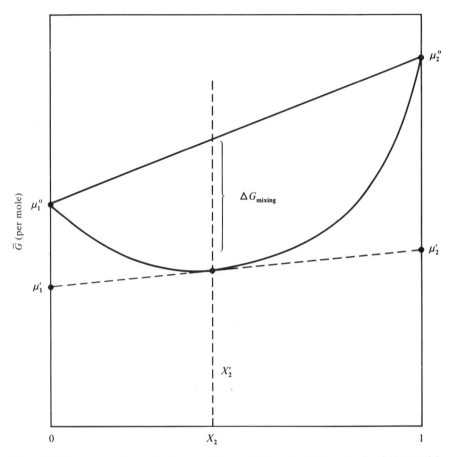

Figure 5-7. Free energy diagram for binary solution 1–2. Here μ_1° and μ_2° are the chemical potentials of the phases of compositions pure 1 and pure 2, respectively; they are equal to the free energy per mole of pure 1 and pure 2, respectively. The parabola is the molal free energy of all binary solutions. At any composition (e.g., X_2'), the chemical potentials of 1 and 2 are given by the vertical-axis intercepts of a tangent line to the curve (μ_1' and μ_2' in the figure). The distance between the solution curve and a straight line joining the chemical potentials of the pure phases is equal to the free energy of mixing.

5-6 FUGACITY, ACTIVITY, AND STANDARD STATES

We have now discussed the nature of chemical potentials and how they relate to free energies. We also know that the method of intercepts is a convenient way to relate chemical potentials to solution compositions. We now need some new tools to treat adequately the complexities of real solutions. Once again, our reference point will be the behavior of the ideal gas, with particular attention to equation (5-38).

Because real gases deviate significantly from ideal behavior as pressures rise above 1 bar, equation (5-38) is of limited use in petrologic applications. G. N. Lewis, a physical chemist active early in the twentieth century, invented a function he called the fugacity, from the Latin root meaning "escape." Lewis saw the significance of fugacity as a measure of the relative "escaping tendency" of a real gas from a solution. The fugacity was defined to have two important properties. First, the relationship between the fugacity f and the chemical potential μ_i of component i must take the form of equation (5-38), so that

$$\mu_i = \mu_i^\circ + RT \ln \frac{f_i}{f_i^\circ} \qquad (5\text{-}52)$$

Note that equation (5-52) defines only the *ratio* of f_i to f_i°, and thus it does not uniquely define a value for f_i. The standard-state fugacity f_i° has the same significance as P°, the standard-state pressure.

The second characteristic of fugacity involves the relationship between fugacity and pressure. As a considerable convenience, the fugacity and pressure scales were designed to converge for the case of ideal behavior; the pressure and fugacity of 1 mol of any ideal gas at any temperature and volume are identical. Because all gases behave ideally at sufficiently low pressures, this convergence can be expressed as

$$\lim \frac{f}{P} \to 1 \quad \text{as} \quad P \to 0 \qquad (5\text{-}53)$$

The definitions implicit in (5-52) and (5-53) are sufficient to show how the fugacity of a nonideal gas can be rigorously calculated from measurable quantities. Specifically, we define α as the difference between the measured volume of 1 mol of any pure gas at some P and T and the volume that the gas would have if it behaved ideally:

$$\alpha = \frac{RT}{P} - \bar{V} \qquad (5\text{-}54)$$

Rearranging and integrating between limits P° and P, we obtain

$$\int_{P^\circ}^{P} \bar{V}\, dP = \int_{P^\circ}^{P} \left(\frac{RT}{P} - \alpha\right) dP \qquad (5\text{-}55)$$

Because of (5-35) and (5-38), this expression becomes

$$\mu - \mu^\circ = RT \ln \frac{P}{P^\circ} - \int_{P^\circ}^{P} \alpha\, dP \qquad (5\text{-}56)$$

We can now substitute one of our defining equations for fugacity, (5-52), into the lefthand side of the equation to obtain

$$RT\ln\frac{f}{f^\circ} = RT\ln\frac{P}{P^\circ} - \int_{P^\circ}^{P} \alpha\,dP \qquad (5\text{-}57)$$

which can be rearranged to give

$$RT\ln\frac{f}{P} = RT\ln\frac{f^\circ}{P^\circ} - \int_{P^\circ}^{P} \alpha\,dP \qquad (5\text{-}58)$$

Now we introduce the second defining relation for fugacity by allowing P° to approach zero. Then $RT\ln(f^\circ/P^\circ)$ also approaches zero because, from (5-53), the limit of f°/P° must approach 1 as P° approaches zero. By substitution for α, we now see how to calculate the fugacity of a gas from P–V–T measurements:

$$\ln\frac{f}{P} = -\frac{1}{RT}\int_{0}^{P}\left(\frac{RT}{P} - \bar{V}\right)dP \qquad (5\text{-}59)$$

Thus fugacities can be calculated if the molal volume of a gas is measured at a number of isobars up to the pressure of interest.

As an example of how this can be done, Figure 5-8 shows calculated values for α (equal to $RT/P - \bar{V}$) obtained from molal volumes of supercritical H_2O measured at 100-bar intervals up to 2 kbar at a constant temperature of 500°C. From the calculus, it is apparent that the shaded area under the curve is equal to the integrand in (5-59) evaluated between the limits of 0 and 2 kbar—by graphical integration using trapezoidal summation over 100-bar intervals, the shaded area is equal to 65,557 cm·bar·mol^{-1}. Thus, the ratio f/P (known as the fugacity coefficient, symbolized by Γ) at 2 kbar and 500°C is $\exp[-1/(83.141 \text{ cm}\cdot\text{bar}\cdot\text{K}^{-1}\cdot\text{mol}^{-1} \times 773.15 \text{ K}) \times 65557 \text{ cm bar mol}^{-1}] = 0.361$. The fugacity of H_2O at the stated temperature and pressure is then $(0.361)(2000 \text{ bar}) = 722$ bar. Although useful in concept, graphical integration is not used to derive accurate fugacities. More commonly, the measured volume data is fitted to a polynomial in P and T, and fugacities are derived analytically from these functions with the aid of computers.

It is also possible to calculate the fugacities of gases at conditions where no P–V–T measurements exist by assuming adherence to some appropriate equation of state. Using modified Redlich–Kwong equations (cf Section 2-5) to calculate the compressibility factor Z, and fitting parameters a and b, we compute the fugacity coefficient for a pure gas as*

$$\ln\Gamma = Z - 1 - \ln\left(Z - \frac{bP}{RT}\right) - \frac{a}{RT^{1.5}b}\ln\left(1 + \frac{bP}{ZRT}\right) \qquad (5\text{-}60)$$

* Cf Holloway (1977, p. 173).

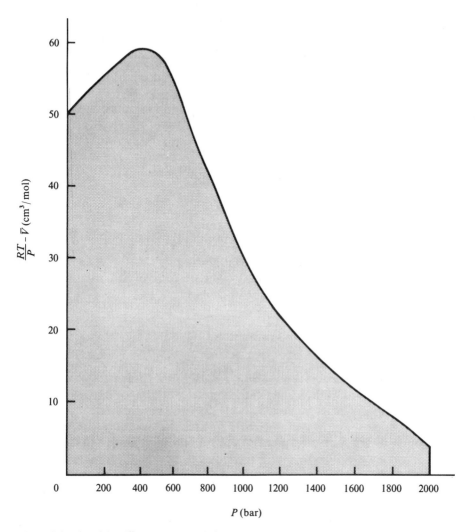

Figure 5-8. Plot of the difference between the ideal and real volumes of H_2O at $500°C$ as a function of pressure. The shaded area is equivalent to the integrand in equation (5-59) and can be used to solve for the fugacity of H_2O at any pressure up to 2000 bar. (data from Burnham et al., 1969*b*)

As an example of the extrapolating power of this approach, Bottinga and Richet (1981) used a modified Redlich–Kwong equation to calculate compressibility factors for pure CO_2 (from which fugacity coefficients may be obtained) to pressures of 50 kbar, based on P–V–T measurements extending no higher than 8 kbar.

Graphical methods are available for estimating fugacity coefficients, but they are all based on the theory of corresponding states, so they must be used with due caution. However, the only data needed to use such graphs are the critical constants of the gas. Such graphs are very similar to those used to estimate compressibility factors, and they are subject to the same uncertainties. One example of a graph for the estimation of fugacity coefficients is reproduced in Figure 5-9.

Thus, fugacities are thermodynamic functions that are directly related to chemical potentials and can be calculated in a straightforward manner from the measured $P-V-T$ properties of a gas. However, fugacities have meaning for solids and liquids as well as gases, because solids and liquids have finite vapor pressures and, whenever an atom or molecule exerts a measurable vapor pressure, a fugacity can always be related to that pressure.

Fugacities are also directly related to activities, one of the most important and essential of the thermodynamic functions. Students commonly have more difficulties manipulating activities than practically any other thermodynamic function. Much of this confusion, however, can be traced to "standard-state" problems and can be avoided by careful attention to all the variables in the problem at hand. It is easy to see the potential source of confusion regarding activities by considering the definition of activity:

$$a_i = \frac{f_i}{f^{\circ}_i}$$ (5-61)

The activity of component i in a phase is equal to the fugacity of component i in the phase divided by the standard-state fugacity of component i in the phase. Thus, activities are dimensionless. Moreover, f°_i is a standard-state fugacity and thus can have any value you choose; there is no such thing as a *unique* activity of a component in a phase even if pressure, temperature, and composition are fixed. This development should be a little frightening—it merits some vigilance. As an example, consider the case of a homogeneous H_2O-CO_2 mixture at a total pressure of 2 kbar and 500°C. Let's say that it is possible to measure the fugacity of H_2O in the mixture, and that its value is 250 bar. If we want to calculate the activity of H_2O in the mixture, we must choose a standard state. Standard states may be either fixed or variable. Examples of fixed standard states are "1 bar," "298.15 K," "5 kbar," "pure Mg_2SiO_4." On the other hand, variable standard states take the form "at the pressure of interest," "at the temperature of interest," or "olivine of measured composition at P and T." Variable standard states appear to contradict the very concept of standard states but, in fact, they are especially useful. Returning to our H_2O-CO_2 example, let's examine three different choices for the standard state of H_2O.

> *Standard state 1*: pure H_2O at P and T (in this case "P and T" is a common shorthand notation meaning "pressure of interest" and "temperature of interest"). This is a typical example of combining fixed (pure H_2O) and variable (P and T) standard states, which is a common practice. Having made this choice, the standard-state fugacity for H_2O is equal to the

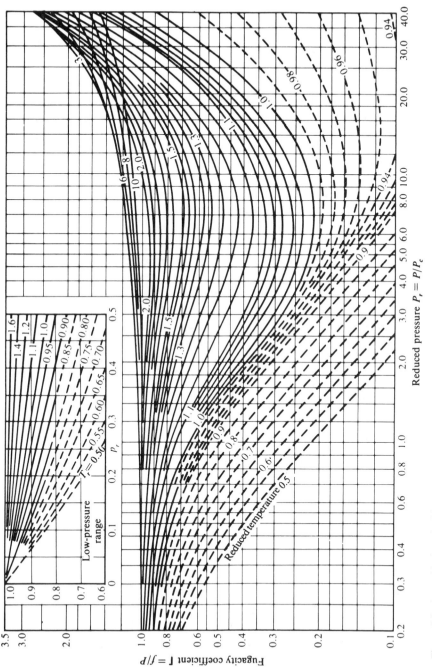

Figure 5-9. A graphical method for determination of the fugacity coefficients of pure gases, based on the principle of corresponding states. Lines are plotted in terms of the reduced pressure ($P_r = P/P_c$), and the contours are isotherms of reduced temperature ($T_r = T/T_c$). (Dickerson, 1969, p. 279)

fugacity of pure H_2O at $P = 2$ kbar and $T = 500°C$, the pressure and temperature of interest. As we saw in Figure 5-8, this value is 722 bar. Accordingly, from (5-61), the activity of H_2O is 250 bar/722 bar = 0.346.

Standard state 2: pure H_2O at 1 bar and T. We have chosen to fix standard-state pressure at 1 bar. To calculate activities, we need to know the fugacity of H_2O at 1 bar and 500°C. Fugacity coefficients for steam at 1 bar are very nearly (but not exactly) equal to one at such low pressures. For this reason, it is convenient to invent a standard state that is exactly equal to one (regardless of temperature); for such a state, the activity at P and T would be numerically equal to the fugacity at P and T. We can obtain the desired result with one additional stipulation to our standard-state definition.

Standard state 2′: pure ideal-gas H_2O at 1 bar and T. By including the (hypothetical) stipulation that H_2O be an ideal gas in its standard state, we are assured that the activity of H_2O will be exactly 250 bar/1 bar = 250.

Standard state 3: H_2O in the solution at P and T. Here all three standard states are variable, and the standard-state fugacity is equal to the fugacity of H_2O in the specific solution under consideration at the P and T of interest. The activity of H_2O is 250 bar/250 bar = 1. Any standard state that defines an activity of unity is particularly useful for measuring deviations in the properties of a solution away from the specific composition of interest. As you can well imagine, a virtually infinite number of standard states, and a correspondingly virtually infinite number of activities, can be defined for this specific solution. Fortunately, only a very few of these possibilities make any sense, so the trick is to pick the standard state that most easily lends itself to solution of your specific geochemical problem.

Next we must consider the vastly important standard-state connection that exists between activity and chemical potential. By substitution of (5-61) into (5-52), we can express the fundamental relation as

$$\mu_i = \mu_i^\circ + RT \ln a_i \qquad (5\text{-}62)$$

Comparison of (5-39) with (5-62) demonstrates the one-to-one correspondence between the activity and the mole fraction of an *ideal* solution:

$$a_{i,\text{ideal}} = X_i \qquad (5\text{-}63)$$

providing that the standard-state chemical potential in (5-39) refers to the composition of pure i at the pressure and temperature of interest.

Because chemical potentials are the driving forces that determine the distribution of components between phases of variable composition in a system, activities can be thought of as parameters that monitor the effective "availability" of component i for reaction in the system. Herein lies their fundamental importance and, because of their reliance on standard-state definitions, their potential pitfall. You can avoid numerous problems in the future by remembering that standard states are implicit in *every* definition of activity (even though no superscript $°$ ever appears with the activity symbol), and that those standard states are tied to the standard-state chemical potential. As an example, consider three different expressions for (5-62), each corresponding to one of the three standard states chosen previously:

standard state 1:

$$RT \ln 0.346 = \mu_{H_2O}^{\text{in soln, 2 kbar, 500°C}} - \mu_{H_2O}^{\text{pure H}_2\text{O, 2 kbar, 500°C}}$$

standard state 2':

$$RT \ln 250 = \mu_{H_2O}^{\text{in soln, 2 kbar, 500°C}} - \mu_{H_2O}^{\text{ideal gas, 1 bar, 500°C}}$$

standard state 3:

$$RT \ln 1 = \mu_{H_2O}^{\text{in soln, 2 kbar, 500°C}} - \mu_{H_2O}^{\text{in soln, 2 kbar, 500°C}}$$

Each activity in this example is actually a measure of the difference in chemical potential between the chosen state and the reference state. Incorrect answers to problems are unconditionally guaranteed if care is not taken to connect the chosen activity with its correct standard-state chemical potential.

Because the whole concept of standard states and their relationship to activities is so important in geochemistry and petrology, we urge readers to delve into the abundant literature on the subject. Many excellent references are available (e.g., Anderson, 1970a, 1970b; Rock, 1967; Whitfield, 1975b).

5-7 EXCESS FUNCTIONS AND ACTIVITY COEFFICIENTS

The ideal-solution model is useful insofar as it provides a clearly defined reference state for solution behavior. Nonideal or real solutions can then be evaluated with respect to the degree to which they deviate from ideality, as was shown at the beginning of this chapter for the vapor pressures of water–dioxane solutions (Figure 5-1). When this comparison between the real and the ideal state is applied to free energy, we have the excess free energy of a solution defined as

$$G_{\text{excess}} = G_{\text{real}} - G_{\text{ideal}} \tag{5-64}$$

The excess free energy can be resolved into contributions from excess enthalpy and excess entropy:

$$G_{\text{excess}} = H_{\text{excess}} - TS_{\text{excess}} \tag{5-65}$$

The excess enthalpy is a measure of the heat exchanged as a result of mixing the pure end-members to form the solution, whereas the excess entropy is a measure of all those energetic effects resulting from a nonrandom distribution of species in solution. If $S_{\text{excess}} = 0$, and there is no significant volume change on mixing, then $G_{\text{excess}} = H_{\text{excess}}$, and the solution is called a *regular* solution.

One of the most important problems of real solutions is the necessity to relate all nonideal thermodynamic parameters to the composition of the solution, because composition can be directly measured to a high degree of precision. The important link is provided by the activity function. In addition to being defined as a fugacity ratio, the activity of component i in solution is also related to the mole fraction of component i through the *rational* activity coefficient* λ_i. The equation is simply

$$a_i = X_i \lambda_i \qquad (5\text{-}66)$$

The activity coefficient is an all-purpose term that contains all those interaction effects between i and the other components in the solution that were neglected in the ideal solution. The activity coefficient of i must depend on the chemical compositions and amounts of all *other* components present, as well as on temperature and pressure. For this reason, the activity of i measures the "availability" of component i for reaction in the system, but the activity coefficient measures the degree to which the amount X_i of i in solution reflects that availability. Note that the activity coefficient of component i in the ideal solution is, by comparison with (5-63), equal to unity. By substitution of (5-66) into the fundamental expression (5-62) for the dependence of chemical potential on activity, we obtain

$$\mu_i = \mu_i^\circ + RT \ln X_i + RT \ln \lambda_i \qquad (5\text{-}67)$$

The total free energy of the solution is found by summing the chemical potentials of all components in solution multiplied by their appropriate mole fractions, so that

$$G_{\text{real}} = \sum_i X_i \mu_i^\circ + RT \sum_i X_i \ln X_i + RT \sum_i X_i \ln \lambda_i \qquad (5\text{-}68)$$

By comparison with (5-50) and (5-64), it is clear that the excess free energy is simply the sum of the activity coefficients of all components, when both temperature and mole fractions of all components are taken into account:

$$G_{\text{excess}} = RT \sum_i X_i \ln \lambda_i \qquad (5\text{-}69)$$

We can obtain further insight into these relations by concentrating on the way that the chemical potential of a single component can vary with composition. Figure 5-10 plots μ_i against $\ln X_i$ and may be viewed as a graphical representation of equation (5-67). At the lefthand abscissa, $\ln X_i = 0$ (or $X_i = 1$); the origin represents pure i. As X_i decreases, $\ln X_i$ tends toward larger negative

---

* The *practical* activity coefficient γ_i is defined for aqueous solutions. See Section 7-2.

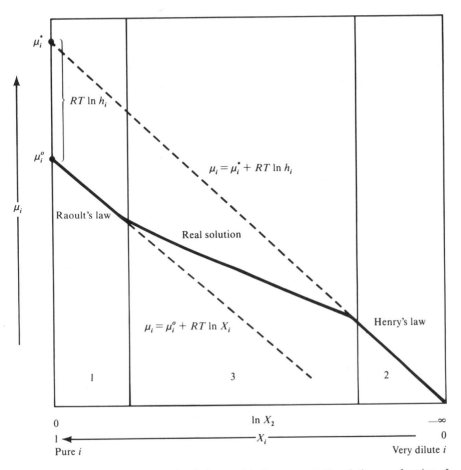

Figure 5-10. Schematic plot of the chemical potential of component i in solution as a function of $\ln X_i$. Here μ_i^o is the chemical potential of pure i at the pressure and temperature of the diagram. The three areas of solution behavior indicated in the figure are discussed in the text. (Adapted from Powell, 1978, p. 47.)

values; i.e., the righthand abscissa represents infinitely dilute i. Component i behaves like a solvent near the left side of the diagram and like a solute near the right side.

For a real solution, it is possible to imagine three distinct types of behavior, depending on the amount of i present.

1. Component i is a solvent, and X_i is very close to unity. Because the mole fractions of all the other species are extremely small, these additional species are dispersed throughout a matrix of i "molecules." So long as enough i is present to overwhelm all the other species, interactions between i and the other species present are insignificant compared with

interactions between i "molecules" alone. When this is true $\lambda_i = 1$, and the solution exhibits ideal behavior, represented by Raoult's law. For all those compositions, μ_i is given by a straight line having slope RT and intercept (at $X_i = 1$) of μ_i°; μ_i° is the chemical potential of pure i at the P and T of interest. This is the graphical expression of equation (5-39) for ideal solutions.

2. Component i is a solute, and X_i is very small. This situation is the exact opposite of case 1. Now component i is totally overwhelmed by all other species present—for this reason, X_i is not sensitive to the amount of i present but is a constant (h_i) depending only on pressure and temperature. Under these extremely dilute conditions, solutions follow Henry's law, and h_i is called the Henry's-law constant for component i. The constant h_i is composition-independent, so that

$$\mu_i = \mu_i^\circ + RT \ln X_i + RT \ln h_i \qquad (5\text{-}70)$$

The final term is a constant at fixed temperature, and it can be regarded as adding to the μ_i° intercept. Defining

$$\mu_i^* = \mu_i^\circ + RT \ln h_i \qquad (5\text{-}71)$$

we have, for solutions that obey Henry's law,

$$\mu_i = \mu_i^* + RT \ln X_i \qquad (5\text{-}72)$$

The chemical potential of i in this dilute range is once again given by a straight line with slope RT but with intercept μ_i^* (Figure 5-10). The intercept μ_i^* is equivalent to the chemical potential that component i would have if X_i could equal unity in the solution. For many solutions, this might well be a hypothetical standard state. If i represents the substitution of a small amount of Cu^{2+} in the octahedral layer of a biotite, then $\mu_{Cu\text{-}bi}^*$ represents the chemical potential of the hypothetical (and physically unobtainable) component copper biotite, $KCu_3Si_3AlO_{10}(OH)_2$.

3. The third region in Figure 5-10 includes all values of X_i intermediate between Raoult's-law and Henry's-law behavior. In this range, interactions between i and all other species present are significant, and λ_i depends on composition as well as pressure and temperature. Accordingly, the dependence of λ_i on X_i must be determined experimentally or theoretically.

5-8 PRINCIPLES OF SOLUTION SEPARATION: IMMISCIBILITY AND EXSOLUTION

We have seen that all ideal-solution models share the concept that the ions or molecules in solution do not interact preferentially as a result of mixing. Viewed in detail on the atomic scale, this assumption is never true. Nonetheless, there are two good reasons for using ideal solution models.

1. Some solutions are formed from species or components having similar physicochemical properties. Such solutions may behave closely enough to ideality that such models are useful predictors of solution behavior.
2. Insufficient data may be available for calculation of nonideal models, so that an ideal solution model may be appealed to as a "last resort."

Both cases must be tested by observation. We may calculate solution properties as though they were ideal and then test the model by observing solution behavior—if the predictions fit the observations, then the assumption of ideality was justifiable. If the assumption appears inappropriate, then we must search for a method to calculate activity coefficients that will measure the deviation from ideality.

Sometimes ideal models are clearly inappropriate. The most extreme example of this situation occurs when repulsive interactions between species in solution become so great that an initially homogeneous solution separates into two solution phases of distinctly different composition in a process known as *exsolution*. The occurrence of thermodynamically stable, coexisting, immiscible solutions is a very common natural phenomenon—an oil-and-vinegar salad dressing serves as a familiar, if pedestrian, example. Geochemical examples are to be found in (1) exsolution lamellae in perthitic feldspar, pyroxenes, oxides, and many other minerals, (2) immiscibility of metal sulfide droplets in silicate melts, and (3) immiscibility of supercritical CO_2 and H_2O at temperatures below $270°C$.

This important process is best understood by analyzing $G–X$ curves of real solutions in detail. In keeping with past practices, we restrict our discussion to binary solutions, but the same principles apply to systems with larger numbers of components. As a first step, we write equation (5-68) for two components:

$$\bar{G} = X_1\mu_1^\circ + X_2\mu_2^\circ + RT(X_1 \ln X_1 + X_2 \ln X_2) + RT(X_1 \ln \lambda_1 + X_2 \ln \lambda_2) \quad (5\text{-}73)$$

Substituting G_{excess} for the last term in (5-73), and differentiating the resulting equation with respect to X_2, we obtain

$$\left(\frac{\partial \bar{G}}{\partial X_2}\right) = \mu_2^\circ - \mu_1^\circ + RT \ln \frac{X_2}{X_1} + \left(\frac{\partial G_{\text{excess}}}{\partial X_2}\right) \quad (5\text{-}74)$$

The second derivative is

$$\left(\frac{\partial^2 \bar{G}}{\partial X_2^2}\right) = \frac{RT}{X_1 X_2} + \left(\frac{\partial^2 G^{\text{excess}}}{\partial X_2^2}\right) \quad (5\text{-}75)$$

Because λ_1 and λ_2 depend on composition, the derivative functions of the excess free energy cannot be expressed in specific terms. If the solution is ideal, $G_{\text{excess}} = 0$, and the second derivative of the Gibbs-free-energy term in (5-75) is always positive. The free energy curve for the solution is then concave upward for every X_2, in agreement with our previous experience about the shape of ideal-solution curves (cf Figure 5-7). However, for a real solution, G_{excess} can be either positive or negative, depending on whether interactions between solution components are predominately repulsive or attractive. Thus, for some combination of T and X_2, the second derivative of excess free energy may be sufficiently negative to cancel the positive term in (5-75), and inflection points may appear in the $G-X$ curve. When this happens, exsolution is possible.

Figure 5-11(a) shows a $G-X$ curve for hypothetical binary system 1–2 in which the free energy of the solution is resolved into ideal and excess contributions. For the sake of this example, the excess free energy is taken to be positive (relative to the mechanical mixture) for all compositions; the second derivative of the excess contribution is always negative. When both ideal and excess contributions are added together, the total $G-X$ curve for the real solution (Figure 5-11b) shows a maximum near the center of the curve. Note the straight line that is drawn mutually tangent to both minima in the solution curve. According to the tangent rule, phases having compositions X_2^a and X_2^b are in equilibrium because $\mu_1^a = \mu_1^b$ and $\mu_2^a = \mu_2^b$. Every possible mixture of phases with compositions X_2^a and X_2^b (hereafter referred to as phases a and b) has a lower free energy than does any homogeneous solution of equivalent composition, as is shown by the arrows in Figure 5-11(b) pointing down from the solution curve to the a–b mixture line. At this temperature, any homogeneous solution within the shaded area could spontaneously separate into the two phases a and b.

The compositions of immiscible phases a and b change systematically with temperature. The ideal solution curve becomes more negative as temperature increases (cf Figure 5-6). In contrast, the excess-free-energy curve changes very slowly with temperature. The relative contribution of the excess-free-energy term increases as temperature falls, causing the minima in the solution curve to move farther apart. As a result, the compositions of coexisting solutions a and b become more widely separated at low temperatures. On the other hand, as temperature increases, the minima in the solution curve and the compositions of coexisting phases a and b approach one another. When the minima merge, every point on the solution curve has a free energy lower than that of any mixture of phases on the join, and a single homogeneous solution is stable for every composition. The locus of coexisting, immiscible solution compositions defines the miscibility gap, or *solvus*, for the solution (Figure 5-11c). The crest of the solvus is defined in terms of both a critical temperature T_c and a critical composition X_c. These principles apply equally to immiscibility in liquids, in supercritical fluids, and in crystalline solutions. The main difference between these cases lies in the kinetic barriers present for solids, because exsolution in crystals can occur only by diffusion of ions in the solid state. Diffusion may be so slow that the equilibrium compositions may not be reached in millions of years, especially at low temperatures.

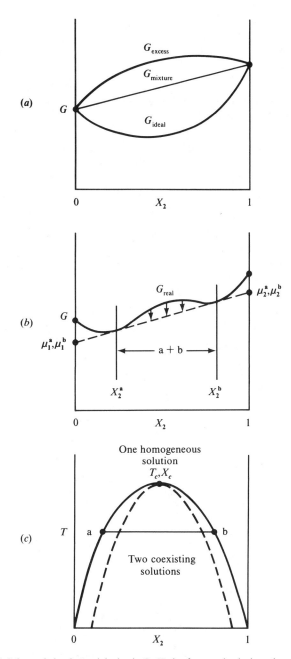

Figure 5-11. (a) Schematic isothermal, isobaric G–X plot for a real solution, showing contributions from mechanical mixing, ideal mixing, and excess mixing. (b) Schematic G–X plot for the real solution of part a ($G_{\text{real}} = G_{\text{ideal}} + G_{\text{excess}}$). A tangent to both minima on the solution curve gives the compositions of coexisting immiscible phases a and b, because $\mu_1^a = \mu_1^b$ and $\mu_2^a = \mu_2^b$. (c) Solvus (solid line) and spinodal (dashed line) for immiscible solutions. Here T_c and X_c are the critical temperature and critical composition, respectively, of the solvus; a and b are compositions of coexisting phases, derived as in part (b). (Adapted from Gordon, 1968.)

Introducing a maximum into the $G-X$ curve produces some interesting and unexpected complexities. Figure 5-12 zeros in on a small compositional range in the vicinity of X_2^a. The scale of Figure 5-11(b) has been exaggerated to emphasize the inflection point located at $X_2^{IP_a}$ on the curve; a complimentary inflection point is located at $X_2^{IP_b}$ near the composition corresponding to minimum X_2^b. Pay special attention to compositions between X_2^a and $X_2^{IP_a}$ (located between the vertical dashed lines in Figure 5-12). According to the tangent construction, the two-phase assemblage a + b is stable within this range (cf Figure 5-11b). Nonetheless, the figure shows that a homogeneous solution has a free energy lower than that of any two-phase mixture in this range. Surprisingly, classical thermodynamics predicts that a homogeneous solution in this range can never

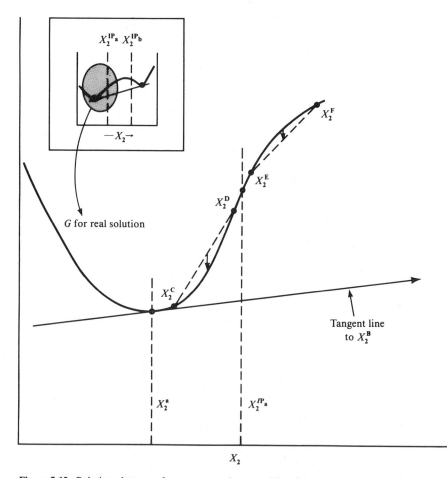

Figure 5-12. Relations between free energy and composition for compositions near the spinode ($X_2^{IP_a}$). The inset shows the region of Figure 5-11(b) that has been enlarged here.

separate into two phases because formation of local areas in the solution with composition slightly different than the homogeneous solution (a necessary first step in phase separation) would cause a local increase in free energy! On an atomic scale, however, random shifts in the positions of atoms, which cannot be predicted by classical interpretations of the second law, may result in local areas of compositional inhomogeneity, leading to macroscopic phase separation. It is important to emphasize that this situation exists only for compositions between the minimum and the inflection point (there will, of course, be two such regions for every temperature where exsolution is possible). In contrast, for all compositions between the two inflection points, exsolution always results in lowering the free energy of the solution. This point is emphasized by comparing the free energy of mixture $X_2^E - X_2^F$ (to the right of the inflection point) with that of mixture $X_2^C - X_2^D$ (to the left of the inflection point); all compositions located to the right of the inflection point in Figure 5-12 will exsolve spontaneously.

The locus of inflection points in the $G-X$ curve when plotted as a function of temperature defines the *spinodal* for the solution (the dashed curve in Figure 5-11c). Compositions that lie between the spinodal and each limb of the solvus can never exsolve, even though a two-phase assemblage has a free energy lower than that of a single homogeneous phase.

SUMMARY

- Solutions are formed when one or more pure substances are dissolved in another substance to produce a homogeneous phase.

- Partial molal quantities record the variation in extensive property Y with the number of moles of component i in solution; they are defined by the partial derivative

$$\left(\frac{\partial Y}{\partial n_i}\right)_{P,T,n_j} = \bar{y}_i$$

where n_j represents the numbers of moles of all other components present. For any solution,

$$Y = \sum_i n_i \bar{y}_i$$

- Partial molal quantity $\bar{y}$ may be determined graphically according to the method of intercepts if $\bar{Y}$ (per mole) is plotted against the mole fraction of component i.

- The activity a is a measure of the availability of a chemical component or species for reaction in a system. Because of the definition

$$a_i = \frac{f_i}{f_i^\circ}$$

(where f_i is the fugacity of i in the system and f_i° is the standard state fugacity), the activity is a dimensionless number whose value depends entirely on the choice of standard state.

- Standard states may be either fixed (e.g., 1 bar, 298.15 K) or variable (P and T of interest) and must be defined with great care. The relationship between the activity of i and the chemical potential of i in solution is

$$\mu_i = \mu_i^\circ + RT \ln a_i$$

where μ_i° is the chemical potential of i in the chosen standard state. Thus, a_i and μ_i° are forever linked by your choice of standard state.

- The fugacities of pure gases can be calculated from P–V–T data or can be estimated from an equation of state. Various modifications of the Redlich–Kwong equation of state have yielded estimates of the fugacities of gases at very high pressures.

- The ideal solution obeys Raoult's law, which states that the vapor pressure of i in equilibrium with a solution is a linear function of the mole fraction of i in the solution. This law leads to the following equations governing the behavior of ideal solutions:

$$\bar{V}_{\text{ideal}} = \sum_i X_i \bar{V}_i = \sum_i X_i \bar{v}_i$$

$$\bar{H}_{\text{ideal}} = \sum_i X_i \bar{H}_i = \sum_i X_i \bar{h}_i$$

$$\bar{S}_{\text{ideal}} = \sum_i X_i \bar{S}_i - R \sum_i X_i \ln X_i$$

$$\bar{G}_{\text{ideal}} = \sum_i X_i \mu_i^\circ + RT \sum_i X_i \ln X_i$$

$$a_{i,\text{ideal}} = X_i$$

Both ΔV_{mixing} and ΔH_{mixing} are zero for ideal solutions.

- Real solutions follow Raoult's law only for low solute concentrations. The activity coefficient λ_i measures the deviation of a solution from ideal behavior and is related to concentration by $a_i = X_i \lambda_i$. The activity coefficient depends on P, T, and the concentrations of all species present in solution, and its measurement must be based on properties of the solution.

- The excess free energy is defined as the difference between the real and the ideal free energy of the solution:

$$G_{\text{excess}} = G_{\text{real}} - G_{\text{ideal}}$$

The excess free energy is related to the activity coefficients by

$$\bar{G}_{\text{excess}} = RT \sum_i X_i \ln \lambda_i$$

- Thus, the free energy of a real solution is

$$\bar{G}_{real} = \sum_i X_i \mu_i = \sum_i X_i \mu_i^\circ + RT \sum_i X_i \ln X_i + RT \sum_i X_i \ln \lambda_i$$

PROBLEMS

Problems 1–4 refer to the following gaseous solution: hypothetical gases 1 and 2 form an ideal binary solution; at 1000 K, the free energies of formation from the elements are $-50\ kJ \cdot mol^{-1}$ for species 1 and $-40\ kJ \cdot mol^{-1}$ for species 2.

1. Calculate ΔG_{mixing} for the solution at 0.1 increments of X_2. Plot your results.

2. Calculate $\bar{G}$ for the ideal solution at 0.1 increments of X_2. Plot your results.

3. Use the method of intercepts to find μ_1 and μ_2 in the solution at $X_2 = 0.2$.

4. Check the accuracy of your results for Problem 3 by calculating μ_1 and μ_2 at $X_2 = 0.2$ (equation 5-39).   (Ans: $\mu_1 = -51.86\ kJ,\ \mu_2 = -53.38\ kJ$)

5. If the difference between the free energies of H_2O at 1000 bar, 800°C and at 0.01 bar, 800°C is $101.186\ kJ \cdot mol^{-1}$, calculate the fugacity of H_2O at 1000 bar and 800°C.   (Ans: 842 bar)

6. Calculate the fugacity of CO_2 at 1 kbar and 500°C, using the following table of measured densities. Use graphical integration (trapezoidal approximation). Be sure to convert the measured densities to molar volumes.   (Ans: 1238 bar)

Density of CO_2 at 500°C

P (bar)	Density $(g \cdot cm^{-3})$	P (bar)	Density $(g \cdot cm^{-3})$
25	0.0171	350	0.2253
50	0.0340	400	0.2536
75	0.0509	450	0.2802
100	0.0677	500	0.3068
150	0.1009	600	0.3565
200	0.1339	700	0.4020
250	0.1658	800	0.4438
300	0.1962	900	0.4824
		1000	0.5165

7. Compare your answer for Problem 6 with the estimated value based on reduced variables (Figure 5-9). Use critical constants for CO_2 given in Table 2-1.

8. With reference to the standard states

        A: pure H_2O at P and T

        B: pure H_2O at 1 bar and T

calculate the activity of H_2O at 1000 bar and 800°C for the following conditions:

a) pure H_2O assuming standard state A (use your answer from Problem 5);

b) pure H_2O assuming standard state B;

c) assuming standard state A for an H_2O–CO_2–CH_4 mixture in which H_2O mixes ideally and $X_{H_2O} = 0.25$;

d) same temperature, pressure, and composition as in part c, but the activity coefficient for H_2O in the mixture is 1.25.

6

CALCULATION OF ACTIVITIES IN GAS MIXTURES, MINERALS, AND SILICATE MELTS

In calculating equilibria among minerals, or among minerals and natural waters, the data that are used in, or result from, the equations, are the activities. Thus, the handling of mineral relations through thermochemical calculations requires translation back and forth between concentrations and activities.

R. M. GARRELS and C. L. CHRIST (1965)

The purpose of this chapter is to show how the generalized relations presented in Chapter 5 can be used to determine activities in gas mixtures, crystalline solutions, and silicate melts. This is a vast and rapidly growing subject, so we can do no more than indicate some of the problems involved and pick out a few highlights that seem important from our viewpoint. We have included in this chapter many references that should encourage you to pursue in depth those subjects of special interest to you.

Throughout this chapter, we assume a standard state of some fixed P and T for calculation of activities—thus, all pure phases having the same composition as that indicated by the standard-state choice will have unit activity. Our goal is to relate activities to the composition of the solution, according to the governing relationship $a_i = X_i \lambda_i$ (equation 5-66). For ideal solutions, $\lambda_i = 1$ and $a_i = X_i$. The only problem is to calculate the mole fraction of i. This is straightforward for gas mixtures, because activities in such mixtures are usually defined in terms of the molecular species actually present. If the relative amount of a given species can be calculated, then the mole fraction can be determined as simply the ratio of moles of species i to total moles present. In contrast, the determination of mole fractions in crystalline solutions is somewhat arbitrary, because it depends on how components in the solution are chosen. We have no obvious guidelines to follow because discrete molecular species are not present in the solid state. Finally, the problem is even worse for silicate melts, where the choice of components and the calculation of their mole fractions is arbitrary. These problems are compounded for the more relevant case of nonideal solutions because we are also faced with the problem of determining the activity coefficients of all components present. The most general approach to this problem has been to assume some nonideal model of solution behavior and then to calculate activity coefficients based on that model. The model can be tested by seeing how well it predicts some property of the solution such as a volume or composition that can be checked by measurement. We begin our discussion with the relatively straightforward gas mixtures and then systematically proceed to the increasingly murky cases of minerals and melts.

6-1 CALCULATION OF ACTIVITIES IN GAS MIXTURES

Mixtures of gas species (e.g., CO_2-H_2O or H_2-H_2O) are commonly encountered in igneous and metamorphic petrology. Because of the high pressures that are characteristic of these environments, it is clear that nonideal behavior should prevail in the gas phase. Thus, the challenge is to determine the fugacities of all gas species that may be present in a specific mixture as functions of temperature, pressure, and composition. These data can be obtained by (1) measuring the partial molal volumes of gas species in specific mixtures at high temperature and pressure and (2) comparing these values with the volumes that would be anticipated from an ideal mixing model. Such measurements are extremely difficult and have always challenged the ingenuity and patience of experimentalists. Unfortunately, relatively little such $P-V-T$ data are currently available for gas mixtures of geologic significance, especially at pressures greater than 1000 bar.

Because of this lack of data, there has long been a tendency to assume a model based on the ideal mixing of *real* gases. From a molecular point of view, this model can be interpreted as neglecting all interactions between *unlike* molecules, while accounting for all interactions between like molecules. Therefore, this model should be most reasonable at high concentrations of like molecules. Calculation of fugacities in such mixtures is straightforward. Because of ideal mixing, we have

$$\mu_i = \mu_i^\circ + RT \ln X_i \qquad (5\text{-}39)$$

which, when combined with the equation relating fugacity to chemical potential, gives

$$\mu_i = \mu_i^\circ + RT \ln \frac{f_i}{f_i^\circ} \qquad (5\text{-}52)$$

If μ_i° is the same in both equations (the chemical potential of pure i at P and T), it is apparent that

$$f_i = X_i f_i^\circ \tag{6-1}$$

which is known as the *Lewis–Randall rule* for calculation of fugacities in ideal gas mixtures. Implicit in (6-1) is the understanding that $\lambda_i = 1$ for every X_i.

According to the Lewis–Randall rule, the fugacity of component i in a mixture is equal to the fugacity of pure i at P and T multiplied by the mole fraction of i. If the fugacity of CO_2 is 3500 bar at some P and T of interest, then the fugacity of CO_2 in a gaseous mixture in which $X_{CO_2} = 0.1$ is 0.1×3500 bar $= 350$ bar. Because this model ignores interactions between unlike molecules, the fugacity of CO_2 under these conditions is 350 bar regardless of the identity of other components present—i.e., it is equally valid for CO_2–H_2O, CO_2–CH_4, CO_2–H_2, etc. and need not be restricted to binary solutions.

As you may well imagine, ideal mixing of real gases is an appealing alternative because it allows calculation of fugacities in very complex natural fluids, given only knowledge of the fugacity coefficients for the pure components you intend to mix. Nonetheless, be forewarned—engaging in this practice does involve risks. Ignoring interactions between unlike molecules is unrealistic except in dilute solutions and can lead to significant errors in calculated fugacities, especially at low temperatures and/or high pressures. Unfortunately, it is not possible to predict *a priori* the magnitude of these errors.

Because of the obvious shortcomings of the Lewis–Randall rule, there has been a tendency in recent years to calculate activities of species in nonideal gas mixtures by assuming adherence both to some equation of state and to a set of mixing rules. As was the case for pure gases, the Redlich–Kwong equation has proved most popular. Recalling the basic equation

$$P = \frac{RT}{\overline{V} - b} - \frac{a}{T^{1/2} \overline{V}(\overline{V} + b)} \tag{2-11}$$

the problem becomes how to adapt constants a and b in (2-11) to accommodate a mixture. The most common approach (e.g., Prausnitz, 1969, p. 153) has been to follow the lead of van der Waals, who suggested

$$a_{\text{mixture}} = \sum_i \sum_j X_i X_j a_{ij} \tag{6-2}$$

and

$$b_{\text{mixture}} = \sum_i X_i b_i \tag{6-3}$$

where indices i and j refer to all species present. For instance, for binary system 1–2, equation (6-2) becomes

$$a_{\text{mixture}} = a_1 X_1^2 + a_2 X_2^2 + 2a_{12} X_1 X_2 \tag{6-4}$$

where a_{12} is the cross coefficient and is generally unknown. A traditional solution has been to take the cross coefficient as the geometric mean; i.e., $a_{12} = (a_1 a_2)^{1/2}$, but this approach is not always satisfactory. It is better to formulate a cross coefficient based on some actual measure of interaction between species 1 and 2. For instance, de Santis et al. (1974) derived an expression for H_2O–CO_2 mixtures that accounts for associative formation of H_2CO_3 molecules, and their equation for calculating $a_{H_2O-CO_2}$ in mixtures has been used by most subsequent researchers in this system. The actual algorithms required to calculate fugacities in gas mixtures using any of the modified Redlich–Kwong equations are overpowering in length and complexity. For examples and details of how these

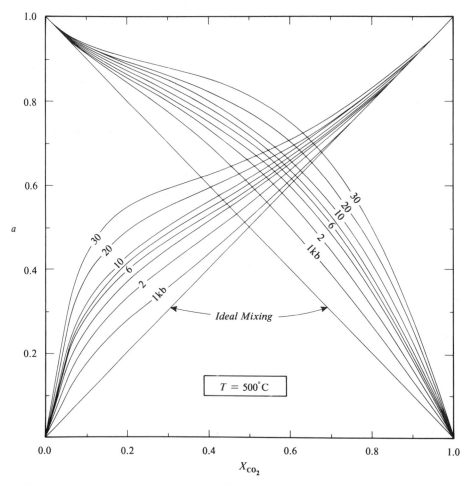

Figure 6-1. Activity–composition diagram for H_2O–CO_2 mixtures at 500°C and 800°C. Contours show total pressure in kbar ($P_{total} = P_{H_2O} + P_{CO_2}$). Calculated from modified Redlich–Kwong equations (Kerrick and Jacobs, 1981).

equations are derived and solved, see Kerrick and Jacobs (1981) for H_2O–CO_2 mixtures and Jacobs and Kerrick (1981) for H_2O–CO_2–CH_4 mixtures.

The activity of the ith species in a nonideal gas mixture is simply obtained for specific values of X_i by dividing the fugacity calculated for that mixture by the fugacity of pure i at the same pressure and temperature. Results of such calculations for H_2O–CO_2 mixtures are shown in Figure 6-1 at 500°C and 800°C for pressures up to 30 kbar. Note that the mixtures become more ideal at high temperature, in agreement with our expectations developed for pure gases in Chapter 2. It bears repeating that these figures are based on calculation, not on measurement. Greenwood (1973) published measurements of activities in H_2O–CO_2 mixtures, but the maximum experimental pressure in that study was 500 bar.

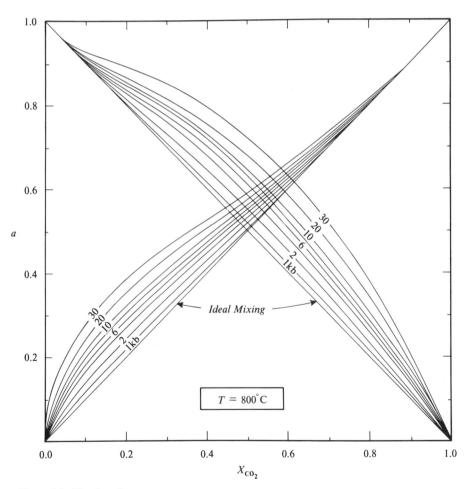

Figure 6-1. (Continued)

6-2   CALCULATION OF ACTIVITIES IN IDEAL CRYSTALLINE SOLUTIONS

The crystalline state is characterized by a three-dimensional periodic arrangement of atoms or ions in a repeatable unit cell motif. Crystalline solutions differ from gaseous or liquid solutions in that mixing occurs on crystallographically distinct sites in the crystal structure. This consideration requires some extra-thermodynamic information on specific site occupancy in crystals in order to derive more realistic thermodynamic expressions. We begin with a brief review of the way that compositions in crystalline solutions are commonly described, and then we shall develop the ideal solution model.

Consider a binary solid solution in which one and only one ion substitutes for another. Variations in solution composition could be described simply in terms of the mole fraction of one of the ions involved in the substitution: e.g., X_{Na^+} for alkali feldspar, $(Na,K)AlSi_3O_8$, or $X_{Mg^{2+}}$ for olivine, $(Mg,Fe)_2SiO_4$. This practice is not generally satisfactory, however, because most natural mineral solutions show much more complicated substitutions. One traditional way around the problem is to resolve fairly complex compositional variations into "end members," each of which represents the stoichiometric composition of an idealized (and sometimes hypothetical) component present in solution. Examples of such end-member components are $Mg_2Si_2O_6$ in pyroxene, $Fe_6Si_4O_{10}(OH)_8$ in chlorite, and FeS in sphalerite. In many solid solutions, simultaneous substitution in more than one site is possible, adding to complexity. Unfortunately, the end-member approach is not always very satisfactory either. Complicated mineral groups such as amphiboles cannot be handled in this way because the number of proposed end members far exceeds the number of possible independent compositional variables.

Definition of components is a matter of convenience and is always arbitrary; in the case of crystalline solutions it may also be somewhat misleading. As an example, consider a pyroxene with solution in two different sites: an eightfold M2 site that can accommodate either Ca^{2+} or Na^+, and a sixfold M1 site containing Mg^{2+}, Fe^{2+}, Fe^{3+}, and Al^{3+}—e.g., $(Ca,Na)(Mg,Fe^{2+},Fe^{3+},Al)Si_2O_6$. One might be tempted to describe such a pyroxene in terms of end members $CaMgSi_2O_6$ (diopside), $CaFe^{2+}Si_2O_6$ (hedenbergite), $NaFe^{3+}Si_2O_6$ (acmite), and $NaAlSi_2O_6$ (jadeite). But do these compositions tell us anything about the thermodynamics of the solution? Two very different possibilities come to mind— either the cations could mix independently on each of the two sites, or substitution in the M1 site could be linked to substitution in the M2 site. Such linkage results in short-range ordering of ion pairs in the crystal, a very different case from independent (and therefore random) mixing. If short-range ordering is present, the mixing model may be easily understood in terms of end members. According to this model, substitution of a trivalent cation (Fe^{3+} or Al^{3+}) for a divalent cation (Fe^{2+} or Mg^{2+}) would require simultaneous substitution of Na^+ for Ca^{2+} in order to maintain local charge balance.

TABLE 6-1

CALCULATION OF THE MOLE FRACTION OF $NaAlSi_3O_8$ (ab)
FROM CHEMICAL ANALYSIS OF PLAGIOCLASE FELDSPAR

Oxide	Weight % oxide	Cation	Moles cation per 8 oxygen anions*
SiO_2	64.6	Si	2.86
Al_2O_3	22.0	Al	1.15
CaO	2.8	Ca	0.13
Na_2O	9.3	Na	0.80
K_2O	1.3	K	0.07
	100.0		

$$Ca + Na + K = 1.00$$
$$X_{ab} = 0.80/1.00 = 0.80$$

* See Deer et al. (1966, p. 515) for a review of the method of calculating cation proportions in minerals from a chemical analysis.

Alternatively, either the acmite or jadeite components could be substituted for the hedenbergite or diopside components in the pyroxene solution. Whether such closely coupled substitution exists in a given crystal is a question that may be answered through X-ray or spectral measurements, but its answer is certainly not obvious *a priori*. This model has been shown to be appropriate for those cases in which independent mixing on sites would result in unacceptable local charge imbalances, and it has been called the local charge balance (LCB) model. A familiar example is the coupled substitution $Ca^{2+} + Al^{3+} = Na^+ + Si^{4+}$ in the plagioclase feldspars. In those cases where the LCB model is appropriate, the activity of the end-member component in an ideal solution may be equated with the mole fraction of that component.

For instance, one could obtain a chemical analysis of a plagioclase feldspar crystal by electron microprobe—such an analysis is shown in Table 6-1. The weight percentages are converted to molecular proportions in the usual way. When these values are recalculated in terms of eight oxygen anions, it is possible to calculate the mole percentage of the $NaAlSi_3O_8$ (ab) end member. Adapting the LCB model, the activity of the ab component in plagioclase solution is equal to X_{ab}, or 0.80 in the example shown in Table 6-1.

As compositional variations in real crystals become more complex (both with respect to the number of ions competing for the same site and with respect to the number of sites exhibiting solution), the idea of substitution by end members becomes much less reasonable. Indeed, it is seldom apparent how best to define appropriate end-member compositions. Fortunately, there is a way to approach such solutions if we resort to the *mixing-on-sites* (MOS) model, which assumes that all ions occupying a given crystallographic site mix independently (and randomly, in the ideal case) on that site. The ideal model implies that ions in the solution show no preferential interactions as a result of mixing—i.e., in addition to random occupancy, we assume that there is no heat of mixing.

The key to quantifying this model lies in the calculation of the entropy change that results from mixing cations randomly on the same crystallographic site. The entropy produced in such ideal mixing must be the same as the configurational entropy term that was derived from the fundamental tenents of statistical thermodynamics—i.e.,

$$\Delta S_{\text{ideal mixing}} = -R \sum_j n_j \sum_i X_{i,j} \ln X_{i,j} \qquad \text{(cf 3-129)}$$

in which n_j is the total number of sites (per formula unit) on which mixing takes place, and $X_{i,j}$ is the mole fraction of the ith ion in the jth site. When mixing takes place on a single site, the equation simplifies to

$$\Delta S_{\text{ideal mixing}} = -nR \sum_i X_i \ln X_i \qquad \text{(6-5)}$$

Because this is the simplest case, we begin by considering single-site mixing only.

It is clear that the value of n in (6-5) depends on how you choose to write one formula unit of the phase in question. For $Mg^{2+} = Fe^{2+}$ substitution in olivine, $n = 2$ for the customary formula $(Mg,Fe)_2SiO_4$ because that formula is written with two exchangeable sites. With the entropy relation established, the relationship between the activity and mole fraction is, according to this model,

$$\boxed{a_{i,\text{ideal}} = X_i^n} \qquad \text{(6-6)}$$

where the exponent once again refers to the number of exchangeable sites. For a complete derivation, see Appendix C. It should be clear from equation (6-6) that the activity of an ion in the MOS model depends on the number of atoms in the formula unit (which determines the exponent) as well as on composition.

We now consider some examples.

Mixing on a single site

If mixing occurs on one site only, it may be convenient to normalize the mineral formula to contain only one exchangeable site, which is equivalent to ensuring that the exponent in (6-6) is equal to unity. For example, consider the possibility of substituting fluoride for hydroxyl in the magnesian mica phlogopite (ph), $KMg_3Si_3AlO_{10}(OH,F)_2$. If we write the phlogopite formula as $K_{1/2}Mg_{3/2}Si_{3/2}Al_{1/2}O_5(F,OH)$, then only one hydroxyl site is involved in the exchange, and the ideal activity of fluoride in that site is equal to the mole fraction of fluoride in the site: $a_F = X_F$. Once the stoichiometry of the site is decided upon, it is essential that all thermodynamic parameters used in subsequent calculations agree with that stoichiometry. For instance, if S_{298} for hydroxyl phlogopite written as the half-cell formula (two hydroxyl sites) is $319.66 \text{ J} \cdot \text{K}^{-1} \cdot \text{mol}^{-1}$, then the entropy that must be used in all calculations based on one hydroxyl site is

$159.83 \, \text{J} \cdot \text{K}^{-1} \cdot \text{mol}^{-1}$. Alternatively, if the two-site formula is used, then $a_F = X_F^2$, because two sites are contributing to the exchange.

Some applications may require the calculation of activities in terms of end-member components—for example $KMg_3Si_3AlO_{10}F_2$, Fph, or $KMg_3Si_3AlO_{10}(OH)_2$, OHph—rather than activities of specific ions in sites. As we have seen, the ideal-mixing model requires that the activity of the component in question be equal to the mole fraction of the component. Thus, in order to apply the MOS model in terms of components rather than individual ions substituting on sites, the problem becomes one of how to calculate the mole fraction of a component in such a mineral. If mixing occurs on one site only, the problem is straightforward. Continuing with the phlogopite example, and defining the Fph component in terms of one exchangeable site, we have for ideal mixing

$$\mu_{Fph}^{ph} = \mu_{Fph}^{\circ} + RT \ln X_F^{ph} \tag{6-7}$$

and

$$a_{Fph}^{ph} = X_F^{ph} = X_{Fph}^{ph} \tag{6-8}$$

Alternatively, if the two-site formula were used, we would have

$$\mu_{Fph}^{ph} = \mu_{Fph}^{\circ} + 2RT \ln X_F^{ph} \tag{6-9}$$

which leads to

$$a_{Fph}^{ph} = (X_F^{ph})^2 = X_{Fph}^{ph} \tag{6-10}$$

Note carefully the difference between the mole fraction of the *component* Fph and that of the fluoride *ion*, and remember that the standard-state chemical potential in (6-9) is exactly twice the standard-state chemical potentials in (6-7). This difference exactly compensates for the differences between the activities and the mole fractions in both cases.

Multisite mixing

When mixing occurs randomly on more than one crystallographically equivalent site, a separate summation of ions must be taken for each site, and equation (3-127) applies directly.

For instance, consider garnets of generalized formula $X_3Y_2Si_3O_{12}$ in which mixing can occur independently on the eightfold X sites and the sixfold Y sites. Assuming for sake of example that mixing is restricted to Al^{3+} and Fe^{3+} in the octahedral site and Mg^{2+}, Fe^{2+}, and Ca^{2+} in the cubic site, then the entropy of ideal mixing will be (Ulbrich and Waldbaum, 1976)

$$\Delta S_{ideal} = -3R(X_{Mg^{2+}} \ln X_{Mg^{2+}} + X_{Fe^{2+}} \ln X_{Fe^{2+}} + X_{Ca^{2+}} \ln X_{Ca^{2+}})$$
$$-2R(X_{Al^{3+}} \ln X_{Al^{3+}} + X_{Fe^{3+}} \ln X_{Fe^{3+}}) \tag{6-12}$$

The activities of cations in individual sites are calculated as before: $a_{Mg^{2+}} = X^3_{Mg^{2+}}$; $a_{Fe^{3+}} = X^2_{Fe^{3+}}$; etc. However, it may be desirable to calculate the activities of garnet components in solution. The components and their compositions are pyrope (py), $Mg_3Al_2Si_3O_{12}$; almandine (alm), $Fe_3Al_2Si_3O_{12}$; grossular (gr), $Ca_3Al_2Si_3O_{12}$; and andradite (an), $Ca_3Fe_2Si_3O_{12}$. Consider pyrope as an example. The chemical potential of pyrope in garnet (gt) will contain mixing contributions from both Mg^{2+} in the cubic site and Al^{3+} in the octahedral site:

$$\mu^{gt}_{py} = \mu^{\circ}_{py} + 3RT \ln X_{Mg^{2+}} + 2RT \ln X_{Al^{3+}} \qquad (6\text{-}13)$$

Combining terms, we obtain

$$\mu^{gt}_{py} = \mu^{\circ}_{py} + RT(\ln X^3_{Mg^{2+}} + \ln X^2_{Al^{3+}}) = \mu^{\circ}_{py} + RT \ln (X^3_{Mg^{2+}} X^2_{Al^{3+}}) \qquad (6\text{-}14)$$

which means that

$$a^{gt}_{py} = X^3_{Mg^{2+}} X^2_{Al^{3+}} = X^{gt}_{py} \qquad (6\text{-}15)$$

Once again, note that the activity and the mole fraction of a component depend on the stoichiometry chosen. The mole fraction of an ion in a site is, of course, independent of stoichiometry.

Similarly,

$$a^{gt}_{alm} = X^3_{Fe^{2+}} X^2_{Al^{3+}} = X^{gt}_{alm} \qquad (6\text{-}16)$$

$$a^{gt}_{gr} = X^3_{Ca^{2+}} X^2_{Al^{3+}} = X^{gt}_{gr} \qquad (6\text{-}17)$$

$$a^{gt}_{an} = X^3_{Ca^{2+}} X^2_{Fe^{3+}} = X^{gt}_{an} \qquad (6\text{-}18)$$

If substitution for Si^{4+} occurs in the tetrahedral site, we would have

$$a^{gt}_{py} = X^3_{Mg} X^2_{Al} X^3_{Si} \qquad (6\text{-}19)$$

Example.   Calculate the activities of py, alm, gr, and an from the structural formula of garnet given in Table 6-2, assuming the MOS model for all sites. Note that Al is present in both tetrahedral and octahedral sites, which is very common in silicate minerals. Having calculated the mole fraction of each ion in its appropriate site, we obtain the activity of each component from equations (6-15) through (6-18). For instance,

$$a^{gt}_{py} = (0.126)^3(0.978)^2(0.994)^3 = 1.879 \times 10^{-3} = X^{gt}_{py}$$

Similarly, you should find

$$a^{gt}_{alm} = X^{gt}_{alm} = 0.424$$

$$a^{gt}_{gr} = X^{gt}_{gr} = 1.321 \times 10^{-4}$$

$$a^{gt}_{an} = X^{gt}_{an} = 6.684 \times 10^{-8}$$

The procedure just described will not work if a component contains two ions that occupy the same site in fixed stoichiometric proportions. This situation

TABLE 6-2
STRUCTURAL FORMULA OF GARNET

Site	Ion	Number of ions	Mole fraction of ion in site	Total ions per site
Tetrahedral	Si^{4+}	2.983	0.994	
	Al^{3+}	0.017	0.006	3.000
Octahedral	Al^{3+}	1.947	0.978	
	Fe^{3+}	0.044	0.022	1.991
Cubic	Mg^{2+}	0.382	0.126	
	Fe^{2+}	2.316	0.767	
	Mn^{2+}	0.167	0.055	
	Ca^{2+}	0.156	0.052	3.021

arises in many common mineral groups; for example, both Si and Al occupy the tetrahedral sites in phlogopite in the ratio 3:1. If all real biotites showed precisely a 3:1 tetrahedral Si:Al ratio, it would be possible to ignore tetrahedral occupancies altogether because there would be no mixing in the tetrahedral site that could contribute to the activity of a biotite component. In reality, real biotites may show quite different tetrahedral ratios. To account for this, you should treat Si and Al occupancies in the usual way and then multiply the calculated activity or mole fraction by a normalization factor that cancels out the tetrahedral contribution whenever the cation occupancy in the tetrahedral site is exactly equal to the ideal stoichiometry. For phlogopite, the mixing contribution from the stoichiometric formula is $X_{Si}^3 X_{Al} = (0.75)^3(0.25) = 0.1055$. Thus, if we multiply the calculated activity of phlogopite in all real biotites by $(0.1055)^{-1} = 9.481$, then a biotite that happens to contain Si and tetrahedral Al in exactly a 3:1 ratio will not add any mixing effects to the activity of phlogopite as a result of mixing in the tetrahedral site.

Example. Calculate the activity of fluor-phlogopite $(KMg_3Si_3AlO_{10}F_2)$ in the biotite described in Table 6-3.

$$a_{Fph} = [X_{K^+} X_{Mg^{2+}}^3 (X_{Si^{4+}}^3 + X_{Al^{3+}})_{tet} X_{F^-}^2] \times 9.481 = 0.00232$$

In the preceding example, note that the only cation proportions used to calculate the activity of the phlogopite component are those that actually appear in the stoichiometric formula for phlogopite. Although the octahedral layer in the biotite contains Al^{3+}, Ti^{4+}, Fe^{3+}, Fe^{2+}, and Mn^{2+} in addition to Mg^{2+}, these cations have no effect whatever in the calculation of phlogopite activity. The reason is that we have chosen an ideal solution model, so these cations serve the

TABLE 6-3
STRUCTURAL FORMULA OF A BIOTITE

Site	Ion	Number of ions	Mole fraction of ion in site*	Total ions per site
Tetrahedral	Si^{4+}	5.790	0.724	
	Al^{3+}	2.210	0.276	8.000
Octahedral	Al^{3+}	0.074	——	
	Ti^{4+}	0.474	——	
	Fe^{3+}	0.105	——	
	Fe^{2+}	1.960	——	
	Mn^{2+}	0.017	——	
	Mg^{2+}	3.210	0.550	5.840
Interlayer	Ca^{2+}	0.260	——	
	Na^+	0.199	——	
	K^+	1.540	0.770	1.999
Anion	OH^-	3.228	——	
	F^-	0.519	0.135	
	Cl^-	0.105	——	3.852

* Mole-fraction values have been computed only for those ions needed to solve the example given in the text.

purpose only of diluting the Mg^{2+} in that site (with proportional diminishing of the activity of Mg^{2+}), and this dilution is faithfully recorded by the mole fraction of Mg^{2+} in the site.

6-3 CALCULATION OF ACTIVITIES IN NONIDEAL CRYSTALLINE SOLUTIONS

Ideal solution models have been widely proposed for many different mineral solutions. However, many years of experimental investigations combined with detailed studies of natural mineral assemblages have repeatedly shown that miscibility gaps are very common in almost every major mineral group. The presence of such gaps requires a nonideal model, as shown in Chapter 5. Many different approaches have been taken to calculate the excess free energy of crystalline solids (see Grover, 1977, pp. 73–74, for a listing of some of these), but undoubtedly the most popular models for mineralogists are those based on Margules equations. These equations, originally derived to explain the vapor pressures of binary solutions, were popularized by J. B. Thompson, Jr., and many of his graduate students at Harvard. Derivations of the fundamental Margules equations appear in a paper by Thompson (1967).

The essential feature of the Margules model is that any excess thermodynamic function can be written as a power series expressed in terms of the mole fractions of the components involved. For instance, for binary system 1–2, the excess free energy can be written as

$$G_{\text{excess}} = A + BX_2 + CX_2^2 + DX_2^3 \qquad (6\text{-}20)$$

where A, B, C, and D are constants.

The simplest example of non-ideal behavior occurs when G_{excess} (and all other nonideal thermodynamic functions) are symmetrical about $X_1 = X_2 = 0.5$. This situation applies to equation (6-20) if $A = D = 0$ and $B = -C$. Defining $W_G = -C = B$, we have

$$G_{\text{excess}} = W_G X_2 (1 - X_2) = W_G X_1 X_2 \qquad (6\text{-}21)$$

where W_G depends on pressure and temperature but not composition. If G_{excess} is large and positive relative to the free energy of ideal mixing, then a solvus that is symmetrical about $X_1 = X_2 = 0.5$ will be the result (cf Figure 5-11). To convince yourself that this statement follows from equation (6-21), be sure to solve Problems 3 through 8 at the end of this chapter.

Most real immiscible mineral solutions, however, have strongly asymmetric solvi. Thus, if we wish to pursue a Margules model, we can obtain more realistic behavior when D in (6-20) is nonzero. In this case, it can be shown (e.g., Thompson, 1967, p. 352) that

$$G_{\text{excess}} = (W_{G_1} X_2 + W_{G_2} X_1) X_1 X_2 \qquad (6\text{-}22)$$

where W_{G_1} and W_{G_2} are once again functions of pressure and temperature only. Moreover, they are related to the Henry's-law constants (cf Figure 5-10) by

$$W_{G_1} = \mu_1^* - \mu_1^\circ = RT \ln h_1 \qquad (6\text{-}23)$$

$$W_{G_2} = \mu_2^* - \mu_2^\circ = RT \ln h_2 \qquad (6\text{-}24)$$

Table 6-4 lists the pertinent equations governing the calculation of chemical potentials and activity coefficients in a two-parameter binary Margules solution (cf Grover, 1977, pp. 75–77). Thus, the calculation of activities in such a solution is automatic, providing that W_{G_1} and W_{G_2} are known as functions of temperature. Both these parameters can be obtained if the compositions of coexisting phases in equilibrium across a miscibility gap in the system 1–2 are known over a range of different temperatures.

As an example, consider the dioctahedral micas muscovite (ms), $KAl_2Si_3AlO_{10}(OH)_2$, and paragonite (pg), $NaAl_2Si_3AlO_{10}(OH)_2$, which form partially immiscible solutions at all subsolidus temperatures. Eugster et al. (1972) calculated interchange energies $W_{G_{\text{ms}}}$ and $W_{G_{\text{pg}}}$ from experimental measurements of the compositions of the coexisting muscovite and paragonite solid solutions, assuming that the excess free energies of the micas can be fit to a Margules equation. From the interchange energies, G_{excess} and ΔG_{mixing} ($= \Delta G_{\text{ideal mixing}} + G_{\text{excess}}$) can be calculated at any temperature from equation (6-22) (see Figure 6-2). The solvus compositions can be obtained by setting the first derivative of G_{excess} with respect to X_2 equal to zero (see expression in Table 6-4). This procedure is equivalent to plotting the pairs of minima in the

TABLE 6-4

SOME IMPORTANT EQUATIONS FOR A TWO–PARAMETER BINARY MARGULES SOLUTION

$$G_{\text{excess}} = X_1 X_2 (W_{G_1} X_2 + W_{G_2} X_1)$$

$$\mu_1 = \mu_1^\circ + RT \ln X_1 + X_2^2 [W_{G_1} + 2(W_{G_2} - W_{G_1})X_1]$$

$$\mu_2 = \mu_2^\circ + RT \ln X_2 + X_1^2 [W_{G_2} + 2(W_{G_1} - W_{G_2})X_2]$$

$$RT \ln \lambda_1 = (2W_{G_2} - W_{G_1})X_2^2 + 2(W_{G_1} - W_{G_2})X_2^3$$

$$RT \ln \lambda_2 = (2W_{G_1} - W_{G_2})X_1^2 + 2(W_{G_2} - W_{G_1})X_1^3$$

$$\left(\frac{\partial G}{\partial X_2} \right) = (\mu_2^\circ - \mu_1^\circ) + RT \ln \frac{X_2}{X_1} + W_{G_1} X_2 (3X_1 - 1) - W_{G_2} X_1 (3X_2 - 1)$$

$$\left(\frac{\partial^2 G}{\partial X_2^2} \right) = \frac{RT}{X_1 X_2} - 2[W_{G_1}(3X_2 - 1) + W_{G_2}(3X_1 - 1)]$$

Note: The interchange energies W_{G_1} and W_{G_2} depend on both P and T.

solution curve in Figure 6-2 against temperature. Because the spinodal compositions represent inflection points in the G–X curve, they are calculated by setting the second derivative of G_{excess} equal to zero. Both the solvus and spinodal curves are shown in Figure 6-3. The activities of both ms and pg were calculated from $W_{G_{\text{ms}}}$ and $W_{G_{\text{pg}}}$ (Table 6-4) and were plotted as a function of composition for 1200 K (above the critical solvus temperature) and for 1000 K (below the critical temperature) in Figure 6-4. Note the dashed lines on the 1000 K isotherms. Compositions a and b, which lie on the extreme ends of these dashed lines, are the equilibrium compositions of coexisting muscovite and paragonite solid solutions (cf Figure 6-3). The dashed lines represent the activities of the ms and pg components in a *metastable* homogeneous mica solution. At equilibrium, the most stable configuration for all compositions between a and b is a two-phase mixture of a paragonite solid solution ($X_{\text{ms}} = 0.18$) and a muscovite solid solution ($X_{\text{ms}} = 0.62$). The activities of both ms and pg components in these solutions are, of course, fixed.

Margules equations have been used to model exsolution in a number of other important mineral systems—e.g., alkali feldspars (Thompson and Waldbaum, 1969), pyroxenes (Lindsley et al., 1981), and Fe–Ti oxides (Andersen and Lindsley, 1981).

An alternative method for calculation of the interchange energies of ionic crystalline solutions is the quasi-chemical model of Guggenheim. This model is deeply rooted in statistical thermodynamics and is strictly applicable only to ionic crystals. Furthermore, the model considers only coulombic interactions between nearest-neighbor ions only; longer-range interactions are completely neglected. From a knowledge of the volumes of ions that are exchanged on

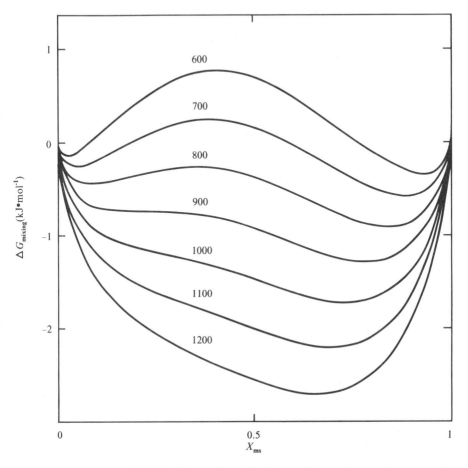

Figure 6-2. Free energy of mixing ($G_{\text{ideal}} + G_{\text{excess}}$) for binary solutions of muscovite–paragonite (ms–pg) micas, calculated for several isotherms at a total pressure of 2.07 kbar. At this pressure, $W_{G_{\text{ms}}} = 13408 + 0.71T$, and $W_{G_{\text{pg}}} = 18210 + 1.653T$. (Calculated from Eugster et al., 1972)

forming a solution and of the compositions of the coexisting phases, and with a good grounding in statistical thermodynamics, one can calculate interchange energies that have the same theoretical significance as those that appear in Margules equations. Once these interchange energies are obtained, the calculation of excess free energies, solvus, spinodal, and activity coefficients can proceed as before. If you want a taste of the equations involved and want to see an example of the quasi-chemical model in action, consult the paper by Green (1970), who examines the halite–sylvite (NaCl–KCl) system and concludes that calculations based on the quasi-chemical model yield a good representation of the solvus in that system.

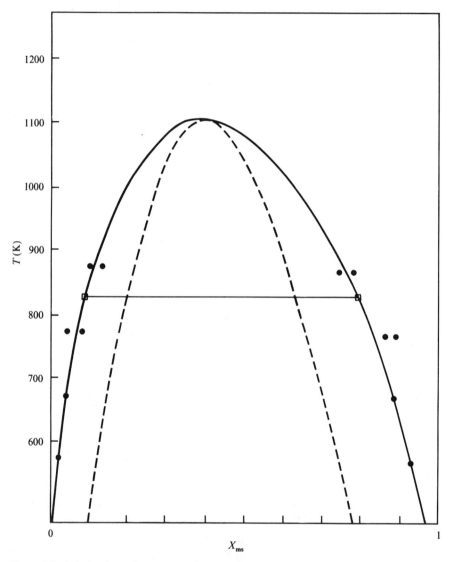

Figure 6-3. Calculated miscibility gap (solid line) and spinodal (dashed line) for muscovite–paragonite solid solutions. Dots are the experimentally obtained compositions of coexisting muscovite and paragonite solid solutions at 2.07 kbar. (Eugster et al., 1972)

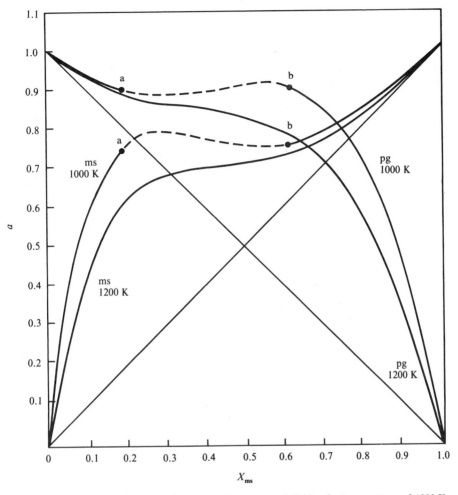

Figure 6-4. Activities in the muscovite–paragonite system at 2.07 kbar for temperatures of 1000 K and 1200 K. The temperature 1200 K is above the critical point of the solvus; 1000 K is below the critical point. Thus, the dashed lines on the 1000 K isotherm give the activities of components for the *metastable* homogeneous solution. Stable immiscible compositions at 1000 K are $X_{ms}^a = 0.18$, $X_{ms}^b = 0.62$. Note that, at these composition, $a_{ms}^a = a_{ms}^b$ and $a_{pg}^a = a_{pg}^b$. (Calculated from data referenced in the text)

6-4   THERMODYNAMIC PROPERTIES OF SILICATE MELTS

Since the early twentieth century, most of the major advances in igneous petrology have come as a result of field, petrographic, and chemical studies of rocks on the one hand, and laboratory measurements of the melting behavior of synthetic and natural rock systems on the other. Thermodynamic investigations of silicate melts tended to lag behind complementary efforts in gases, aqueous systems, and minerals; only recently have considerable strides been made in this area. The principal reason is that, until recently, critical high-temperature thermodynamic data for silicate melts or glasses have been scant, primarily because of the experimental difficulties inherent in making such measurements. Because of space limitations, we can only highlight some of the problems involved. For more rigorous and critical reviews, see Navrotsky (1981) and Bottinga et al. (1981).

In order to achieve a fundamental understanding of the thermodynamic properties of silicate melts, we must know how atoms behave in these liquids— are they disordered as in gases or ordered as in crystals? Are the predominant species in solution ionic or molecular? The latter question can be answered in favor of ions, because silicate melts have been shown to conduct electrical currents much as molten salts do, indicating that ionic species must be present. Early data suggested that the cations are mobile and apparently carry the electrical charges, whereas anions are present as a relatively stationary framework (e.g., Bockris et al., 1952).

The question of order in silicate melts has been answered by combined efforts of Raman spectroscopy (e.g., Sharma et al., 1978) and X-ray radial-distribution analysis (e.g., Taylor and Brown, 1979) performed on silicate glasses. These studies have shown that aluminosilicate glasses (and presumably the melts from which they were quenched) contain $(Si,Al)O_4$ tetrahedra that may be polymerized by sharing of corner oxygens into chains and frameworks similar to those found in silicate minerals. They differ from mineral structures in that (1) any periodicity in the polymerized silicate anions is short-range rather than long-range, and (2) metal cations do not occupy specific sites. The discovery of mineral-like structural units in silicate melts is a very significant development that has formed the basis for empirical models of the thermodynamic behavior of such melts.

Before considering some of these models, let's start with some familiar thermodynamic equations that we have already discussed and see how they might apply to high-temperature melts. Consider equilibrium between a crystalline solid (s) and liquid (l) in binary system 1–2. The equilibrium condition for component 1 is

$$\mu_1^s = \mu_1^l \tag{6-25}$$

The chemical potentials of 1 in liquid and solid are given by

$$\mu_1^l = \mu_{1,l}^\circ + RT \ln a_1^l \tag{6-26}$$

and

$$\mu_1^s = \mu_{1,s}^\circ + RT \ln a_1^s \tag{6-27}$$

Because of (6-25), we have

$$\mu_{1,l}^\circ - \mu_{1,s}^\circ = RT \ln \frac{a_1^s}{a_1^l} \tag{6-28}$$

Assuming a standard-state composition for both melt and solid of pure 1, the lefthand side of equation (6-28) is equal to the standard free-energy change for the reaction solid 1 → liquid 1, which is the same as the free energy of fusion ($_f\Delta G$) for crystalline phase 1. Furthermore, $_f\Delta G = _f\Delta H - T \,_f\Delta S$, where $_f\Delta H$ and $_f\Delta S$ are the *enthalpy* and *entropy of fusion*, respectively. We can make one simplification by assuming that both the enthalpy and entropy of fusion are independent of temperature, which is equivalent to saying that the heat-capacity difference (ΔC_P) between solid and melt does not change with temperature. Moreover, at the melting point, melt and solid must be in equilibrium, so $_f\Delta G = 0$ and $_f\Delta S = _f\Delta H/T_f$, where T_f is the melting point. Substituting into (6-28), we obtain

$$\ln \frac{a_1^s}{a_1^l} = \frac{_f\Delta H_1}{R} \left(\frac{1}{T} - \frac{1}{T_{f,1}} \right)$$
(6-29)

which is the general equation governing the depression of the melting point of one phase by the addition of a second component to the melt. This is the familiar principle that operates when ethylene glycol (antifreeze) is added to water in your car's radiator—the freezing point of ice is significantly reduced in the binary system ethylene glycol–H_2O. We can make the model as simple as possible by assuming that both the solid and the melt phases behave ideally. Replacing activities by mole fractions, we have

$$\ln \frac{X_1^s}{X_1^l} = \frac{_f\Delta H_1}{R} \left(\frac{1}{T} - \frac{1}{T_{f,1}} \right)$$
(6-30)

What does such ideal behavior imply? If the crystalline phase in question does not change composition ($X_1^s = 1$ for all melt compositions), then the liquidus* composition depends only on the enthalpy of fusion and the melting temperature for the pure solid. The implication is that the liquidus composition is independent of the second component, so the depression of the melting point for phase 1 should be identical in every binary system, providing that the compositions are calculated on a mole-fraction basis. The latter conversion from activities to mole fractions is a nontrivial exercise, however, because there is no obvious way to determine the mole fraction of a component in a melt. As always, we are free to choose any components that seem convenient, but the results (in this case, the calculated degree of lowering of the liquidus) will depend on that choice.

For example, consider the melting of anorthite ($CaAl_2Si_2O_8$) in the binary systems anorthite–leucite ($KAlSi_2O_6$), anorthite–silica (SiO_2), and anorthite–diopside ($CaMgSi_2O_6$). The composition of anorthite does not change significantly in the diopside and leucite melts, and it shifts to only slightly more silica-rich compositions in the SiO_2 melt. Assuming unit activity of the anorthite component in all three melts is not an unreasonable

---

* The liquidus is defined as the surface representing all liquid *compositions* that are saturated with (in equilibrium with) a given crystalline phase.

approximation, at least insofar as a test of ideal mixing is concerned. Because framework structural units are present in aluminosilicate melts, it makes sense to define components in the melt based on a constant number of oxygens, reflecting the unifying frameworks present. For reasons that soon will become apparent, it also is reasonable to adopt the anorthite structural unit as a component model and to calculate all other compositions in terms of the eight oxygens present in one formula unit of anorthite. The three binary joins then become $Ca_2Al_2Si_2O_8 - 1.33\ KAlSi_2O_6$, $CaAl_2Si_2O_8 - Si_4O_8$, and $CaAl_2Si_2O_8 - 1.33\ CaMgSi_2O_6$. Ludington (1979) has compiled experimental data on the anorthite-saturated liquidus in each of the three systems, and it is an easy matter to plot these liquidus temperatures against the mole fraction of anorthite in each melt, normalized to eight oxygen atoms (Figure 6-5). The figure shows that the observed liquidus curves are very different from each

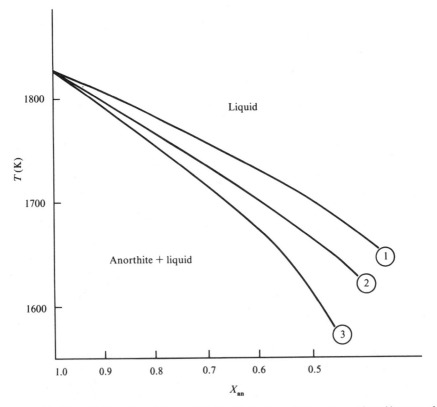

Figure 6-5. Observed depression of the anorthite liquidus in three silicate systems, plotted in terms of mole fraction $CaAl_2Si_2O_8$. Line 1: $Ca_2Al_2Si_2O_8 - 1.33\ KAlSi_2O_6$. Line 2: $Ca_2Al_2Si_2O_8 - Si_4O_8$. Line 3: $Ca_2Al_2Si_2O_8 - 1.33\ CaMgSi_2O_6$. The data indicate that the anorthite component does not mix ideally in these melts. (Data from Ludington, 1979, p. 81)

other, which implies that the assumption of ideal mixing of the anorthite component in all these melts—as defined by equation (6-30)—is inappropriate, at least with the given choice of components.

With this in mind, it is reasonable to ask how to determine directly the excess thermodynamic properties of components in silicate melts. Because of the excessive temperatures required to melt silicate minerals, a convenient approach is to measure the mixing properties of silicate glasses by dissolving them in an appropriate solvent such as lead borate. Under these conditions, measurements can be made in the range 700–800°C. This technique has been pursued with great success by A. Navrotsky and her students and colleagues (see Chapter 11).

The technique involves measurement of the heats of solution of silicate glasses in the borate solvent. If a given solution were ideal, the heat of solution should be a linear function of composition. The degree to which the measurements deviate from the projected ideal values is a measure of ΔH_{mixing} and H_{excess}. Heat-capacity data is further required to verify that these enthalpies do not change substantially between the measurement temperatures and magmatic temperatures.

For instance, Navrotsky et al. (1980) measured the heats of mixing of glasses in the system albite–anorthite–diopside (ab–an–di) in a molten PbO–B_2O_3 flux at 712°C. Results (Figure 6-6) show that these excess enthalpies do not vary with

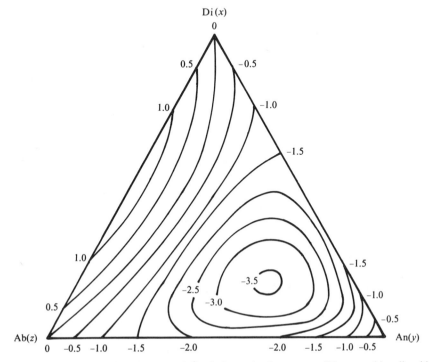

Figure 6-6. Heats of mixing (in cal·mol^{-1}) of glasses in the system albite–anorthite–diopside (ab–an–di) at 710°C. (Navrotsky et al., 1980)

composition in any simple or predictable way; note that the heats of mixing are positive on the join ab–di and negative on the ab–an and di–an joins. These data suggest that any fundamental model that would explain these results and would permit extrapolation to more complicated systems would be extremely complex. Unfortunately, these data are not sufficient by themselves to calculate activities of ab, an, and di in the melt because it is the excess free energies and not the enthalpies that are needed. However, one may resort to one of several simplified melt models to obtain the partial molal entropies, which can be combined with the measured excess enthalpies to yield the chemical potentials required to calculate the activities of all melt components. One important additional constraint is the measured entropies of ab and an glasses (e.g., Robie et al., 1978). Once all these factors are taken into account, it is possible to solve for the temperatures and compositions of melts in equilibrium with a crystalline phase. Navrotsky (1981) computed pyroxene saturation surfaces in the system ab–an–di, based on the enthalpy data referenced earlier and using a number of different entropy models. Some of the calculated diagrams are encouragingly close to the compositions of the observed liquidus surface.

Many attempts have been made to design models, both fundamental and empirical, that explain the observed melting relations of silicate minerals. Fundamental models exist to explain simple binary molten salts, and they have been adapted to silicate melts by specifying arbitrarily the nature and degree of polymerization of silicate anions in the melt (Fraser, 1977, pp. 307–312). Because of the many simplifications inherent in these models, they have not been very successful in explaining the observed behavior of the very complex silicate melts.

On the other hand, empirical models have met with much greater success. The most noteworthy of these is the *quasi-crystalline* model developed by Burnham (1981). This model assumes the existence of structural units in the melt that are either identical or very similar in both structure and composition to the crystalline phases that appear on the liquidus; the evidence supporting the existence of such units has already been mentioned. The second and most far-reaching premise of the model is that the aluminosilicate components in the melt mix ideally, given proper choice of components. The evidence for this assertion comes from measurements of the solubility of water in $NaAlSi_3O_6$ (albite) melt. Burnham (1975) showed that the activity of H_2O in the melt is related to the mole fraction of H_2O according to

$$a_{H_2O}^{melt} = K(X_{H_2O}^{melt})^2 \quad \text{for } X_{H_2O} \leq 0.5 \tag{6-31}$$

where K is the Henry's-law constant for the dissociated solute.

This Henrian behavior was explained as the combination of the hydrolysis reaction

$$H_2O_{(g)} + O_{(melt)}^{2-} = 2OH_{(melt)}^- \tag{6-32}$$

with an exchange reaction of Na^+ ions that occupy nonspecific structural sites, providing charge balance on the aluminosilicate framework, with protons present in the melt. The overall reaction (cf Burnham, 1979) is written

$$H_2O_{(g)} + O^{2-}_{(melt)} + Na^+_{(melt)} = OH^-_{(melt)} + ONa^-_{(melt)} + H^+_{(melt)} \qquad (6\text{-}33)$$

This proposed mechanism, combined with the empirical results of equation (6-31), was used by Burnham as the basis for calculating components in aluminosilicate melts on the basis of eight oxygens (i.e., albitelike units). The reaction of H_2O with the anhydrous melt depolymerizes the silicate framework, causing a marked decrease in viscosity. It was further shown that the Henry's-law constant in (6-31) is the same in $NaAlSi_3O_8$, $KAlSi_3O_8$, and $CaAl_2Si_2O_8$ melts, which means that the activity of H_2O in a generalized *feldspar* melt depends only on X_{H_2O}, for constant temperature and pressure. As shown by Burnham et al. (1978), the implication is that all three feldspar components mix ideally with each other in hydrous melts. From this premise, it is possible to study proposed ideal mixing of other melt components in an aluminosilicate "solvent." The trick once again is a proper choice of the nonfeldspar components. Burnham (1981) calculated activities of components in hydrous melts from experimentally measured liquidus temperatures and an internally consistent set of standard-state free energies. In those cases where deviations from Raoult's law were noted (i.e., $X_i^{melt} - a_i^{melt} \neq 0$ within experimental uncertainty), it was proposed that homogeneous reactions occur in the melt phases, producing new species with distinctly different stoichiometries. The implication is that calculations based on a new set of components more accurately reflecting the actual speciation present would restore Raoultian behavior to the aluminosilicate melt. This model will undoubtedly be subject to further testing in the future.

A very different approach to the calculation of activities in silicate melts is based on the premise that activities of some components in a melt may be fixed or "buffered" by melt–crystal equilibrium. For instance, the activity of SiO_2 in melt is fixed at constant temperature by the coexistence of olivine (ol) and orthopyroxene (px):

$$\begin{array}{ccc} Mg_2SiO_4 + & SiO_2 & \rightleftharpoons Mg_2Si_2O_6 \\ \text{olivine} & \text{in melt} & \text{pyroxene} \end{array} \qquad (6\text{-}34)$$

(Carmichael et al., 1970). At equilibrium,

$$\mu^{px}_{Mg_2Si_2O_6} - \mu^{ol}_{Mg_2SiO_4} - \mu^{melt}_{SiO_2} = 0 \qquad (6\text{-}35)$$

Substituting the familiar standard chemical potential and activity terms for each of the three chemical potentials in (6-35) and defining $\Delta G^\circ = \mu^\circ_{Mg_2Si_2O_6} - \mu^\circ_{Mg_2SiO_4} - \mu^\circ_{SiO_2}$ for the standard-state compositions pure Mg-orthopyroxene, Mg-olivine, and silica glass at 1 bar and T, we have

$$\ln a^{melt}_{SiO_2} = \frac{\Delta G^\circ}{RT} + \ln a^{px}_{Mg_2Si_2O_6} - \ln a^{ol}_{Mg_2SiO_4} \qquad (6\text{-}36)$$

According to (6-36), the activity of SiO_2 in a magma can be calculated at any temperature if the melt is precipitating both olivine and orthopyroxene of known composition, providing that an appropriate model exists for converting mineral compositions to activities. Lower silica activities are defined by equilibrium with mineral phases less saturated with silica such as kalsilite, leucite, and nepheline. By judicious choice of reactions, one can also define activities of components other than silica (e.g., Nicholls, 1977, p. 331). The buffering concept is important in petrology and will be discussed in following chapters.

6-5 CRYSTAL–MELT EQUILIBRIA IN SILICATE SYSTEMS: $\bar{G}$–X PLOTS AND PHASE DIAGRAMS

In the first half of the twentieth century, the melting relations in many important mineral systems were determined empirically at 1 atm. If all the thermodynamic properties of silicate melts had been known, the temperatures and compositions of all crystal–melt equilibria could have been calculated; because we are very far removed from that goal, most of our knowledge about these systems is still based on the measured phase diagrams. There are many important thermodynamic principles to be learned here. In particular, we shall illustrate how equilibrium conditions in crystal–melt systems can be read from $\bar{G}$–X diagrams.

We restrict our attention to the binary system at constant pressure (e.g., 1 bar). The phase rule requires that the number of degrees of freedom in a binary system be $f = 4 - p$. If pressure cannot vary, one of those degrees of freedom is not available and $f(\text{isobaric}) = 3 - p$. For crystal–melt equilibria, we shall ignore the possible existence of a gas phase. If S represents any crystalline phase and L represents any melt phase, then the following assemblages are possible:

Assemblage type	f	p	Possible assemblages
Invariant	0	3	2S + L, 2L + S, 3S
Univariant	1	2	S + L, 2S, 2L
Divariant	2	1	S, L

The maximum number of possible degrees of freedom is two, corresponding to temperature and phase composition. These two variables will be the coordinate axes of our phase diagram.

It is now possible to show each of the divariant, univariant, and invariant assemblages on a $\bar{G}$–X diagram, remembering that the temperature of each $\bar{G}$–X diagram is fixed. The purpose of this exercise is to show generalized relationships, so each of the following figures should be regarded as a cartoon.

1. $f = 2$. A single homogeneous solid solution or melt is stable over a range of compositions (Figure 6-7(a)).
2. $f = 1$. One degree of freedom is removed, and the compositions of all phases must be fixed because the temperature is already fixed by the $\bar{G}$–X section. The following four possibilities exist.

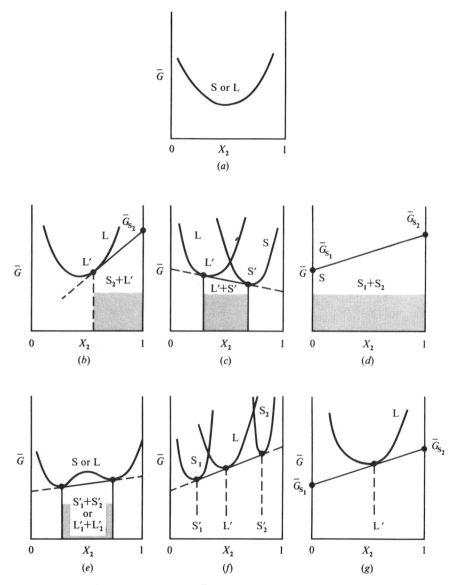

Figure 6-7. Schematic isothermal-isobaric $\bar{G}-X$ diagrams for a binary system in which solids (S) and liquids (L) are the only possible phases. See text for discussion.

a) Melt + solid of fixed composition (Figure 6-7(b)). The equilibrium requirement is $\mu_i^S = \mu_i^L$. Because the solid composition is fixed, μ_i^S is equal to $\bar{G}_{S_2}$. Thus, because μ_i^L must also equal $\bar{G}_{S_2}$, the composition of the melt phase in equilibrium with pure S_2 must be given by a tangent to the melt curve, one end of which is connected to $\bar{G}_{S_2}$. All compositions within the shaded area in Figure 6-7(b) are represented by a two-phase mixture of pure S_2 and a melt of composition L'.

b) Melt + solid solution (Figure 6-7(c)). Each phase is represented by a parabola on the $\bar{G}$–X plot. We still require $\mu_i^L = \mu_i^S$, so the equilibrium compositions of melt and solid must be given by the mutual tangent to the two curves. The reason for this should be clear. The intercept of this tangent with the $X_2 = 0$ sideline defines the condition $\mu_1^{S'} = \mu_i^{L'}$, while the intercept with the $X_2 = 1$ sideline shows $\mu_2^{S'} = \mu_2^{L'}$. Both conditions are required for equilibrium between two binary phases of variable composition. The shaded area in Figure 6-7(c) defines a two-phase region of S' + L' having the specified compositions. All compositions to the left of the shaded area will be represented by a single homogeneous liquid phase, whereas all compositions to the right of the shaded area will be a single homogeneous solid-solution phase.

c) Two stoichiometric solids (Figure 6-7(d)). Because no solid solution is possible, all compositions are represented by a two-phase mixture of the pure solids, $S_1 + S_2$.

d) Two immiscible solids or two immiscible melts (Figure 6-7(e)). This graph is equivalent to the solution curve with a maximum caused by a large excess free energy term, and it has already been discussed. The relation between the chemical potentials is the same as in Figure 6-7(c).

3. $f = 0$. Because no degrees of freedom are available, these invariant assemblages can occur only at one unique temperature (the isobarically invariant temperature). In other words, we are not free to choose the temperature of the invariant $\bar{G}$–X section. The compositions of all phases must likewise be fixed. The most significant case is the equilibrium of two solids + melt. If the chemography of the phases is

$$\underset{\vdash}{S_1'} \qquad \underset{+}{L} \qquad \underset{\dashv}{S_2'}$$

then the implied *univariant* reaction is $S_1 + S_2 = L$. Because only solids can be stable on the lefthand side of the reaction, the isobarically invariant coexistence of $S_1 + S_2 + L$ represents the lowest temperature at which liquid can be stable on the binary join; this temperature is called the *eutectic*. If, on the other hand, the chemography of the phases is

$$\underset{\vdash}{S_1'} \qquad \underset{+}{S_2'} \qquad \underset{\dashv}{L}$$

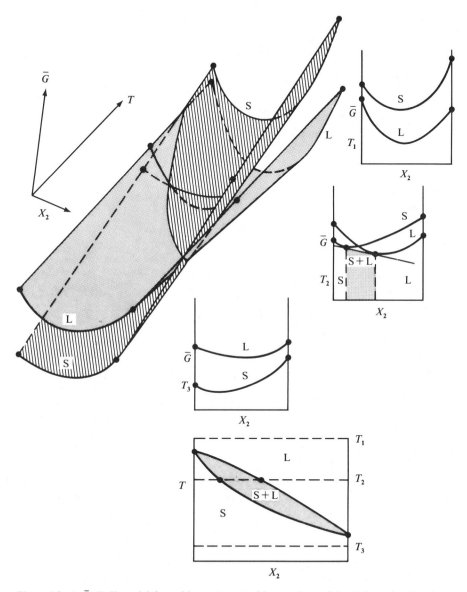

Figure 6-8. A $\bar{G}$–T–X model for a binary system with complete solid solution, showing three isothermal sections and the T–X phase diagram with the isotherms indicated on it. In the $\bar{G}$–T–X model, the shaded surface is the liquid (L) phase; the ruled surface is the solid (S) phase.

then the corresponding univariant reaction is $S_2 = S_1 + L$. Liquids are stable at temperatures both above and below the reaction point, which is also known as a *peritectic*. The invariant equilibrium condition is $\mu_i^{S_1} = \mu_i^{S_2} = \mu_i^{L'}$. Accordingly, the tangent line on the G–X plot must touch the compositions of all three phases. The eutectic can be recognized because the equilibrium liquid composition lies between the compositions of the two solids. The peritectic is distinguished by the liquid curve being positioned on the outside of the solid curves. Figure 6-7(f) shows the eutectic configuration when both solids show small degrees of solid solution, and Figure 6-7(g) represents a eutectic involving two pure solid phases S_1 and S_2.

The T–X phase diagram can be derived theoretically from a three-dimensional $\bar{G}$–T–X model that contains the free energy surfaces of the liquid and solid phases. At every temperature, the model is examined for the phase or phases with the lowest free energy according to the principles shown in Figure 6-7, and the compositions of those phases are projected onto the T–X plane. The simplest example is the melting diagram for complete solid solution. This system contains two free energy surfaces—one for the solid solution, and one for the melt. These surfaces are similar in appearance—each is a cylindrical trough that plunges downward from high to low temperature. Because of the effect of temperature on the free energy of mixing, the troughs are deeper at high temperature than they are at low temperature (Figure 6-8). Moreover, at very high temperature, the liquid trough must lie completely below the solid trough because a melt phase is stable at every composition. At very low temperature, the configuration must be reversed. Thus, at intermediate temperatures the troughs intersect, and equilibrium between liquid and solid is possible. Figure 6-8 shows three isothermal sections taken through the G–T–X model and shows how these sections are related to the T–X phase diagram. This diagram is typical of complete solid solution; this situation has been observed for the plagioclase feldspars and the (Mg,Fe) olivines.

A binary eutectic can be demonstrated by making two small but significant changes in the three-dimensional model of Figure 6-8. First, we let the melting points of the pure solids be more nearly equal, which will cause two intersections of the S and L curves for some compositions. Second, if the shaded surface representing the solid contains a large excess free energy contribution at lower temperatures, two discrete crystalline phases will form in place of one homogeneous phase at low temperatures. Figure 6-9 shows five isothermal sections through this model along with the corresponding T–X diagram (Figure 6-9(f)). The dashed line on the T–X section represents the top of the metastable solvus. Note that a maximum is present in the free energy curve for the solid below the metastable critical temperature (isotherms T_1 through T_3), whereas the maximum is absent (signifying a homogeneous solution) at temperatures above the top of the metastable solvus (isotherms T_4 and T_5). The eutectic corresponds to isotherm T_2. Based on the concept of intersecting surfaces, there is only one temperature

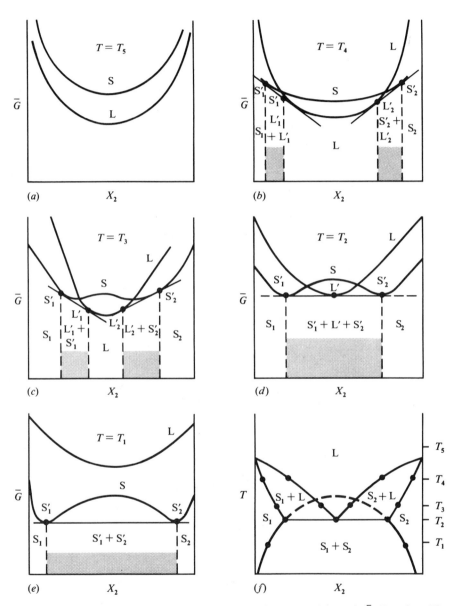

Figure 6-9. A binary eutectic showing partial solid solution. $(a-e)$ Schematic $\bar{G}-X$ sections. The dots on the phase diagram and the $\bar{G}-X$ sections indicate the compositions of coexisting phases for the particular isotherm. Shaded areas correspond to two-phase regions.

for which a single tangent can be drawn to both solid and liquid curves. This is an alternative view of the isobaric invariant condition predicted from the phase rule. This type of eutectic diagram has been recognized in the alkali feldspar system at a constant water pressure of 5 kbar.

In addition to the historical impact that phase diagrams representing silicate melt–crystal equilibrium have had on the evaluation of igneous petrology (e.g., see Morse, 1980), they are also valuable sources of thermodynamic data. As demonstrated in this section, the observed compositions of coexisting liquids and solids are indirect indicators of how the chemical potentials of these phases behave as a function of temperature. Unfortunately, as we have seen in this chapter, the chemical potentials of the melt phase depend on temperature and composition in a very complex fashion, so the phase diagrams themselves cannot be used to determine the free energy surface of the silicate-melt phase. Perhaps the major thermodynamic value of the phase diagrams is that they can be used as a check on proposed models of melt behavior. Once a model is proposed for the calculation of the free energies of silicate melts, these data can be combined with thermodynamic data for appropriate mineral phases, and equilibrium melt compositions can be calculated as a function of temperature. By comparing these calculated phase boundaries to the observed ones, one can assess the validity of the proposed solution model.

SUMMARY

- The Lewis–Randall rule for the ideal mixing of real gases is $f_i = f_i^\circ X_i$ where f_i is the fugacity of species i in the mixture, and f_i° is the fugacity of pure i at the pressure and temperature of interest. The rule works best at high temperatures, at low total pressures, and for dilute solutions of i.

- Because relatively few data are available on the compressibility of gas mixtures of geologic interest, the activities of gas species in real mixtures are most commonly calculated from one of several modified Redlich–Kwong equations of state. The equations require specific assumptions about mixing rules, but they can be used to predict activities at very high pressures.

- The activity of the ith ion in the crystallographic site of a mineral is given by $a_i = X_i^n$ if the ion mixes randomly and ideally in that site, where n is equal to the number of sites in one formula unit of the mineral.

- The activity of a component in a mineral is obtained by summing contributions from ions in all crystallographic sites for which mixing is possible. If ideal mixing occurs on all sites, $a_i = \Pi_i X_i^{n_i}$ where each of the i ions in the summation is the mole fraction of the ion present in that site relative to the composition of the pure component. The mole fraction of the component in the ideal solution is equal to the activity calculated according to the summation.

- Margules equations are one method for calculating the activities of components in nonideal crystalline solutions. They are based on excess free energy parameters that depend only on pressure and temperature. These parameters can be retrieved from measurements of the compositions of coexisting phases in equilibrium across a miscibility gap.

- Certain silicate-melt compositions are characterized by a striking degree of short-range order of structural units. However, fundamental models that have been proposed to explain thermodynamic properties of molten salts have not been very successful when applied to the more complex silicate melts. Measurements of the heats of mixing in silicate glasses, observation of immiscible liquids in some compositions, and comparison of liquidus compositions in phase diagrams indicate that most silicate melts cannot be treated with ideal models.

- Crystal–melt equilibria in binary systems can be interpreted in terms of schematic isothermal-isobaric $\bar{G}$–X diagrams.

PROBLEMS

1. Assuming the validity of the Lewis–Randall rule for gas mixtures, calculate the fugacity of CO_2 in a mixture of the following composition:

$$X_{CO_2} = 0.25$$
$$X_{CH_4} = 0.05$$
$$X_{CO} = 0.05$$
$$X_{H_2} = 0.10$$
$$X_{H_2O} = 0.55$$

The total pressure of the mixture is 1 kbar, and the temperature is 500°C. The fugacity of pure CO_2 under these conditions is found in the answer to Problem 6 of Chapter 5 (*Ans:* 309.5 bars). Calculate the fugacity of methane in the mixture using the reduced-variables plot of Figure 5-9 to estimate the fugacity of pure methane (*Ans:* 78 bars). If the activity of CO_2 in the mixture is 0.35 (standard state: pure CO_2 at P and T), calculate the true fugacity of CO_2 in the nonideal mixture (*Ans:* 433.3 bars).

2. With reference to the structural formula for the biotite given in Table 6-3, calculate the activities of the following components assuming the MOS model for all ions in all sites:

a) $KMg_3Si_3AlO_{10}(OH)_2$ (*Ans:* 0.089)

b) $KFe_3^{+2}Si_3AlO_{10}F_2$ (*Ans:* 5.48 × 10⁻⁴)

c) $K_{1/2}Mg_{3/2}Si_{3/2}Al_{1/2}O_5F$ (*Ans:* 0.048)

Problems 3 through 8 refer to a symmetrical binary solution 1–2 for which the excess free energy is represented by a one-parameter Margules equation (cf equation (6-21)). The temperature dependence of W_G is given by

$$W_G = 16000 + 1.2T$$

(W_G in joules, T in kelvins).

3. Prepare a table similar to Table 6-4 in which you set $W_{G_1} = W_{G_2} = W_G$ and solve for the chemical potentials, activity coefficients, and first and second derivatives of the solution curves for the one-parameter binary Margules solution.

4. Calculate G_{excess} and ΔG_{mixing} at 0.1 intervals of X_2 for 800 K, 1000 K, and 1200 K. Then plot ΔG_{mixing} vs X_2 for each of the three isotherms. Are homogeneous solutions stable at all three temperatures?   (*Ans:* No)

5. Prove that the critical temperature of the solvus in any such solution is given by $T_c = W_G/2R$, and then calculate the critical temperature for this solution.   *Hint:* At the critical temperature, the first and second derivatives of G with respect to X_2 are equal, and $X_1 = X_2 = 0.5$.   (*Ans:* 1037 K)

6. Prove that the spinodal curve in this system is given by $X_1 X_2 = RT/2W_G$. Calculate the spinodal for this system, and plot the results in terms of temperature and X_2.

7. Calculate the activity coefficients λ_1 and λ_2 as a function of composition at 0.1 intervals of X_2 for 1200 K, 1000 K, 800 K, and 600 K. Convert these values to activities of components 1 and 2 in the solution, and plot these activities as a function of X_2 for each of the four isotherms.   *Note:* Compare your results with Figure 6-4.

8. Because this system is symmetrical about $X_1 = X_2 = 0.5$, when two phases A and B are in equilibrium across the miscibility gap, the compositions of the coexisting phases are related by $X_1^A = (1 - X_1^B)$ and $X_2^B = (1 - X_2^A)$. These special conditions lead to the activities of *both* components being equal in *both* phases; i.e., $a_1^A = a_1^B = a_2^A = a_2^B$. Use this condition along with the four graphs you drew in Problem 7 and with the critical temperature you calculated in Problem 4 to calculate the solvus for the system. Plot your results on the same graph.

9. Consider the binary system $CaSiO_3$–$CaAl_2Si_2O_8$ and phases pseudowollastonite ($CaSiO_3$) and anorthite ($CaAl_2Si_2O_8$).

Phase	$_f\Delta H$ (kJ)	T_f (K)
Anorthite ($CaAl_2Si_2O_8$)	81.00	1830
Pseudowollastonite ($Ca_{8/3}Si_{8/3}O_8$)	73.09	1817

Calculate the liquidus surface for the system at 1 bar assuming (1) no solid solution between pseudowollastonite and anorthite, and (2) ideal behavior in the melt phase. Make all calculations on a mole-percentage basis, normalized to eight oxygens. Compare your calculated eutectic temperature and composition with the observed values of 1307°C and 47% anorthite (by weight).   (*Ans: T* (calc) = 1605 K,   44% anorthite by weight)

10. Calculate the activity of SiO_2 in a melt that is in equilibrium at 1 bar and 1400 K with olivine (in which $X_{Mg} = 0.85$ in each of two octahedral sites) and orthopyroxene (in which there are two octahedral sites: $X_{Mg} = 0.89$ in the M1 site and 0.52 in the M2 site).

Assume ideal mixing on all of the octahedral sites for both olivine and orthopyroxene. For the reaction as written, $\Delta G°$ is -9.372 kJ at 1400 K.   (*Ans*: 0.29)

11. The peritectic equivalent of Figure 6-9 is schematically drawn here:

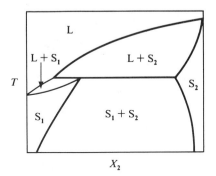

Draw $G-X$ cartoons for isotherms above and below the peritectic temperatures. Make these sketches carefully, including tangent lines wherever they are needed to show the compositions of coexisting phases.

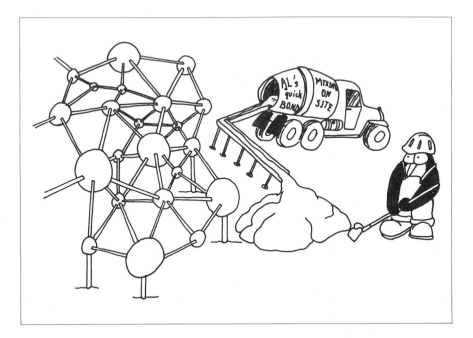

7

ELECTROLYTE THEORY

Natural waters are, as Lavoisier pointed out, the "rinsings of the earth," and during their interminable cycling they interact with the air, with the rocks, and with the biota to form solutions of great complexity and variability.

M. WHITFIELD (1979)

Few rocks have ever formed in the total absence of water. Water is abundant in the earth's atmosphere, on the earth's surface, and to great depths within the earth. During its journey from the atmosphere through the hydrologic cycle, water picks up a multitude of soluble and insoluble constituents. Soluble particles exist everywhere, created by the breakdown of rocks, by biological activity, and by human perturbations of natural systems. The physical and chemical properties of water containing dissolved substances can be markedly different from those of pure water, because of the presence of ionic conductors or electrolytes. The theory of electrolyte solutions not only is essential to our understanding of the behavior of dissolved substances in natural waters, but it also forms an integral part of the thermodynamics of solutions.

Electrolyte theory contrasts with the principles of thermodynamics in two important ways: it is *theory*, and it is *microscopic*. Thermodynamics is not considered to be theory (see McGlashan, 1979) because it deals with well tested quantitative relationships between observable bulk properties of matter. These properties are macroscopic and need no assumptions regarding the underlying microscopic picture. The study of electrolytes involves theory because it is possible mathematically to derive thermodynamic functions from basic assumptions about the microscopic behavior of charged particles. Unfortunately, the theoretical development of electrolytes has not proceeded far enough to formulate a single framework that is adequate for all species under all conditions. This chapter will show that ionic interactions can be described from many different points of view and that these various viewpoints usually necessitate a semiempirical approach to the problem. The empirical nature should be considered not a serious disadvantage of electrolyte theory but rather an inevitable outcome of trying to give quantitative meaning to the movement of atomic-sized particles whose time in contact may be on the order of picoseconds and whose structure is only slightly better defined than that of the elusive electron by Heisenberg's uncertainty principle.

The overall goal of this chapter is to derive meaningful values for the chemical potential of aqueous electrolyte solutions from measurable quantities. The fundamental equation, from Chapter 5, is:

$$\mu_i = \mu_i^\circ + RT \ln a_i \qquad (5\text{-}62)$$

where i may refer to aqueous ions. Two questions are immediately apparent: how do we define the standard-state chemical potential of an electrolyte, and how do we measure the activity of an electrolyte? Before we answer these questions, we should first look at the general behavior of single-salt solutions in terms of thermodynamic concepts already discussed.

7-1 ACTIVITY–COMPOSITION RELATIONS FOR A SINGLE ELECTROLYTE

Consider a simple aqueous system of solid halite crystals in water at 25°C and 1 bar total pressure. The saturation equilibrium reaction is written

$$NaCl_{(s)} \rightleftarrows Na^+_{(aq)} + Cl^-_{(aq)} \qquad (7\text{-}1)$$

There are two phases (halite and solution) and four chemical species (halite, water, $Na^+_{(aq)}$, and $Cl^-_{(aq)}$) if we ignore the dissociation of water. The compositions of all species can be described conveniently in terms of the two components NaCl and H_2O. The other method of finding the number of components, subtracting the number of independent chemical restrictions or equations from the total number of species, gives $4 - 1 = 3$. Because only two components are sufficient, there must be one more restriction that applies to electrolyte solutions. The missing equation is the electrical charge balance, or the condition of electrical neutrality:

$$\sum_i m_i z_i = 0 \qquad (7\text{-}2)$$

where m_i and z_i are the molality and the charge, respectively, of the ith ionic species. The charge balance is a fundamental concept in aqueous chemical calculations. It is also identical to the sum of ionic concentrations on an equivalent weight basis if negative signs are applied to anions.

Because we want to know how the activity of a soluble salt such as halite can be related to its measured concentration, we would like to have measurements from $X_{NaCl} = 0$ to $X_{NaCl} = 1$. Unfortunately, our saturated solution is only 6.144 molal, or $X_{NaCl} = 0.1$. Any binary mineral–H_2O system suffers from the same problem at low temperatures—that is, limited solubility prevents the measurement of activity over a wide range of mole fraction. For the solution to remain in the liquid state at any mole fraction, the temperature of the system must be no lower than the melting point of the pure salt (800.5°C in the case of NaCl). Most other salts of geologic interest have high melting points. Pitzer (1980) has described two systems for which the activity of water has been measured over the full compositional range from fused salt to dilute aqueous solution. These systems, shown in Figure 7-1, are $(Ag,Tl)NO_3-H_2O$ measured at 98°C, and $(Li,K)NO_3-H_2O$ measured at 119°C for the region of high mole fraction of salt and at 100°C for the remainder of the mole-fraction range. The Ag/Tl and Li/K mole ratios were fixed at about 1.0. The curves for these binary systems look very similar to those for nonelectrolytes and fluid systems such as CO_2-H_2O (compare Figure 6-1). The dashed line in Figure 7-1 shows the ideal-solution behavior where $a_{H_2O} = X_{H_2O}$. A relatively simple equation derived for nonelectrolytes fits the data quite well, as shown in Figure 7-2 (see Pitzer, 1980), and thereby avoids the pitfalls of traditional electrolyte theory. However, such a method has been little explored and is of little use in geochemistry without much more development and refinement. Both Henry's law and Raoult's law apply, and they provide a basis for standard states, reference states, and quantitative relationships between activity and concentration. For simplicity, we shall consider H_2O as component 1 and $(Li,K)NO_3$ as component 2. As the mole fraction of water approaches unity, its activity converges with ideal behavior. This convergence is

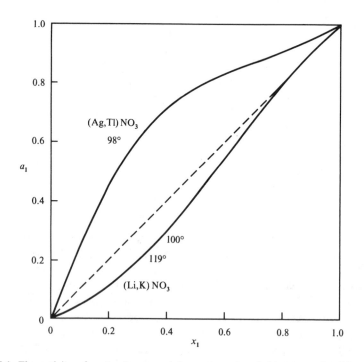

Figure 7-1. The activity of water $(a_1 = a_{H_2O})$ for water–salt solutions over the full range of composition $(X_1 = X_{H_2O})$. (From Pitzer, 1980)

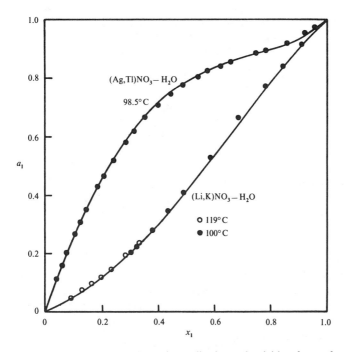

Figure 7-2. Comparison of calculated and experimentally observed activities of water for water–salt solutions over the full range of compositions. (From Pitzer, 1980)

an expression of Raoult's law:

$$\lim_{X_1 \to 1} a_1 = X_1 \tag{7-3}$$

For the same range of mole fraction, the salt behaves according to Henry's law:

$$\lim_{X_2 \to 0} a_2 = k_2 X_2 \tag{7-4}$$

At the other end of the mole-fraction range, we see the same pattern, except that here it is component 2 that follows Raoult's law and component 1 that follows Henry's law.

7-2 STANDARD STATES AND ACTIVITY COEFFICIENTS

The solute activity in aqueous solutions is a complicated function that depends upon the temperature, the pressure, and the compositions and amounts of all species in solution. As we emphasized for nonelectrolyte systems, activity has no meaning until the standard state is defined. The temperature, pressure, and concentration units are defined by the standard state, and deviations from ideality are defined by the activity coefficient. Some individual judgement must be used in choosing the appropriate standard state for the particular problem to be solved. In Chapter 5, we established that the activity a_2 of a solute component 2 replaces the mole fraction for real solutions:

$$\text{Ideal solution:} \quad \mu_2 = \mu_2^\circ + RT \ln X_2 \qquad \text{(cf 5-39)}$$

$$\text{Real solution:} \quad \mu_2 = \mu_2^\circ + RT \ln a_2 \qquad \text{(cf 5-62)}$$

We also showed that the deviation from ideality is defined by the *rational* activity coefficient:

$$\lambda_2 = \frac{a_2}{X_2} \qquad \text{(cf 5-66)}$$

Aqueous solutions are usually measured in terms of molarity rather than mole fraction. The deviation from ideality in terms of molarity M is the *practical activity coefficient*:

$$\gamma_2 = \frac{a_2}{M_2} \qquad \text{(7-5a)}$$

Alternatively, the practical activity coefficient can be defined on a molality scale by substituting $X_2 = n_2/(n_{H_2O} + n_2)$ and noting that $X_2 \approx n_2/n_{H_2O}$ for dilute solutions. If the concentration n_2/n_{H_2O} is multiplied by the conversion factor $(1000 \text{ g} \cdot \text{kg}^{-1})/(18.016 \text{ g} \cdot \text{mol}^{-1})$, or $55.51 \text{ mol} \cdot \text{kg}^{-1}$, then the mole fraction of the aqueous species is transformed to molality m, and

$$\gamma_2 = \frac{a_2}{m_2} \qquad \text{(7-5b)}$$

These definitions of practical activity coefficients lead to an inconsistency in units for activity. The original definition of activity as the relative fugacity, f_i/f_i°, gives it dimensionless units, consistent with the mole-fraction scale. But a definition in terms of molality or molarity means that either (1) the activity is taken as dimensionless, and the activity coefficient has units of reciprocal concentration, or (2) the activity takes on concentration units, and the activity coefficient is dimensionless. Both possibilities can be found in standard textbooks. Because the activity coefficient is really a correction factor demonstrating the degree of nonideality, it seems more appropriate to consider it dimensionless. Then it would be consistent with the fugacity coefficient, which is dimensionless. There is no convention that is generally acceptable, and it makes little difference which is chosen. The value will always be the same for any given concentration scale.

The standard state for an aqueous solute (component 2) in an ideal solution should be defined for $X_2 = 1$, because then

$$RT \ln X_2 = 0$$

and

$$\mu_2 = \mu_2^\circ$$

from equation (5-39). However, this state is unattainable in most geochemical systems, because mineral solubilities reach saturation values at much lower concentrations. As we pointed out earlier, few mineral-water solutions have mole fractions much above $X = 0.1$, and many of the more common minerals reach saturation at $X < 0.01$. Therefore, the only practical choice for the standard state is one based on Henry's law using the molality concentration scale. This approach not only simplifies the problem of having the solute standard state in the same physical form as the solution itself, but it greatly simplifies the calculations. This state has been adopted for most solutes, and it usually is described as a "hypothetical" 1 molal (1 m) solution at 298.15 K and 1 bar. The solution is hypothetical because any real solution would not have $a = 1$ when $m = 1$, due to a nonunity value for the activity coefficient at any measurable concentration. These general relationships are illustrated schematically in Figure 7-3. The dashed line represents the ideal solution (where $a = m$ and $\gamma = 1$), and the solid line is the actual activity curve. The standard state is chosen so as to simplify the calculations. Any real solution, of course, would have $a \neq 1$ at $m = 1$, and therefore the solute standard state is hypothetical. There *is* a real solution for which $a = 1$ that should not be confused with the standard state. This activity will occur at some $m > 1$ where the real solution line crosses $a = 1$.

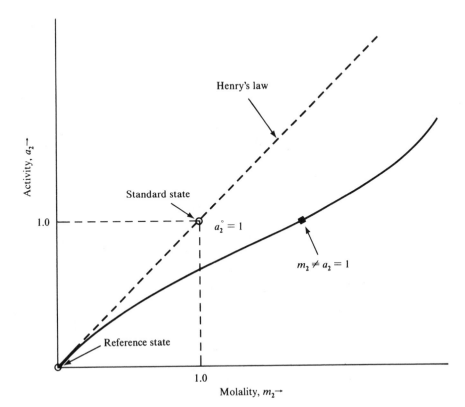

Figure 7-3. Solute standard state based on a molality scale and Henry's law.

TABLE 7-1
STANDARD STATES AND REFERENCE STATES FOR AQUEOUS SOLUTIONS

Substance	Concentration scale	Standard state*	Reference state	Chemical potential†
Water	Mole fraction, X_{H_2O}	Ideal pure liquid: $a_{H_2O} = 1$ $\lambda_{H_2O} = 1$	Ideal pure liquid: $a_{H_2O} = 1$ $\lambda_{H_2O} = 1$	$\mu_{H_2O} = \mu_{H_2O}^{\circ} + RT \ln \lambda_{H_2O} X_{H_2O}$
Solute				
a. Single solute	Molal, m_i	Hypothetical ideal solution of unit molality: $a_i = 1$ $\gamma_i = 1$	Infinitely dilute solution: $\gamma_i \rightarrow 1$ as $m_i \rightarrow 0$	$\mu_i = \mu_i^{\circ} + RT \ln \gamma_i m_i$
b. Multiple solutes	Molal, m_i	Hypothetical ideal solution of unit molality: $a_i = 1$ $\gamma_i = 1$	1. Infinitely dilute solution of all solutes: $\gamma_i \rightarrow 1$ as $\sum m_i \rightarrow 0$ 2. Constant ionic medium solution: $\gamma_i \rightarrow 1$ as $m_i \rightarrow 0$ where $\sum_j m_j$ = constant for $i \neq j$	$\mu_i = \mu_i^{\circ} + RT \ln \gamma_i m_i$

* Fixed-temperature and fixed-pressure standard states of 298.15 K and 1 bar are assumed, although other choices can be made.

† The standard-state potential μ° is expressed in mole-fraction units for water and in molal units for the solutes.

In addition to the standard state, we must define a *reference state* that is accessible to experimental measurement or extrapolation from such measurements. *The reference state for an aqueous solute is the limiting condition whereby the activity approaches the molal concentration as the concentration becomes infinitely dilute*—that is, a physically realizable limiting behavior in which the limit is the ideal solution ($\gamma = 1$). This state is really an expression of Henry's law (7-4) and can be simply stated as

$$\lim_{m_i \to 0} \frac{a_i}{m_i} = 1 \qquad (7\text{-}6)$$

Because the limiting reference state is pure water, we find that the reference states for the solute and the solvent are the same, and this fact simplifies the experimental measurements.

Natural waters contain a variety of solutes, and they are classified as mixed electrolyte solutions. Some natural waters maintain a rather constant composition—e.g., surface seawater has very nearly the same concentration of major and minor ions almost everywhere in the world. For such solutions, it is convenient to define a reference state of infinite dilution for solute component i in a solution of constant salt concentration. In a manner analogous to equation (7-6), we then obtain

$$\lim_{\substack{m_i \to 0 \\ \text{all } m_j = \text{constant} \\ (i \neq j)}} \frac{a_i}{m_i} = 1 \qquad (7\text{-}7)$$

This method of measuring solute activities, known as the *constant ionic medium method*, is used to measure stoichiometric equilibrium constants (see Chapter 8). This method is not suitable, of course, if the medium interacts with the component being measured. It is also advisable not to apply data gathered in this way to natural systems that vary considerably in medium composition. Estuaries and oceanic depth profiles vary enough in composition that this approach would not be advisable.

Table 7-1 summarizes the standard states and reference states that are used for both solvent and solute.

7-3 CONSEQUENCES OF IONIC DISSOCIATION

Pure water is a very poor conductor of electricity. The conductivity of water can be markedly enhanced by the presence of dissolved solutes, especially those solutes that are predominantly ionic in bonding character and highly soluble. This observation led Arrhenius, in 1887, to postulate that certain salts, which we call electrolytes, dissociate when dissolved in water to form charged particles, or ions. The formation of electrolytes is now accepted as a fact because of the large number of conductivity measurements that can be explained on this basis. Those electrolytes that cause large increases in conductivity—such as $NaCl$, H_2SO_4, and $Mg(NO_3)_2$—are called strong electrolytes, whereas those that make small increases—such as H_3BO_3—are called weak electrolytes.

The phenomenon of electrolytic dissociation has important consequences for the representation of solute activities. If we examine the observed Henry's-law limiting slopes for three real electrolytes (NaCl, CaCl$_2$, and AlCl$_3$), we find that they are essentially zero (see Figure 7-4) which is not the expected behavior. Furthermore, as the charge on the cation increases, the activity curve becomes progressively flatter. Now let's see what form these activities take if full dissociation is assumed. For a solute AB that fully dissociates

$$AB_{(aq)} \rightarrow A^+_{(aq)} + B^-_{(aq)} \tag{7-8}$$

and the chemical potentials are

$$\mu_{AB_{(aq)}} \equiv \mu_{A^+_{(aq)}} + \mu_{B^-_{(aq)}} \tag{7-9}$$

because chemical potentials are additive terms for the component ions that make up the total solute. Therefore, substituting (7-9) into (5-62), we obtain

$$\mu_{AB_{(aq)}} = \mu^\circ_{A^+_{(aq)}} + \mu^\circ_{B^-_{(aq)}} + RT \ln a_{A^+_{(aq)}} + RT \ln a_{B^-_{(aq)}} \tag{7-10}$$

and

$$\mu_{AB_{(aq)}} = \mu^\circ_{A^+_{(aq)}} + \mu^\circ_{B^-_{(aq)}} + RT \ln \left[a_{A^+_{(aq)}} a_{B^-_{(aq)}} \right] \tag{7-11}$$

For an ideal solution, $\gamma_{A^+_{(aq)}} = \gamma_{B^-_{(aq)}} = 1$, and

$$\mu_{AB_{(aq)}} = \mu^\circ_{A^+_{(aq)}} + \mu^\circ_{B^-_{(aq)}} + RT \ln \left[m_{A^+_{(aq)}} m_{B^-_{(aq)}} \right] \tag{7-12}$$

But, for a uni-univalent solute, we know that

$$m_{AB_{(aq)}} = m_{A^+_{(aq)}} = m_{B^-_{(aq)}} \tag{7-13}$$

Therefore,

$$\mu_{AB_{(aq)}} = \mu^\circ_{AB_{(aq)}} + RT \ln m^2_{AB_{(aq)}} \tag{7-14}$$

and we have the important relation

$$a_2 \propto m^2_2 \quad \text{and} \quad \frac{a}{m} \neq \text{constant} \tag{7-15}$$

or, more generally, for any salt that fully dissociates into v^+ positive ions and v^- negative ions for a total of $v = v^+ + v^-$ ions we find

$$\boxed{a_2 \propto m^v} \tag{7-16}$$

When we replot the data from Figure 7-4 in terms of equation (7-16), we find that the limiting slopes are no longer zero (see Figure 7-5).

One further consequence of ionic dissociation is inherent in equations (7-9) and (7-10). It must be kept in mind that, although we may write the chemical potentials for individual ions, *these potentials cannot be individually measured experimentally*. Each ion is always related to a stoichiometric amount of an oppositely charged ion by the principle of electroneutrality. We need an

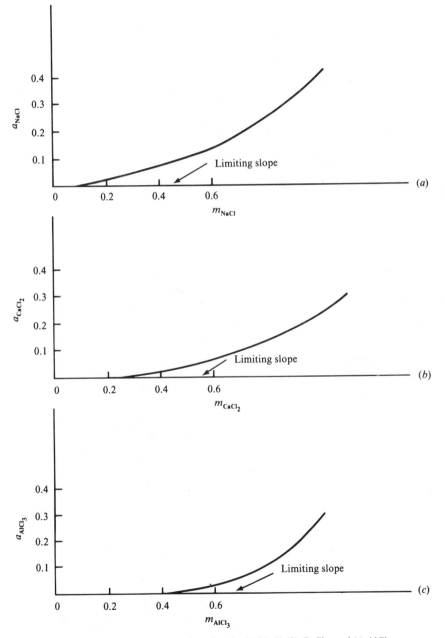

Figure 7-4. Activity–concentration plots for (a) NaCl, (b) $CaCl_2$, and (c) $AlCl_3$.

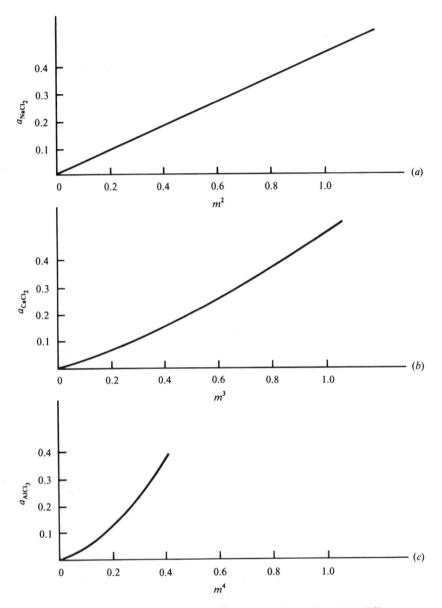

Figure 7-5. Plots of activity versus m^ν for (a) NaCl, (b) CaCl$_2$, and (c) AlCl$_3$.

expression that recognizes the existence of separate ions but allows us to make real measurements of solution activities. To meet this requirement, the concepts of mean ionic chemical potentials and mean ionic activities have been developed. For any strong electrolyte (one that fully dissociates) of general stoichiometry $A_{v^+}B_{v^-} \rightarrow v^+A^{z+} + v^-B^{z-}$, the chemical potentials of the undissociated and dissociated ions are represented by

$$\mu_{A_{v^+}B_{v^-}} = v^+\mu_{A^{z+}} + v^-\mu_{B^{z-}} \qquad (7\text{-}17)$$

Using $v = v^+ + v^-$, we can take a mean value of the ionic chemical potential represented by

$$\frac{v^+\mu_{A^{z+}} + v^-\mu_{B^{z-}}}{v} = \mu_{\pm A_{v^+}B_{v^-}} \qquad (7\text{-}18)$$

The symbol $\mu_{\pm}$ represents the mean ionic chemical potential. Substituting equation (5-62), we have

$$v\mu_{\pm} = v\mu_{\pm}^{\circ} + vRT\ln a_{\pm} \qquad (7\text{-}19)$$

where $a_{\pm}$ is the mean ionic activity. Because activity and molality are related by equation (7-5b), then we also have the mean ionic activity coefficient $\gamma_{\pm}$ and the mean ionic molality $m_{\pm}$. The relationship between the solute component parameters and the mean ionic parameters are simply

$$\mu_2 = v\mu_{\pm} = v^+\mu_+ + v^-\mu_- \qquad (7\text{-}20)$$

$$a_2 = a_{\pm}^v = a_+^{v^+}a_-^{v^-} \qquad (7\text{-}21)$$

$$\gamma_2 = \gamma_{\pm}^v = \gamma_+^{v^+}\gamma_-^{v^-} \qquad (7\text{-}22)$$

$$m_2 = m_{\pm}^v = m_+^{v^+}m_-^{v^-} = (v^+m_2)^{v^+}(v^-m_2)^{v^-} = m_2(v^+)^{v^+}(v^-)^{v^-} \qquad (7\text{-}23)$$

The quantity $(v^+)^{v^+}(v^-)^{v^-}$ will be unity for symmetrical electrolytes such as NaCl, $MgSO_4$, and $AlPO_4$ where the stoichiometries are always 1:1. This quantity will equal 4 for uni-divalent and di-univalent electrolytes such as Na_2SO_4 and $CaCl_2$, and it will equal 27 for tri-univalent and uni-trivalent electrolytes such as $AlCl_3$ and Na_3PO_4. These relationships are important to keep in mind when it becomes necessary to convert between solute activities, solute molalities, and mean ionic values. For example, an extrapolated mean activity coefficient for $Al_2(SO_4)_3$ was needed by Nordstrom (1982) to calculate a mineral solubility. The molality was the measured solute molality, and a mean ionic molality was required for the solubility calculation. For $Al_2(SO_4)_3$,

$$m_{\pm Al_2(SO_4)_3} = [(2)^2(3)^3]^{1/5}m_2 = (108)^{1/5}m_2$$

and

$$\gamma_{\pm Al_2(SO_4)_3} = [\gamma_{Al^{3+}}^2 + \gamma_{SO_4^{2-}}^3]^{1/5}$$

The general relationships between mean ionic functions and solute functions are summarized in Table 7-2. This table can be used as a guide when looking up

TABLE 7-2
IMPORTANT FUNCTIONS FOR AQUEOUS ELECTROLYTES

Function	Sucrose	Electrolyte					
		NaCl	CaCl$_2$	AlCl$_3$	MgSO$_4$	Al$_2$(SO$_4$)$_3$	A$_{\nu_+}$B$_{\nu_-}$
a_2	a_{sucrose}	$a_+ a_-$	$a_+ a_-^2$	$a_+ a_-^3$	$a_+ a_-$	$a_+^2 a_-^3$	$a_+^{\nu^+} a_-^{\nu^-}$
$a_\pm$	——	$(a_+ a_-)^{1/2}$	$(a_+ a_-^2)^{1/3}$	$(a_+ a_-^3)^{1/4}$	$(a_+ a_-)^{1/2}$	$(a_+^2 a_-^3)^{1/5}$	$(a_+^{\nu^+} a_-^{\nu^-})^{1/\nu}$
$m_\pm$	——	m_2	$4^{1/3} m_2$	$(27)^{1/4} m_2$	m_2	$(108)^{1/5} m_2$	$[(\nu^+)^{\nu^+} (\nu^-)^{\nu^-}]^{1/\nu} m_2$
$\gamma_\pm$	——	$a_\pm/m_\pm$ $= (\gamma_+ \gamma_-)^{1/2}$	$a_\pm/m_\pm$ $= (\gamma_+ \gamma_-^2)^{1/3}$	$a_\pm/m_\pm$ $= (\gamma_+ \gamma_-^3)^{1/4}$	$a_\pm/m_\pm$ $= (\gamma_+ \gamma_-)^{1/2}$	$a_\pm/m_\pm$ $= (\gamma_+^2 \gamma_-^3)^{1/5}$	$a_\pm/m_\pm$ $= (\gamma_+^{\nu^+} \gamma_-^{\nu^-})^{1/\nu}$

Source: Modified from Klotz and Rosenberg (1974).

activity coefficients (usually given as mean values) and converting to mean activities from molalities (always given as a solute molality).

Conversion to mean ionic functions also has the same effect on the solute activity curve as did the plots of a_2 versus m_2^v in Figure 7-5. By converting to mean ionic activities, we incorporate the effect of ionic dissociation because

$$a_\pm = a_2^{1/v} \propto m_2 \qquad (7\text{-}24)$$

Now we can plot the mean ionic activity or the mean ionic activity coefficient as a direct function of molality for any solute. As an example, the real and ideal behavior of $CaCl_2$ has been replotted in Figure 7-6 along with the activity of

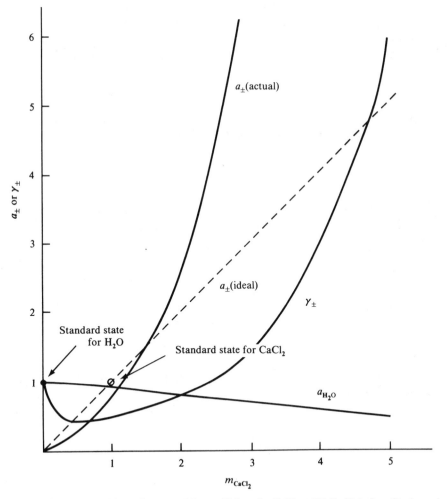

Figure 7-6. Mean activities and mean activity coefficients for $CaCl_2$ at 298 K. (Data from Staples and Nuttall, 1977)

water. Standard states for both the solute (at $a = m = \gamma = 1$) and the solvent (at $a = \gamma = X = 1$) are shown. The mean activity coefficient initially decreases rapidly at low molality, goes through a minimum at $m \simeq 0.5$, and then increases above unity rapidly at $m > 2.4$. The mean ionic activity at first shows negative deviation from the Henry's-law limiting slope but rapidly increases to a positive deviation due to the increasing activity coefficient.

7-4 INCOMPLETE DISSOCIATION OF AN ELECTROLYTE

In 1926, Bjerrum introduced the concept of ion pair formation. He proposed that oppositely charged ions attract each other and that this electrostatic attraction can overcome the disruptive forces of thermal energy for some brief period of time. The concept of ion pairs greatly aided the interpretation of conductivity data for aqueous solutions. In fact, the conductance method is one of the primary methods used to determine dissociation constants of electrolytes. Other techniques include electrochemical, spectroscopic, and solubility measurements.

Ion pair formation is another type of deviation from ideality that helps to explain the need for activity coefficients. An advantage of this concept is that ion pairs can be treated as equilibrium aqueous species that behave according to the law of mass action. For example, when calcite dissolves in water, it is postulated that bicarbonate and carbonate ions pair up with calcium ions:

$$Ca^{2+}_{(aq)} + HCO^{-}_{3(aq)} \rightleftarrows CaHCO^{+}_{3(aq)}$$

$$Ca^{2+}_{(aq)} + CO^{2-}_{3(aq)} \rightleftarrows CaCO^{\circ}_{3(aq)}$$

where $CaCO^{\circ}_{3(aq)}$ represents a neutral ion pair—not crystalline $CaCO_3$ nor the aqueous solute, $CaCO_{3(aq)}$, which is the total stoichiometric amount. These reactions effectively reduce the individual ion concentrations, and thereby modify the activities, of Ca^{2+}, HCO^{-}_3, and CO^{2-}_3. It now becomes necessary to modify our definitions of activities and activity coefficients.

Returning to the chemical potential of a uni-univalent solute, which forms a certain fraction θ of ion pairs having a chemical potential of $\mu'_{\pm}$, we see that

$$\mu_2 = \theta\mu'_{\pm} + (1 - \theta)\mu'_+ + (1 - \theta)\mu'_- \tag{7-25}$$

where μ'_+ is the chemical potential of the remaining free cation, and μ'_- is the chemical potential of the remaining free anion. Because the chemical potential of the solute must be identical at equilibrium to the total or stoichiometric chemical potential,

$$\mu_2 = \mu_+ + \mu_- = \theta\mu'_{\pm} + (1 - \theta)(\mu'_+ + \mu'_-) \tag{7-26}$$

Also, at equilibrium, the Gibbs free energy ΔG_r for the ion association reaction is given by

$$\Delta G_r = \mu'_{\pm} - \mu'_+ - \mu'_- = 0 \tag{7-27}$$

or

$$\mu'_{\pm} = \mu'_{+} + \mu'_{-}$$

By substituting (7-27) into (7-26), we find that the sum of the stoichiometric chemical potentials is equal to the sum of the chemical potentials for the individual free ions:

$$\mu_{+} + \mu_{-} = \mu'_{+} + \mu'_{-} \qquad (7\text{-}28)$$

If the chemical potential equation (5-62) is substituted for (7-28), the result is

$$\mu^{\circ}_{+} + RT \ln a_{+} + \mu^{\circ}_{-} + RT \ln a_{-} = \mu^{\circ\prime}_{+} + RT \ln a'_{+} + \mu^{\circ\prime}_{-} + RT \ln a'_{-}$$

where a'_{+} and a'_{-} represent the remaining free-cation activity and free-anion activity, respectively, after ion pairing. Because the standard state refers to an ideal solution at unit molality where no ion pairing exists, the standard-state chemical potentials are equivalent and cancel each other. This derivation leads to the equivalency of the mean stoichiometric activity and the mean free-ion activity:

$$a_{\pm}^{2} = a_{+}a_{-} = a'_{+}a'_{-} = (a'_{\pm})^{2} \qquad (7\text{-}29)$$

or

$$a_{\pm} = a'_{\pm}$$

The mean stoichiometric (or total) activity coefficient $\gamma_{\pm}$ is not equal to the mean free-ion activity coefficient $\gamma'_{\pm}$ because, in a uni-univalent electrolyte,

$$m_{2} = m_{+} = m_{-}$$

$$m'_{+} = (1 - \theta)m_{+}$$

$$m'_{-} = (1 - \theta)m_{-}$$

Substituting into equation (7-29), we obtain

$$(\gamma_{+}m_{2})(\gamma_{-}m_{2}) = [\gamma'_{+}(1 - \theta)m_{2}][\gamma'_{-}(1 - \theta)m_{2}] \qquad (7\text{-}30)$$

$$\gamma_{+}\gamma_{-} = (1 - \theta)\gamma'_{+}\gamma'_{-}$$

$$\gamma_{\pm}^{2} = (1 - \theta)^{2}(\gamma'_{\pm})^{2} = \alpha^{2}(\gamma'_{\pm})^{2}$$

$$\gamma_{\pm} = \alpha\gamma'_{\pm} \qquad (7\text{-}31)$$

where $\alpha = 1 - \theta$, the fraction of free ions.

The relationships (7-29) and (7-31) are important in the utilization of activity coefficient data to calculate mineral solubilities or aqueous ion pair reactions as shown in the following chapters. Very often, it is necessary to calculate the mean free-ion activity of a natural water. Data compilations of activity coefficients (e.g., Robinson and Stokes, 1955) are given in terms of the mean stoichiometric activity coefficients. Therefore, the $\gamma_{\pm}$ values from the literature are not equivalent to $\gamma'_{\pm}$ values needed in some thermodynamic calculations unless α is known. Alternatively, if $\gamma_{\pm}$ values are converted to $a_{\pm}$, then these values are equivalent to $a'_{\pm}$ values by equation (7-29).

Equations (7-29) and (7-31) can also be seen as justification for calculating the ion pair distribution of an aqueous solution when free-ion activity coefficients are being used. A simple example should make this point clear. Solutions of Na_2SO_4 exhibit some association between the anion and the cation by the reaction

$$Na^+_{(aq)} + SO^{2-}_{4(aq)} \rightleftarrows NaSO^-_{4(aq)}$$

At high concentrations, this association becomes more and more dominant. If we calculate individual free-ion activity coefficients (see Section 7-5) assuming no association, and combine these to calculate the mean free-ion activity coefficient, we get a curve like the dashed line shown in Figure 7-7. This curve compares very poorly with the experimental measurements. If we consider ion association and include it in the calculations, however, then the agreement with experiment is quite good. The difference between the curves is the factor α, which indicates the amount of ion pairing.

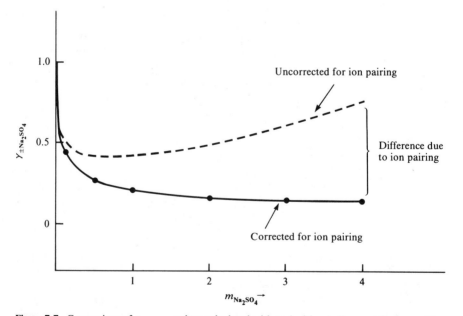

Figure 7-7. Comparison of $\gamma_{\pm Na_2SO_4}$ values calculated with and without allowance for ion pairing. Large dots represent actual measurements.

7-5 THEORETICAL EXPRESSIONS FOR MEAN ACTIVITY COEFFICIENTS

In 1923, Debye and Hückel published a paper in which they derived a relationship between the concentration of a fully dissociated electrolyte and its mean activity coefficient. This publication was a major achievement because it (1) gave very successful fits to electrolyte data at low concentrations, (2) provided a means of reliable extrapolation to infinite dilution (the reference state), and (3) was a simple equation derived from electrostatic theory. One of the clearest derivations for the Debye–Hückel equation is given by Bockris and Reddy (1970). Nearly every successful representation of electrolyte activity coefficients uses some form of the basic Debye–Hückel equation.

Before the Debye–Hückel (D–H) equation was introduced, it was already known from experimental and theoretical considerations that the logarithm of the activity coefficient is related to the square root of the concentration in dilute solutions:

$$\log \gamma = \alpha m^{1/2} \tag{7-32}$$

where α is simply a constant, independent of the solute. An important concept that aided the development of the D–H equation was the *ionic strength*, introduced by Lewis and Randall in 1921. The ionic strength measures the total concentration of charge in a solution as defined by the equation

$$I = \tfrac{1}{2}\sum_i m_i z_i^2 \tag{7-33}$$

where m_i is the molality of the ith species of z_i charge. The factor $\tfrac{1}{2}$ makes the ionic strength equivalent to the molality for uni-univalent electrolytes. A 1 m solution of NaCl, $CaCl_2$, or $AlCl_3$ will have ionic strength of

$$I = \tfrac{1}{2}[(1)(1)^2 + (1)(1)^2] \ = 1 \ m \quad \text{for 1 } m \text{ NaCl}$$

$$I = \tfrac{1}{2}[(1)(2)^2 + (2)(1)^2] \ = 3 \ m \quad \text{for 1 } m \text{ CaCl}_2$$

$$I = \tfrac{1}{2}[(1)(3)^2 + (3)(1))^2] = 6 \ m \quad \text{for 1 } m \text{ AlCl}_3$$

For a mixed solution containing 1 $mol \cdot kg^{-1}$ of each electrolyte,

$$I = \tfrac{1}{2}[(1)(1)^2 + (1)(2)^2 + (1)(3)^2 + (6)(1)^2] = 10 \ m$$

The ionic strength is a particularly useful parameter for the calculation of activity coefficients in natural waters, which are highly variable mixed-electrolyte solutions. Now we shall see how the ionic strength is useful in calculating activity coefficients.

From electrostatic theory and the consideration of only long-range attractive forces between ions or clusters of ions of opposite charge, a limiting law was derived by Debye and Hückel for an ith ion in very dilute solutions:

$$\log \gamma_i = -A z_i^2 \sqrt{I} \tag{7-34}$$

where A is a solvent parameter that depends on the density, the dielectric constant, and the temperature of the pure solvent. Equation (7-34) has the same form as equation (7-32) but includes the ion charge and a solvent parameter that has a theoretical basis. The mean activity coefficient can be derived from (7-34):

$$\log \gamma_{\pm} = -A|z_+z_-|\sqrt{I} \tag{7-35}$$

Although this limiting law was an important breakthrough, it represents activity-coefficient data only at very low concentrations ($I < 0.01$ for univalent ions; $I < 0.005$ for divalent ions; $I < 0.001$ for trivalent ions). Equation (7-35) lacks individuality and cannot account for the wide range of activities between ions of the same charge. In addition to long-range interactions involving charges of opposite sign, there are short-range interactions that become increasingly dominant as the concentration of the electrolyte increases. These short-range forces can be handled in several different ways. Debye and Hückel considered the attractive forces operating from an infinite distance up to the distance of closest approach between one ion and another of opposite charge. This distance approximates the hydrated ionic diameter and is sometimes called the "effective" ionic diameter, $\mathring{a}_i$. The complete Debye–Hückel equation is

$$\log \gamma_i = \frac{-Az_1^2\sqrt{I}}{1 + B\mathring{a}_i\sqrt{I}} \tag{7-36}$$

where B is another solvent parameter—also a function of the density, the dielectric constant of water, and the temperature. The A and B parameters are

TABLE 7-3
DEBYE–HÜCKEL SOLVENT
PARAMETERS

$T(^{\circ}C)$	A	$B \times 10^8$
0	0.4911	0.3244
25	0.5092	0.3283
50	0.5336	0.3325
75	0.5639	0.3371
100	0.5998	0.3422
125	0.6416	0.3476
150	0.6898	0.3533
175	0.7454	0.3592
200	0.8099	0.3655
225	0.8860	0.3721
250	0.9785	0.3792
275	1.0960	0.3871
300	1.2555	0.3965

Source: Helgeson and Kirkham (1974)

given in Table 7-3 for $0-300°C$, and $\mathring{a}_i$ parameters are given in Table 7-4. Equation (7-36) is a much better representation of activity coefficients than (7-35) for ionic strengths up to about 0.10 m.

Numerous attempts (with corrections for ion pairs) have been made to modify equation (7-36) so that it is reliable at higher concentrations. These developments originated with the specific ion interaction theory of Brønsted (1922). He suggested that a term linear in the molality can be added to equation (7-32) to give

$$\log \gamma = \alpha m^{1/2} + \beta m \tag{7-37}$$

in which β is a "specific ion interaction parameter." This parameter is different for every solute and, therefore, is a "specific" property of each ion or solute. Guggenheim and Scatchard (for references, see Table 7-5) made important contributions to the specific ion interaction (SII) theory by incorporating the Debye–Hückel equation (7-36), showing that the equation fits a large number of solutes at $I \leq 0.1$ *without the need for invoking ion-pair reactions* and pointing out that β is not a constant but a slowly varying function of m. These developments led Pitzer and Brewer (see Lewis and Randall, 1961) to tabulate a compact summary of electrolyte data by listing $B (= 2\beta/2.303)$ values for all available solutes. The improved form of the mean activity coefficient of a uni-univalent solute now becomes

$$\log \gamma_{\pm} = \frac{-A|z_+ z_-|\sqrt{I}}{1 + \sqrt{I}} + Bm \tag{7-38}$$

It should always be kept in mind that SII equations such as (7-38) are really stoichiometric activity coefficient expressions because the interaction parameter incorporates weak ion-pairing effects.

Many modifications of (7-38) have been tried to extend the range of validity to $I \geq 0.1$; some of these are shown in Table 7-5. The most successful modification is the approach derived by Pitzer (1977, 1979), based on recent advances in the statistical-mechanical treatment of electrolyte solutions. The basic premise is that ionic solutions can be modeled by a virial-type equation of state that can account for all specific ion interactions except strongly bound, covalent association of ions. The general equation becomes

$$\ln \gamma_{\pm} = \frac{-A}{3}|z_+ z_-|f(I) + \frac{2v_+ v_-}{v}B(I)m + \frac{2(v_+ v_-)^{3/2}}{v}Cm \tag{7-39}$$

where

$$f(I) = \frac{\sqrt{I}}{1 + 1.2\sqrt{I}} + 1.67\ln(1 + 1.2\sqrt{I})$$

$$B(I) = 2\beta° + \frac{2\beta'}{\alpha^2 I}[1 - (1 + \alpha I^{1/2} - \tfrac{1}{2}\alpha^2 I)e^{-\alpha I^{1/2}}]$$

where

TABLE 7-4

EFFECTIVE DIAMETERS OF HYDRATED IONS IN AQUEOUS SOLUTIONS

$10^8 a_i$ (cm)	Inorganic Ions: Charge 1
9	H^+
6	Li^+
4–4.5	Na^+, $CdCl^+$, ClO_2^-, IO_3^-, HCO_3^-, $H_2PO_4^-$, HSO_3^-, $H_2AsO_4^-$, $Co(NH_3)_4(NO_2)_2^+$
3.5	OH^-, F^-, NCS^-, NCO^-, HS^-, ClO_3^-, ClO_4^-, BrO_3^-, IO_4^-, MnO_4^-
3	K^+, Cl^-, Br^-, I^-, CN^-, NO_2^-, NO_3^-
2.5	Rb^+, Cs^+, NH_4^+, Tl^+, Ag^+

	Inorganic Ions: Charge 2
8	Mg^{2+}, Be^{2+}
6	Ca^{2+}, Cu^{2+}, Zn^{2+}, Sn^{2+}, Mn^{2+}, Fe^{2+}, Ni^{2+}, Co^{2+}
5	Sr^{2+}, Ba^{2+}, Ra^{2+}, Cd^{2+}, Hg^{2+}, S^{2-}, $S_2O_4^{2-}$, WO_4^{2-}
4.5	Pb^{2+}, CO_3^{2-}, SO_3^{2-}, MoO_4^{2-}, $Co(NH_3)_5Cl^{2+}$, $Fe(CN)_5NO^{2-}$
4	Hg_2^{2+}, SO_4^{2-}, $S_2O_3^{2-}$, $S_2O_8^{2-}$, SeO_4^{2-}, CrO_4^{2-}, HPO_4^{2-}, $S_2O_6^{2-}$

	Inorganic Ions: Charge 3
9	Al^{3+}, Fe^{3+}, Cr^{3+}, Sc^{3+}, Y^{3+}, La^{3+}, In^{3+}, Ce^{3+}, Pr^{3+}, Nd^{3+}, Sm^{3+}
6	$Co(ethylenediamine)_3^{3+}$
4	PO_4^{3-}, $Fe(CN)_6^{3-}$, $Cr(NH_3)_6^{3+}$, $Co(NH_3)_6^{3+}$, $Co(NH_3)_5H_2O^{3+}$

	Inorganic Ions: Charge 4
11	Th^{4+}, Zr^{4+}, Ce^{4+}, Sn^{4+}
6	$Co(S_2O_3)(CN)_5^{4-}$
5	$Fe(CN)_6^{4-}$

	Inorganic Ions: Charge 5
9	$Co(SO_3)_2(CN)_4^{5-}$

	Organic Ions: Charge 1
8	$(C_6H_5)_2CHCOO^-$, $(C_3H_7)_4N^+$
7	$[OC_6H_2(NO_2)_3]^-$, $(C_3H_7)_3NH^+$, $CH_3OC_6H_4COO^-$
6	$C_6H_5COO^-$, $C_6H_4OHCOO^-$, $C_6H_4ClCOO^-$, $C_6H_5CH_2COO^-$, $CH_2{=}CHCH_2COO^-$, $(CH_3)_2CHCH_2COO^-$, $(C_2H_5)_4N^+$, $(C_3H_7)_2NH_2^+$
5	$CHCl_2COO^-$, CCl_3COO^-, $(C_2H_5)_3NH^+$, $(C_3H_7)NH_3^+$
4.5	CH_3COO^-, CH_2ClCOO^-, $(CH_3)_4N^+$, $(C_2H_5)_2NH_2^+$, $NH_2CH_2COO^-$
4	$NH_3^+CH_2COOH$, $(CH_3)_3NH^+$, $C_2H_5NH_3^+$
3.5	$HCOO^-$, H_2-citrate$^-$, $CH_3NH_3^+$, $(CH_3)_2NH_2^+$

TABLE 7-4 (Continued)

	Organic Ions: Charge 2
7	$OOC(CH_2)_5COO^{2-}$, $OOC(CH_2)_6COO^{2-}$, Congo-red anion^{2-}
6	$C_6H_4(COO)_2^{2-}$, $H_2C(CH_2COO)_2^{2-}$, $(CH_2CH_2COO)_2^{2-}$
5	$H_2C(COO)_2^{2-}$, $(CH_2COO)_2^{2-}$, $(CHOHCOO)_2^{2-}$
4.5	$(COO)_2^{2-}$, H-citrate^{2-}

	Organic Ions: Charge 3
5	Citrate^{3-}

Source: Kielland (1937)

β° and β' are specific ion parameters, α is a constant for a single class of electrolytes (e.g., for all di-divalent electrolytes), and C is also a specific ion parameter independent of ionic strength. The B function is analogous to a second virial coefficient, and C is analogous to a third virial coefficient. Although this equation seems cumbersome, it does fit the $\gamma_{\pm}$ of many electrolytes for molalities up to 6 m, and it can easily be adapted for mixed electrolytes by summing the interaction parameters from the single electrolyte data. The general applicability of this method, without the need for a vast array of ion-association constants, holds considerable promise for interpreting mineral equilibrium reactions in saline waters and brines.

Example. Calculate the mean activity coefficient for $CaCl_2$ at $m = 1$ and 25°C using (a) the D–H limiting law, (b) the D–H equation, and (c) the Davies equation.

   a) Debye–Hückel limiting law:

$$\log \gamma_{\pm} = -A|z_+z_-|\sqrt{I} = -0.509 \times 2 \times \sqrt{3} = -1.7632$$

$$\gamma_{\pm} = 0.0172$$

   b) Debye–Hückel equation ($\mathring{a} = \mathring{a}_{Ca^{2+}} + \mathring{a}_{Cl^-}$):

$$\log \gamma_{\pm} = \frac{-A|z_+z_-|\sqrt{I}}{1 + B\mathring{a}\sqrt{I}} = \frac{-0.509 \times 2 \times \sqrt{B}}{1 + 0.328 \times 4.5 \times \sqrt{3}} = -0.495$$

$$\gamma_{\pm} = 0.320$$

   c) Davies equation (See Table 7-5):

$$\log \gamma_{\pm} = -A|z_+z_-|\left(\frac{\sqrt{I}}{1+\sqrt{I}} + 0.3I\right) = -0.509 \times 2 \times \left(\frac{\sqrt{3}}{1+\sqrt{3}} + 0.9\right)$$

$$= 1.5616$$

$$\gamma_{\pm} = 0.0274$$

TABLE 7-5

THEORETICAL EQUATIONS FOR MEAN ACTIVITY COEFFICIENTS

Name	Equation	Reference
Debye–Hückel limiting law	$\log \gamma_{\pm} = -A|z_+ z_-|\sqrt{I}$	Debye and Hückel (1923)
Brønsted	$= -A|z_+ z_-|\sqrt{m} + \beta m$	Brønsted (1922)
Extended Debye–Hückel	$= \dfrac{-A|z_+ z_-|\sqrt{I}}{1 + B\mathring{a}\sqrt{I}}$	Debye and Hückel (1923)
Güntelberg	$= \dfrac{-A|z_+ z_-|\sqrt{I}}{1 + \sqrt{I}}$	Güntelberg (1926)
Brønsted–Guggenheim–Scatchard	$= \dfrac{-A|z_+ z_-|\sqrt{I}}{1 + \sqrt{I}} + \nu\beta m$	Brønsted (1922); Hückel (1925); Scatchard (1936); Guggenheim (1935); Guggenheim and Turgeon (1955)
Davies	$= -A|z_+ z_-|\left(\dfrac{\sqrt{I}}{1 + \sqrt{I}} + 0.3I \right)$	Davies (1938, 1962)
National Bureau of Standards	$= \dfrac{-A|z_+ z_-|\sqrt{I}}{1 + B'\sqrt{I}} + CI + DI^2 + \cdots$	Lietzke and Stoughton (1962); Hamer and Wu (1972); Staples and Nuttall (1977)

Note: I = ionic strength; $A = 1.824928 \times 10^6 \times \rho_0^{1/2}/(\varepsilon T)^{3/2} \, \text{mol}^{-1/2} \cdot (\text{kg H}_2\text{O})^{1/2}$; $B = 50.29117 \times 10^8 \times \rho_0^{1/2}/(\varepsilon T)^{1/2} \, \text{cm}^{-1} \cdot \text{mol}^{-1} \cdot (\text{kg H}_2\text{O})^{1/2}$; $\mathring{a}$ = ion-size parameter; β = empirical fitting parameter now considered to be a function of the ionic strength; B', C, D, ... are all empirical fitting coefficients, specific for each electrolyte and independent of ionic strength; ε is the dielectric constant for water and ρ_0 is the density of water.

The experimental value is $\gamma_\pm = 0.496$. Clearly, the complete Debye–Hückel equation fits better than the limiting law or the Davies equation, although it is still in error by 36%.

7-6 INDIVIDUAL ION ACTIVITY COEFFICIENTS

The geochemical interpretation of natural waters is made considerably easier if individual ionic activities can be measured or calculated. Unfortunately, individual ion properties cannot be independently measured because each ion must be accompanied by an ion of the opposite charge—a circumstance reflected by the principle of electroneutrality. Individual ion activity coefficients can be calculated using any one of a number of equations such as those shown in Table 7-5, but these are mostly valid in rather dilute solutions.

The mean-salt method (Garrels and Christ, 1965) is an old procedure that provides reasonable estimates for individual ion activity coefficients. This method is based on the MacInnes convention, which says that

$$\gamma_{\pm KCl} = \gamma_{K^+} = \gamma_{Cl^-} \tag{7-40}$$

because the ion size and ion mobility of K^+ and Cl^- are nearly equivalent in dilute solutions (MacInnes, 1919). From this assumption, the activities of other ions can be calculated. For example, the activity coefficient of $Ca^{2+}_{(aq)}$ is

$$\gamma_{\pm CaCl_2} = (\gamma_{Ca^{2+}} \gamma^2_{Cl^-})^{1/3} = (\gamma_{Ca^{2+}} \gamma^2_{\pm KCl})^{1/3}$$

and

$$\gamma_{Ca^{2+}} = \frac{\gamma^3_{\pm CaCl_2}}{\gamma^2_{\pm KCl}}$$

or, for $Al^{3+}_{(aq)}$,

$$\gamma_{\pm AlCl_3} = (\gamma_{Al^3} \gamma^3_{Cl^-})^{1/4} = (\gamma_{Al^3} \gamma^3_{\pm KCl})^{1/4}$$

$$\gamma_{Al^{3+}} = \frac{\gamma^4_{\pm AlCl_3}}{\gamma^3_{\pm KCl}}$$

Example. Calculate the activity coefficient of Mg^{2+} at $I = 3$ from the mean activity coefficients. From Robinson and Stokes (1955), we have

$$\gamma_{\pm KCl} = 0.569 \quad \text{at } I = 3 \quad (m = 3)$$

$$\gamma_{\pm MgCl_2} = 0.570 \quad \text{at } I = 3 \quad (m = 1)$$

$$\gamma_{Mg^{2+}} = \frac{(0.570)^3}{(0.569)^2} = 0.572$$

There are some obvious problems with the mean-salt method. For example, suppose we calculate γ_{Na^+} from NaCl and from NaBr solutions. At the same ionic strength, we should get the same value. For $I = 2$, we have

$$\gamma_{Na^+(NaCl)} = \frac{\gamma^2_{\pm NaCl}}{\gamma_{\pm KCl}} = \frac{(0.668)^2}{0.573} = 0.779$$

However, because

$$\gamma_{Br^-} = \frac{\gamma^2_{\pm KBr}}{\gamma_{\pm KCl}}$$

we have

$$\gamma_{Na^+(NaBr)} = \frac{\gamma^2_{\pm NaBr}\gamma_{\pm KCl}}{\gamma^2_{\pm KBr}}$$

$$= \frac{(0.731)^2(0.573)}{(0.593)^2} = 0.871$$

There is a large difference between individual ion activity coefficients calculated by different pathways. This difference reflects specific ion properties at high ionic strengths and points out the disadvantages of the method: (1) the MacInnes assumption is not valid at moderate to high ionic strengths; (2) the activity coefficients are *not* independent of the medium; and (3) the method is not valid for electrolytes that associate. Nevertheless, this approximation does help to extend the usefulness of the Debye–Hückel equation.

Single-ion activity coefficients can be more successfully calculated for a broader range of ionic strengths by using the SII method. Mixed-electrolyte solutions such as natural waters can be modeled with this procedure (see Whitfield, 1975a,b,c, 1979; Harvie and Weare, 1980), but the lack of interaction and mixing parameters for some solutes has made it less generally applicable than the ion-association (IA) model. The SII calculations are much more reliable at high ionic strengths, whereas the IA method is more versatile.

TABLE 7-6
TRUESDELL–JONES
INDIVIDUAL ION ACTIVITY-
COEFFICIENT PARAMETERS

Ion	$\mathring{a}$	b
Ca^{2+}	5.0	0.165
Mg^{2+}	5.5	0.20
Na^+	4.0	0.075
K^+	3.5	0.015
Cl^-	3.5	0.015
SO_4^{2-}	5.0	-0.04
HCO_3^-	5.4	0
CO_3^{2-}	5.4	0

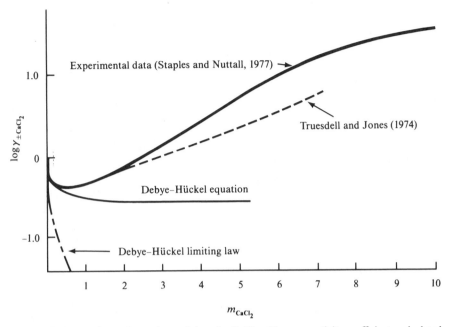

Figure 7-8. Comparison of experimental data for $CaCl_2$ with mean activity coefficients calculated from three theoretical formulas.

A noteworthy compromise has been offered by Truesdell and Jones (1974) to calculate ion activities in natural waters. By combining a simplified form of the Brønsted–Guggenheim equation (see Table 7-5) with the mean-salt method, they were able to derive reasonable single-free-ion activity coefficients for the major ions in natural waters. The activity coefficients for Ca^{2+}, Mg^{2+}, Na^+, K^+, Cl^-, and SO_4^{2-} were calculated by the mean-salt method and then fitted to the equation

$$\log \gamma_i = \frac{-Az_i^2 \sqrt{I}}{1 + B\mathring{a}\sqrt{I}} + bI$$

where $\mathring{a}$ and b are constant for each ion (Table 7-6).

In Figure 7-8, the experimental mean activity coefficient of $CaCl_2$ is shown along with the Debye–Hückel and Truesdell–Jones approximations. These curves show that the Debye–Hückel equation begins to deviate from the experimental curve at about $m = 0.2$ $(I = 0.6)$, whereas the Truesdell–Jones formulation is reliable up to about $2\,m$ $(I = 6)$. Improved equations must include interaction parameters, which usually are empirical fits to the residual deviations (after Debye–Hückel electrostatic interactions have been accounted for). Because more reliable expressions usually involve empirical fitting, the National Bureau of Standards has used a simple power series to fit mean activity coefficients in their evaluations (see Staples and Nuttall, 1977)—e.g.,

$$\log \gamma_\pm = \frac{-A|z_+ z_-|\sqrt{I}}{1 + B'\sqrt{I}} + CI + DI^2 + EI^3 + \cdots$$

where $B', C, D, E, \ldots$ are fitting parameters. This form is convenient for smoothing data and making calculations, but it lacks the flexibility to be applied to mixed-electrolyte solutions.

We shall digress briefly to discuss a question that may have occurred to you: If the SII method works well without any explicit accounting for ion pairs and the IA method works well with ion pairs, is there really ion association or is there not? Unfortunately, there is no completely satisfactory answer to this question. Spectroscopic techniques that demonstrate ion pairing do not give the same amount of association for different techniques. Interionic distances tend to become vague and depend heavily on experimental technique. The problem has been elegantly summed up by Pitzer and Brewer (see Lewis and Randall, 1961).

What do we mean by degree of dissociation?

While we have this question before us and before we return to our purely thermodynamic treatment, it may be of interest to view for a moment the logical implications of such a term as degree of dissociation.

Let us consider the equilibrium in the vapor phase, between diatomic and monatomic iodine, and at such a temperature that on the average each molecule of I_2, after it has been formed by combination of two atoms, remains in the diatomic condition 1 min before it redissociates. During this minute such a molecule will traverse several miles in a zigzag path, and after its dissociation each of its constituents will traverse a similar path before it once more combines with another atom. If we imagine an instantaneous photograph of such a gaseous mixture, with such enormous magnifying power as to show us the molecules as they actually exist at any instant, then by counting the single and double molecules we should doubtless find the same degree of dissociation which is actually determined by physicochemical methods.

On the other hand, if we should choose a condition in which the dissociation and reassociation occurs 10^{13} or 10^{14} times as frequently, the atoms of the dissociated molecules would hardly emerge from one another's sphere of influence before they would once more combine with each other or with new atoms. In such a case the time required in the process of dissociation would be comparable with the total time during which the atoms would remain free, and even our imaginary instantaneous photograph would not suffice to tell us the degree of dissociation. For, first, it would be necessary to know how far apart the constituent atoms of a molecule must be to warrant our calling the molecule dissociated. But such a decision would be arbitrary; and, according to our choice of this limiting distance, we should find one or another degree of dissociation.

Until a problem has been logically defined, it cannot be experimentally solved; and it seems evident in such a case as we are now considering that, just as we should obtain different degrees of dissociation by different choices of the limiting distance, so we should expect to find different degrees of dissociation when we come to interpret different experimental methods.

Now it is generally agreed that ionic reactions are among the most rapid of chemical processes, and it is in just such reactions that we should expect to find difficulty in determining, either logically or experimentally, a really significant value of the degree of dissociation.

On the whole, we must conclude that the degree of dissociation and the concentration of the ions are quantities which cannot be defined without some degree of arbitrariness. If consistency is maintained, no inaccuracy is introduced into a purely thermodynamic treatment. The problem is quite analogous to the treatment of NO_2, where one can arbitrarily choose to treat the vapor as a mixture of monomer and dimer or to take only a single component and to treat the interactions between monomers by a virial equation. Likewise in solutions of electrolytes, we can arbitrarily treat the system as a mixture of undissociated molecules and of ions, or we can treat the electrolyte as being completely dissociated, but taking into account the interaction between ions.

The practical decision of whether or not to recognize undissociated molecules in a strong electrolyte solution may be influenced by the existence of nonthermodynamic properties which distinguish clearly between dissociated and undissociated species. Thus Raman spectra show distinct features characteristic of HNO_3 molecules and NO_3^- ions, and it is possible to use the intensities of these spectral bands to calculate the degree of dissociation of nitric acid in concentrated solutions. In dilute aqueous solution the HNO_3 band disappears, and we know that HNO_3 behaves as a typical strong electrolyte. It should be realized, however, that the selection of the Raman spectrum as the criterion of dissociation is arbitrary, that other experimental phenomena might give different values of the degree of dissociation, and that it is possible to give an exact thermodynamic treatment on the assumption of complete dissociation.

The attempt to find a universally applicable activity-coefficient equation for all solutes at any concentration and any mixture is still being debated and researched. In the following chapters, we shall make use of the equations described in this chapter to show what can be done to compute chemical equilibrium properties of geochemical fluids and to show the limitations that exist in these computations.

SUMMARY

- Aqueous electrolyte solutions are solutions that conduct an electric current by charged particles called ions. In addition to the usual constraints of chemical equilibrium, electrolytes are subject to the principle of electroneutrality:

$$\sum_i m_i z_i = 0$$

- The standard state for aqueous solutes is based on Henry's law; it is an ideal, hypothetical, 1 molal (1 m) solution in which $a_2 = m_2 = \gamma_2 = 1$ at 298 K and 1 bar.

- There are two reference states for aqueous solutes in solution:
  1. the limiting condition in which

$$\lim_{m_i \to 0} \frac{a_i}{m_i} = 1$$

  2. the constant ionic medium reference state

$$\lim_{m_i \to 0} \frac{a_i}{m_i} = 1$$

where m_j = constant for $i \neq j$.

- The dissociation or ionization of electrolytes leads to the need for mean ion chemical potentials $\mu_\pm$, mean ion activities $a_\pm$, and mean ion activity coefficients $\gamma_\pm$.

- Some electrolytes show partial association to form ion pairs. This complication requires a correction factor (the fraction α of free ions) to relate the stoichiometric mean activity coefficient to the mean free-ion activity coefficient.

- The activity coefficient of solutes or individual ions can be calculated from the theoretical equations of Debye and Hückel:

$$\log \gamma_i = \frac{-Az_i^2\sqrt{I}}{1 + B\mathring{a}\sqrt{I}}$$

up to ionic strengths of about 0.1 m. For higher ionic strengths, specific-ion-interaction parameters may be added to this equation. The most successful modification has been the Pitzer equations. A less reliable, but equally convenient, approach used by Truesdell and Jones (1974) gives a useful expression for equilibrium calculations in natural waters.

PROBLEMS

1. Derive the mean ionic molality for the solute $Ca_3(PO_4)_2$ in terms of the solute molality. What is the expression for the mean ion activity coefficient for fully dissociated $Ca_3(PO_4)_2$?

2. Using the Debye–Hückel equation (7-36), calculate the mean activity coefficients for $NaCl$, $MgCl_2$, and $MgSO_4$, and compare these values to the experimental values from Robinson and Stokes (1955) for $I = 0.1, 0.5, 1.0, 2.0, 3.0$ and 4.0. Use Table 7-4 for $\mathring{a}$ values, and average the ion sizes. At what ionic strengths do the deviations become greater than 10% for each salt?

3. Using the mean activity coefficient data for $CaCl_2$ from Robinson and Stokes (1955), plot the difference function

$$\log \gamma_\pm + \frac{A|z_+ z_-|\sqrt{I}}{1 + \sqrt{I}}$$

(refer to equation (7-38)), and show that it is a well behaved, increasing function of molality. What is the interaction parameter B for $CaCl_2$?

4. Using the Truesdell–Jones parameters, plot the individual ion activity coefficients for the major ions in natural waters (Ca^{2+}, Mg^{2+}, Na^+, K^+, Cl^-, SO_4^{2-}, and HCO_3^-) from $I = 0$ to $I = 5$. What can you say about the differences between negative ions and positive ions? between univalent and divalent ions?

5. The major constituents in average surface seawater are shown below in mg/kg

Cl	19350	Ca	411
Na	10760	K	399
SO$_4$	2710	HCO$_3$	142
Mg	1290	Br	67

Density $= 1.0234$ g/cm^3

Calculate (a) the molality of each constituent, (b) the ionic strength of seawater and (c) the single–ion activity coefficients for each constituent using the Truesdell–Jones formulation and assuming that the values of $\mathring{a}$ and b for Br are 3 and 0, respectively.

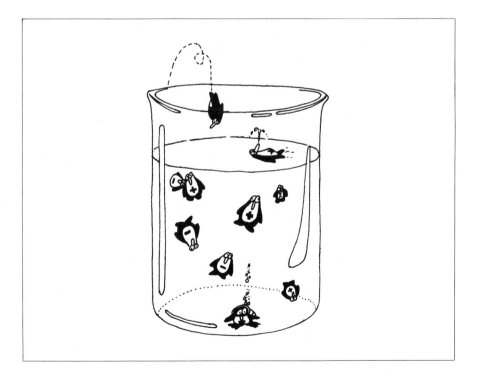

8

MINERAL EQUILIBRIA
I. THE EQUILIBRIUM
CONSTANT

*Two Norwegian investigators of the last century, Guldberg and Waage,
stumbled almost by accident on the key to the numerical handling of chemical
equilibrium.... Working with slow reactions, they were impressed with the
effect on these reactions of changing the amounts of reactants and concluded
that the "driving force" of a reaction depends on the mass of each substance.*

K. KRAUSKOPF (1979)

In preceding chapters, geochemical reactions have been described in terms of
enthalpies, entropies, free energies, and their partial molal quantities such as
chemical potentials. We are now ready to consider heterogeneous multicomponent equilibria in terms of P, T, and compositional variables, especially activities.
Chemical potentials for heterogeneous reacting systems are obtained from
activities, and the resulting relationship, the equilibrium constant, is an essential
feature of most geochemical calculations. Equilibrium constants—that are also
functions of pressure, temperature, and composition—are the main topic of this
chapter. The following two chapters explore various ways of representing
mineral equilibria by diagrams and the additional complication of oxidation-
reduction in geochemistry. These three chapters complete the essential features
of geochemical thermodynamic principles.

8-1 THE EQUILIBRIUM CONSTANT

Calculation of mineral equilibria in systems where all phases have constant
composition can be approached in straightforward fashion through Gibbs free
energies and the dependence of these functions on pressure and temperature.

However, when one or more of the phases in question is a solution, chemical potentials clearly are more convenient descriptors because they incorporate the variation of free energy with composition.

For a general reaction such as

$$a\,A + b\,B \rightleftharpoons c\,C + d\,D \tag{8-1}$$

carried out under isobaric and isothermal conditions, the Gibbs free energy of a system composed of reactant species A and B and product species C and D is, by expansion of equation (5-11),

$$G = n_A\mu_A + n_B\mu_B + n_C\mu_C + n_D\mu_D \tag{8-2}$$

At equilibrium, the free energies of product and reactant assemblages are equal, so we have

$$\Delta G_r = (n_C\mu_C + n_D\mu_D) - (n_A\mu_A + n_B\mu_B) = 0 \tag{8-3}$$

In any chemical reaction, the absolute values of the stoichiometric coefficients are the same as the number of moles of each reacting species. Furthermore, because stoichiometric coefficients of reactants are negative by convention, we have the following important condition relating the chemical potentials in an equilibrium system:

$$\sum_i v_i\mu_i = 0 \tag{8-4}$$

where the index i in the summation refers to all species or phases that appear in the reaction as written. For example, we would write the reaction for gypsum–anhydrite equilibrium as

$$CaSO_4 \cdot 2H_2O_{(s)} \rightleftharpoons CaSO_{4(s)} + 2H_2O_{(l)}$$

and

$$\mu_{anhydrite} + 2\mu_{H_2O} - \mu_{gypsum} = 0 \tag{8-5}$$

If we are dealing with a variable composition phase such as jarosite, $KFe_3(SO_4)_2(OH)_6$, which commonly contains Na^+ and H_3O^+ replacing K^+ and Al^{3+} replacing Fe^{3+}, then the buffer reaction proposed for acid mine-tailings piles (Miller, 1979, 1980)

$$K^+_{(aq)} + 3\,Fe(OH)_{3(s)} + 2\,SO^{2-}_{4(aq)} + 3\,H^+_{(aq)} \rightleftharpoons KFe_3(SO_4)_2(OH)_{6(s)} + 3\,H_2O_{(l)}$$

might be represented by

$$\mu^{jarosite}_{KFe_3(SO_4)_2(OH)_6} + 3\mu_{H_2O} - \mu_{K^+} - 3\mu_{Fe(OH)_3} - 2\mu_{SO_4^{2-}} - 3\mu_{H^+} = 0 \quad (8-6)$$

Note that in this example the equilibrium condition is described with respect to "pure" $KFe_3(SO_4)_2(OH)_6$ in the jarosite phase.

The equilibrium-constant expression is derived by recalling the chemical potential of a species i,

$$\mu_i = \mu^\circ_i + RT \ln a_i \tag{5-62}$$

and substituting this into equation (8-4):

$$\sum_i v_i \mu_i^\circ + RT \sum_i v_i \ln a_i = 0$$

or

$$\sum_i v_i \mu_i^\circ + RT \ln \prod_i a_i^{v_i} = 0 \tag{8-7}$$

where $\prod_i a_i^{v_i}$ is the continuous product of all activities, each raised to the power of its stoichiometric coefficient. The first term on the lefthand side is the sum of the standard-state chemical potentials multiplied by the appropriate stoichiometric coefficients, and thus this term is equivalent to the standard-state Gibbs free energy for the reaction:

$$\sum_i v_i \mu_i^\circ = \Delta G_r^\circ = -RT \ln \prod_i a_i^{v_i} \tag{8-8}$$

For a given standard state at a fixed temperature, there can be only one fixed value of ΔG° and therefore a fixed value of $\prod_i a_i^{v_i}$. This constant product of activities is therefore called the equilibrium constant K, and

$$\boxed{\Delta G_r^\circ = -RT \ln K} \quad \text{or} \quad \boxed{\Delta G_r^\circ = -2.303 RT \log K} \tag{8-9}$$

Equation (8-9) is one of the most important and useful relationships in chemical thermodynamics. Again considering the gypsum–anhydrite equilibria, the equilibrium constant for equation (8-5) is:

$$K = \frac{a_{CaSO_4} a_{H_2O}^2}{a_{CaSO_4 \cdot 2H_2O}} \tag{8-10}$$

Likewise, for the jarosite–ferric hydroxide buffer,

$$K = \frac{a_{KFe_3(SO_4)_2(OH)_6} a_{H_2O}^3}{a_{K^+} a_{Fe(OH)_3}^3 a_{SO_4^{2-}}^2 a_{H^+}^3} \tag{8-11}$$

If possible, it is advantageous to choose standard states such that one or more activities become unity. If one or more of the *phases* involved does not change composition, then we can adopt a "pure phase at P and T" standard state for that phase. When we apply this standard state to stoichiometric gypsum, anhydrite, jarosite, and ferric hydroxide, their activities become unity, and equations (8-10) and (8-11) are reduced to

$$K = a_{H_2O}^2 \tag{8-12}$$

and

$$K = \frac{a_{H_2O}^3}{a_{K^+} a_{SO_4^{2-}}^2 a_{H^+}^3} \tag{8-13}$$

which greatly simplifies the form of the equilibrium constant and highlights the important compositional variables. For example, we see that it is the activity of water that describes the equilibrium reaction between gypsum and anhydrite. In the example of the jarosite–ferric hydroxide buffer, the activity of water is often very close to unity, so that the activities of potassium, sulfate, and hydrogen ions determine the equilibrium between these two iron minerals.

Formulating reactions in terms of equilibrium constants carries with it an important responsibility. Because standard states are implicitly involved in every equilibrium constant, you must clearly define the standard state from several possible choices. Several examples that follow will make this clear. First, however, we shall present the various general forms of the equilibrium constant for different types of phases.

Equilibrium constant for gases

For reactions between pure ideal gases, the equilibrium constant is expressed in terms of the partial pressure P_i. If the gaseous reaction is equivalent to reaction (8-1), the equilibrium constant is

$$K = \frac{(P_C)^c(P_D)^d}{(P_A)^a(P_B)^b} \tag{8-14}$$

For nonideal gases, the activities are represented by the fugacity relative to the standard-state fugacity of 1 bar. Equation (8-14) then becomes

$$K = \frac{(f_C)^c(f_D)^d}{(f_A)^a(f_B)^b} \tag{8-15}$$

Equilibrium constant for solids

Reactions that involve only crystalline phases of fixed composition can be adequately treated without resorting to equilibrium constants, as shown in Chapter 4. However, if one or more of the solids participating in the equilibrium can vary its composition, then equilibrium constants are needed. Consider, for example, the four-phase assemblage plagioclase (plag), garnet (gt), kyanite (ky), and quartz. Because two of these minerals may be solid solutions, we have the following compositions (considering only the most important cations in the solution phases):

plagioclase	$(Ca,Na)(Al,Si)AlSi_2O_8$
garnet	$(Ca,Mg,Fe^{2+})_3(Al,Fe^{3+})_2Si_3O_{12}$
kyanite	Al_2SiO_5
quartz	SiO_2

Suppose all four phases were found in the same metamorphic rock; could they have ever been at equilibrium? First, we could use an electron microprobe to measure the compositions of the feldspar and garnet phases. From the compositional ranges just given, it should be clear that it would be impossible to write a stoichiometric reaction involving all phases. There is no way to balance the Na in plagioclase or the Mg, Fe^{2+}, and Fe^{3+} in garnet. But it is possible to write a reaction involving pure Ca-plagioclase (an) and pure Ca-garnet (gross):

$$3\,CaAl_2Si_2O_8 = Ca_3Al_2Si_3O_{12} + 2\,Al_2SiO_5 + SiO_2$$
$$\text{anorthite} \qquad \text{grossular} \qquad \text{kyanite} \quad \text{quartz} \tag{8-16}$$

Here is where the equilibrium constant comes in. We take equation (8-16) to represent standard-state compositions, and for the reaction we define

$$K_{eq} = \frac{a_{gross}^{gt}(a_{ky}^{ky})^2 a_{SiO_2}^{quartz}}{(a_{an}^{plag})^3} \tag{8-17}$$

If we adopt the complete standard-state declaration of pure phases (defined by equation (8-16)) at P and T, then we may simplify the equilibrium constant to

$$K_{eq} = \frac{a_{gross}^{gt}}{(a_{an}^{plag})^3} \tag{8-18}$$

because quartz and kyanite are always pure and thus have unit activities at every P and T. Equation (8-18) is an extremely convenient form, because now we can investigate equilibrium conditions involving any composition of garnet or plagioclase, so long as we assume that both minerals were in equilibrium with both kyanite and quartz. The only problem is that, although it is easy to measure the compositions of garnet and plagioclase, it is not easy to convert the measured compositions to activities. As discussed in Chapter 6, the activity coefficients that relate these functions vary in complex fashion with respect to temperature, pressure, and composition. One way around the problem is to define a distribution coefficient K_D in terms of the mole fractions of the same components:

$$K_D = \frac{X_{gross}^{gt}}{(X_{an}^{plag})^3} \tag{8-19}$$

The distribution coefficient can now be calculated in straightforward fashion from analytical data by adopting an appropriate ideal-solution model to convert the measured compositions of plagioclase and garnet to mole fractions of anorthite and grossular components. Note that the relationship between the equilibrium constant and the distribution coefficient must be

$$K_{eq} = K_D K_\lambda \tag{8-20}$$

where K_λ is the ratio of the activity coefficients:

$$K_\lambda = \frac{\lambda_{gross}^{gt}}{(\lambda_{an}^{plag})^3} \tag{8-21}$$

Distribution coefficients are functions of temperature and pressure, although they will not behave exactly like the equilibrium constants to which they are related. The unknown factor is the dependence of K_λ on pressure, temperature, and composition, which makes the calculation of K_D from free energies problematical. Nonetheless, distribution coefficients calculated from measured compositions of coexisting minerals are routinely used to extract the temperature or pressure of equilibration of a mineral assemblage. The dependence on temperature and pressure is obtained either from experimental calibration of K_D or by calculation of K_{eq} as a function of P and T from thermodynamic data; this is the foundation for geothermometry and geobarometry based on element partitioning between minerals. If the latter path is chosen, some model such as ideal mixing in one or more phases must be assumed in order to relate K_{eq} to K_D. The literature on the subject of mineral geothermometers and geobarometers is extensive. (For reviews of these applications, see Carmichael et al., 1974, pp. 78–124, and Essene, 1982; see also Problems 5 and 6.) If you are curious to see how the garnet–plagioclase–kyanite–quartz assemblage just discussed relates to pressure and temperature, see Ghent (1976).

Equilibrium constants for aqueous species

Equilibrium constants for aqueous species are a bit more complex because of the wider choice of standard states. We first define the "thermodynamic" equilibrium constant K in terms of activities:

$$K = \frac{(a_C)^c (a_D)^d}{(a_A)^a (a_B)^b} \tag{8-22}$$

The constant for the aqueous reaction involving aluminum hydrolysis,

$$Al^{3+}_{(aq)} + H_2O_{(l)} \rightleftharpoons Al(OH)^{2+}_{(aq)} + H^+_{(aq)}$$

would be written as

$$K = \frac{a_{Al(OH)^{2+}} a_{H^+}}{a_{Al^{3+}} a_{H_2O}} = \frac{m_{Al(OH)^{2+}} \gamma_{Al(OH)^{2+}} m_{H^+} \gamma_{H^+}}{m_{Al^{3+}} \gamma_{Al^{3+}} m_{H_2O} \gamma_{H_2O}} \tag{8-23}$$

Actual measurements of hydrolysis constants often use a pH electrode to measure a_{H^+} and a constant ionic medium that keeps the activity coefficients constant. Equation (8-23) then reduces to:

$$K^{app} = \frac{m_{Al(OH)^{2+}} a_{H^+}}{m_{Al^{3+}}} = \frac{K \gamma_{Al^{3+}}}{\gamma_{Al(OH)^{2+}}} \tag{8-24}$$

assuming unit activity of water. Equation (8-24) is an example of an "apparent" equilibrium constant (sometimes called a conditional, effective, concentration, or stoichiometric constant). The phrase "apparent constant" is usually applied to aqueous reactions involving a_{H^+}. For other aqueous reactions, it would be more appropriate to speak of the "stoichiometric" constant. For example, the ion-pair formation of magnesium sulfate,

$$Mg^{2+}_{(aq)} + SO^{2-}_{4(aq)} \rightleftharpoons MgSO^{\circ}_{4(aq)}$$

would be represented by the stoichiometric association constant

$$K = \frac{m_{MgSO^{\circ}_4}}{m_{Mg^{2+}} m_{SO^{2-}_4}} \tag{8-25}$$

This relationship is a constant only for a fixed and specified composition.

Many important reactions involve equilibrium between a mineral and a saturated aqueous solution. For instance, the dissolution of fluorite,

$$CaF_{2(s)} \rightleftharpoons Ca^{2+}_{(aq)} + 2F^{-}_{(aq)} \tag{8-26}$$

would be represented by the thermodynamic equilibrium constant

$$K = \frac{a_{Ca^{2+}} a^2_{F^-}}{a_{CaF_2}} \tag{8-27}$$

Because fluorite may be a pure phase, we assume a standard state where $a_{CaF_2} = 1$, which gives

$$K_{sp} = a_{Ca^{2+}} a^2_{F^-} \tag{8-28}$$

Because equation (8-28) is a simple product of activities, it is called a solubility-product constant, K_{sp}. The stoichiometric solubility-product constant for fluorite would be

$$K^{stoic}_{sp} = m_{Ca^{2+}} m^2_{F^-} = \frac{K_{sp}}{\gamma_{Ca^{2+}} \gamma^2_{F^-}} = \frac{K_{sp}}{\gamma^2_{\pm CaF_2}} \tag{8-29}$$

The constant-ionic-medium standard state has been widely used for applications in chemical oceanography because seawater approximates a constant composition. It is very important, however, to keep the same standard state for all reactions in the same system. If a constant ionic medium of 1 m is adopted, then it must be used for all aqueous species and solids that react in the system.

In Chapter 7, we pointed out that thermodynamic properties of aqueous species and reactions are often obtained by extrapolating to the reference state of infinite dilution. This method is commonly used to obtain equilibrium constants, whether they be association constants or solubility-product constants. Some stoichiometric (or apparent) constants can be measured over a concentration range dilute enough that a form of the Debye–Hückel equation can be used to extrapolate with some confidence. This procedure will be discussed in Chapter 11. Some types of aqueous reactions, however, cannot be measured at low concentrations without the ionic strength changing during the course of the measurements. This problem makes it impossible to measure the equilibrium with any degree of reliability. Therefore, an "inert" electrolyte is often added at such high concentrations that changes occurring during the measurements are swamped out. This is the reason why the constant-ionic-medium approach has been so popular. Commonly, concentrations or ionic strengths of 1, 2 or 3 m are used. These data cannot be easily extrapolated or corrected to infinite dilution to obtain thermodynamic data, but they are useful in telling us whether or not a particular reaction may be important.

8-2   TEMPERATURE, PRESSURE, AND COMPOSITIONAL DEPENDENCE OF THE EQUILIBRIUM CONSTANT

Temperature dependence of K: calcite solubility

The effect of pressure, temperature, and composition on the equilibrium constant is a simple transformation of the same variables on the Gibbs free energy. Several different approaches can be used to derive the appropriate relationships, and the most straightforward is to consider the temperature dependence first, using simple substitution. Recalling that the standard-state enthalpy and entropy are related to the free energy by

$$\Delta G° = \Delta H° - T \Delta S° \tag{3-76}$$

we substitute this expression into

$$\Delta G_r° = -RT \ln K \tag{8-9}$$

and rearrange to get

$$\ln K = -\frac{\Delta H_r°}{RT} + \frac{\Delta S_r°}{R} \tag{8-30}$$

This equation can be written as

$$\ln K = \frac{A}{T} + B \tag{8-31}$$

which is the equation for a straight line if $\ln K$ is plotted against reciprocal temperature, and if A and B are constants. Because $A = -\Delta H_r°/R$ and $B = \Delta S_r°/R$, the implication is that $\Delta H_r°$ and $\Delta S_r°$ are independent of temperature. This is true only if the heat-capacity terms for products and reactants exactly cancel ($\Delta C_P = 0$), which can never be tacitly assumed. As we shall see, for reactions that involve aqueous species, the assumption is commonly invalid.

TABLE 8-1

STANDARD-STATE THERMODYNAMIC VALUES FOR THE REACTION
$CaCO_3 \rightleftharpoons Ca^{2+} + CO_3^{2-}$

Species	C_P° $(J \cdot mol^{-1} \cdot K^{-1})$	S° $(J \cdot mol^{-1} \cdot K^{-1})$	ΔH_f° $(kJ \cdot mol^{-1})$	ΔG_f° $(kJ \cdot mol^{-1})$
Calcite	83.47	91.71	−1207.37	−1128.842
$Ca_{(aq)}^{2+}$	−27.61	−53.10	−542.83	−553.54
$CO_{3(aq)}^{2-}$	——	−56.90	−677.14	−527.90

Source: Data from Robie et. al. (1978).

Nonetheless, it is customary to plot $\ln K$ for any reaction as a function of the reciprocal temperature. If a small-enough temperature range is chosen, the data often assume a linear dependence. We return to a recurring theme: if heat-capacity data are available, they should always be used. If they are not available, calculations can proceed for certain kinds of reactions as long as the limitations of the $\Delta C_P = 0$ assumptions are kept in mind.

The solubility of calcite is a good example to discuss because it is very well known experimentally. Let's assume that you want to predict the solubility of calcite as a function of temperature from $0°C$ to $100°C$. The first step, if you want to use equation (8-30), would be to find enthalpy and entropy data for the reaction

$$CaCO_{3(s)} \rightleftharpoons Ca_{(aq)}^{2+} + CO_{3(aq)}^{2-} \tag{8-32}$$

Robie et al. (1978) give the values listed in Table 8-1. From these data, we find that at 25°C and 1 bar $\Delta H_r^\circ = -12.60 \ kJ \cdot mol^{-1}$ and $\Delta S_r^\circ = -201.71 \ J \cdot K^{-1} \cdot mol^{-1}$. These values, when substituted into equation (8-30), give us

$$\ln K = \frac{1515}{T} - 24.26 \tag{8-33}$$

or

$$\log K = \frac{657.8}{T} - 10.53$$

Figure 8-1 is a plot of the calcite solubility-product constant according to equation (8-33). Assuming that ΔH_r° and ΔS_r° are independent of temperature is equivalent to assuming $\Delta C_P = 0$. To see just how good this approximation is, we shall carry the calculations forward one more step, assuming $\Delta C_P =$ constant. Heat-capacity data for most minerals are readily available, but very few heat capacities of aqueous ions can be found. As shown in Table 8-1, the heat capacity

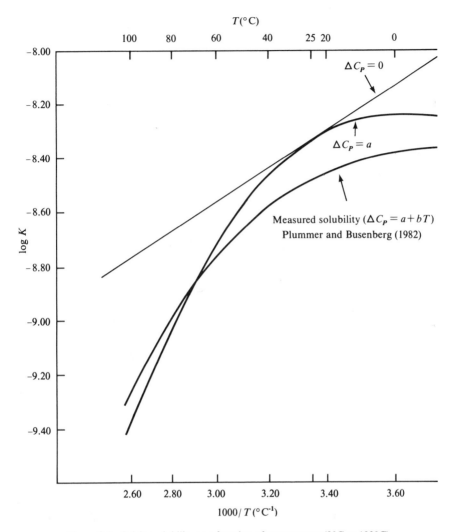

Figure 8-1. Calcite solubility as a function of temperature (0°C to 100°C).

for calcite (Robie et al., 1978; Jacobs et al., 1981) and $Ca^{2+}_{(aq)}$ (Helgeson et al., 1981) are known, but the value for $C^{\circ}_P(CO_3^{2-})$ is unknown. An estimate of $C^{\circ}_P(CO_{3(aq)}^{2-})$ can be obtained from correlations that exist between heat capacities of ions and their entropies (Helgeson et al., 1981; see estimation techniques in chaps. 11 and 12). For oxygenated anions,

$$C^{\circ}_P = -318 + 1.5S^{\circ}_i$$

which gives $C_P^\circ(CO_{3(aq)}^{2-}) = -403.3 \, \text{J} \cdot \text{mol}^{-1} \cdot \text{K}^{-1}$ and

$$\Delta C_{P_r}^\circ = -514.4 \, \text{J} \cdot \text{mol}^{-1} \cdot \text{K}^{-1}$$

Adding the heat-capacity data to the ΔH_r° and ΔS_r° data and recalculating an equation for the equilibrium constant, we obtain

$$\ln K = 390.12 - \frac{16930.9}{T} - 61.869 \ln T \qquad (8\text{-}34)$$

or

$$\log K = 169.40 - \frac{7351.7}{T} - 61.869 \log T$$

Equation (8-34) is plotted in Figure 8-1 for comparison with the linear solubility curve. Large deviations from linearity are quite apparent at temperatures greater than about 20 degrees from the reference temperature (25°C). This comparison points out a very important aspect of calculating mineral solubilities over the range of 0°C to 100°C: *the heat capacities must be taken into account when dealing with reactions involving both solid and aqueous species.*

Finally, it is always a useful exercise to compare the solubility-product constant determined from actual solubility measurements with those calculated from available enthalpy, entropy, and heat-capacity data. Plummer and Busenberg (1982) have made very precise measurements of calcite solubility over the temperature range of 5°C to 90°C. After considerable discussion of and correction for many of the uncertainties that enter into these measurements and the data reduction, they found the solubility-product constant to behave according to the relation

$$\log K = -171.9065 - 0.077993\,T + \frac{2839.319}{T} + 71.595 \log T \qquad (8\text{-}35)$$

Equation (8-35) is plotted in Figure 8-1. Notice that equation (8-35) has one more term than equation (8-34); this extra term results from assuming $\Delta C_P = a + bT$ and was needed to give a better fit to the data. A further point that emerges from this comparison is the considerable discrepancy between the two curves in the region of 25°C. Thermodynamic discrepancies such as these are the subject of Chapter 12, in which we discuss the compilation and evaluation of data. We simply mention here that there can be a larger uncertainty inherent in adding together several independent thermodynamic measurements than is present in a single set of highly reliable solubility measurements. In many cases, it is thermodynamically more expedient to use good solubility data in deriving solubility product constants rather than relying on enthalpy, entropy, free-energy, and heat-capacity data.

Combining the data of Plummer and Busenberg (1982) with the precise heat-capacity data for calcite recently obtained by Jacobs et al. (1981), we can derive the temperature dependence of the aqueous solute $CaCO_3$ at 25°C. We know that

$$\Delta C_{P_r}^{\circ} = C_P^{\circ}(Ca^{2+}) + C_P^{\circ}(CO_3^{2-}) - C_P^{\circ}(\text{calcite})$$

$$= C_P^{\circ}(CaCO_{3(aq)}) - C_P^{\circ}(\text{calcite})$$

From Plummer and Busenberg (1982), we have

$$\Delta C_P^{\circ} = 592.26 - 2.986T$$

From Jacobs et al. (1981), we have

$$C_P(\text{calcite}) = 49.252 + 0.1140T$$

for the range 0°C to 100°C. Therefore,

$$C_P^{\circ}(CaCO_{3(aq)}) = \Delta C_{P_r}^{\circ} + C_P^{\circ}(\text{calcite})$$

$$= 644.51 - 2.872T$$

$$= -211.9 \text{ J} \cdot \text{mol}^{-1} \cdot \text{K}^{-1} \quad \text{at } 25°C$$

To learn how this value compares to the estimated value, see Problem 3.

Pressure dependence of K: silica solubility

The effect of pressure on the free energy was described in chapter 4:

$$\left(\frac{\partial(\Delta G_r)}{\partial P}\right)_T = \Delta V_r \tag{4-4}$$

If we now substitute

$$\Delta G_r^{\circ} = -RT \ln K \tag{8-9}$$

we obtain

$$\left(\frac{\partial(\ln K)}{\partial P}\right)_T = -\frac{\Delta V_r^{\circ}}{RT} \tag{8-36}$$

However, it is very important to keep standard states in mind. If a fixed-pressure standard state is chosen, then there is no pressure effect because

$$\left(\frac{\partial(\ln K)}{\partial P}\right)_T = 0$$

Only for variable-pressure standard states, where the standard-state pressure is equivalent to the pressure on the system, will a pressure dependence be calculable. Integrating equation (8-36), we obtain

$$\ln K_{P_2} - \ln K_{P_1} = -\frac{\Delta V_r}{RT}(P_2 - P_1) \tag{8-37}$$

assuming that ΔV_r does not depend on pressure.

To demonstrate the effect of pressure on mineral equilibria, we now consider an example of great importance in geological systems that is relatively simple chemically: the solubility of silica, which may be written as

$$SiO_{2(s)} + 2\,H_2O \rightleftharpoons H_4SiO_{4(aq)} \tag{8-38}$$

The calculation of the pressure dependence for this equilibrium requires knowledge of the partial molal volumes of solid and aqueous silica. The actual hydration of silica in solution involves two water molecules,

$$SiO_{2(aq)} + 2\,H_2O \rightleftharpoons H_4SiO_{4(aq)}$$

except near the critical region, where four waters of hydration are indicated (Walther and Orville, 1983). For these calculations, we assume that

$$\mu_{SiO_{2(aq)}} = \mu_{H_4SiO_{4(aq)}}$$

There are several silica polymorphs, including α-quartz, β-quartz, cristobalite, chalcedony, and amorphous silica. We shall consider α-quartz—one of the most-common polymorphs and the most stable at 298 K and 1 bar. The necessary molal volume data can be obtained from Walther and Helgeson (1977); V° (α-quartz) = 22.688 cm$^3 \cdot$ mol^{-1}, and V° ($SiO_{2(aq)}$) = 13.6 cm$^3 \cdot$ mol^{-1}. Substituting into equation (8-37), we obtain

$$\ln K_{P_2} = \ln K_{P_1} + \frac{0.109(P_2 - P_1)}{T} \tag{8-39}$$

or

$$\log K_{P_2} = \log K_{P_1} + \frac{0.04747(P_2 - P_1)}{T}$$

Equation (8-39) shows that the effect of increased pressure is to increase the solubility of α-quartz. A plot of equation (8-39) at three temperatures, 25°C, 100°C, and 300°C is shown in Figure 8-2. For comparison, the four-term expression for molal volume containing six fit coefficients from Walther and Helgeson (1977) yields the pressure dependence for quartz solubility shown in Figure 8-2. This comparison demonstrates that a simple calculation, which assumes that the molal volumes are not functions of pressure and temperature, is fairly good at low pressures and for all pressures in the vicinity of 100°C. At low temperatures and at very high temperatures, the assumption of constant molal volume becomes less valid with increasing pressure.

Pressure dependence of K: devolatilization equilibria

We have seen in Chapter 4 that the assumption of constant ΔV_r for reactions involving only solids does not lead to significant errors except at extremely high pressures; the preceding example shows that the same assumption may even be

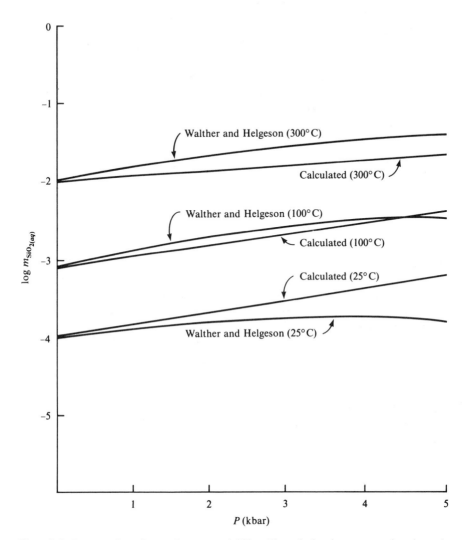

Figure 8-2. Pressure dependence of quartz solubility. The calculated curves are based on the assumption that ΔV_r is independent of pressure.

tolerable for some reactions that involve aqueous species. In contrast, however, many important geologic reactions involve both minerals and gaseous phases such as steam, carbon dioxide, or oxygen. Gases are far more compressible than solid phases are, and their molal volumes are very dependent on both pressure and temperature. For such reactions, the constant-molal-volume assumption is invalid under any circumstances. Although there is no *a priori* reason to treat reactions involving gases differently from reactions involving only solids, the fact that ΔV_r for these reactions is a complex function of pressure and temperature suggests that alternative approaches might be worth pursuing. We shall first consider those mineral–gas reactions in which only one gas species is present and where the total pressure on the system is equal to the pressure on the gas molecules in the system. The geochemically significant cases are dehydration ($P_{total} = P_{H_2O}$) and decarbonation ($P_{total} = P_{CO_2}$). Consider, for example, the dehydration of brucite:

$$
\begin{array}{ccc}
\text{Mg(OH)}_2 = & \text{MgO} & + \text{H}_2\text{O} \\
\text{brucite} & \text{periclase} & \text{water}
\end{array}
\tag{8-40}
$$

The equilibrium constant is

$$
K_{eq} = \frac{a_{\text{MgO}}^{\text{periclase}} \, a_{\text{H}_2\text{O}}^{\text{steam}}}{a_{\text{Mg(OH)}_2}^{\text{brucite}}}
\tag{8-41}
$$

Before deciding on the most appropriate choice for a standard state for the activities in (8-41), it will be helpful to examine the separate contributions of both the minerals and the steam phase to the total free energy of the reaction, and to evaluate how both contributions change with pressure and temperature. We will proceed in two steps.

First, all phases must be raised from the reference condition of 1 bar and 298.15 K to the equilibrium temperature (T_e):

STEP 1:   1 bar, 298.15 K $\to$ 1 bar, T_e

This is familiar ground: if we know the standard entropies and heat capacities of all phases, we can calculate the free-energy change at any temperature. The fundamental relationship is

$$
\Delta G_{r(1,T_e)} = \Delta G_{r(1,298)} - \int_{298}^{T_e} \Delta S(T) \, dT
\tag{4-7}
$$

Alternatively, we can always write

$$
\Delta G_{r(1,T_e)} = \Delta H_{r(1,T_e)} - T_e \Delta S_{r(1,T_e)}
$$

where both ΔH_r and ΔS_r are functions of temperature. The only new twist is that the values for ΔG_r, ΔH_r, ΔS_r, and ΔC_{P_r} which are calculated from standard data between 298.15 K and 373.15 K must include H_2O as the *steam* phase, even though steam is metastable with respect to liquid water in this temperature range. With this in mind, there is no reason to treat the steam phase differently from the mineral phases in the heating step.

The second step raises all phases from 1 bar and the equilibrium temperature to the equilibrium pressure (P_e):

STEP 2:   1 bar, T_e $\to$ P_e, T_e

In this step, the different behavior of solids and gases is apparent. The fundamental equation, however, is equally familiar:

$$\Delta G_{r(P_e, T_e)} = \Delta G_{r(1, T_e)} + \int_1^{P_e} \Delta V_r(P)\, dP \qquad \text{(cf 4-6)}$$

At once, we see that this reaction is fundamentally different from previously studied reactions that involve only solids. Because ΔV_r depends on both P and T, it cannot be moved to the left of the integral sign. For this reason, it is convenient to evaluate the integral in (4-6) in two parts: one term that involves only the solids, and one term that contains the volume of steam:

$$\int_1^{P_e} \Delta V_r(P)\, dP = \int_1^{P_e} \Delta V_s\, dP + \int_1^{P_e} \bar{V}_{H_2O}(P)\, dP \qquad (8\text{-}42)$$

The first term is the familiar $(P-1)\Delta V_s$ pressure correction for a reaction involving solids only. What is the second term? Because the molar volume of steam is known over a wide range of both P and T, it may be possible to evaluate the second integral in (8-42) by analytical methods—clearly, a tedious task. Alternatively, we can evaluate the integral in general terms as before:

$$\left(\frac{\partial \mu_{H_2O}}{\partial P} \right)_T = \bar{V}_{H_2O} \qquad (5\text{-}35)$$

By integrating both sides of (5-35) with respect to P between limits of 1 bar and P_e, we see that the volume integral for H_2O in (8-42) is

$$\int_1^{P_e} \bar{V}_{H_2O}\, dP = \mu_{H_2O}^{P_e, T_e} - \mu_{H_2O}^{1, T_e} = RT \ln a_{H_2O}^{P_e, T_e}$$

To satisfy the lower limit of the integration, the implied standard state for H_2O is 1 bar and T. Under these circumstances, $f_{H_2O}^\circ = 1$ and $a_{H_2O} = f_{H_2O}$. We have now raised both minerals and steam from 1 bar to P_e, and we may write

$$\Delta G_{r(P_e, T_e)} = 0 = \Delta G_{r(1, T_e)} + (P_e - 1)\Delta V_s + RT_e \ln f_{H_2O}^{P_e, T_e} \qquad (8\text{-}43)$$

To obtain the equilibrium constant for reaction (8-40), we now divide both sides of (8-43) by $-RT$:

$$\frac{-\Delta G_{r(1, T_e)}}{RT_e} - \frac{(P_e - 1)\Delta V_s}{RT_e} - \ln f_{H_2O} = 0$$

$$= \ln K_{(1, T_e)} - \frac{(P_e - 1)}{RT_e} \Delta V_s - \ln f_{H_2O}$$

$$(8\text{-}44)$$

We now must choose standard states for the activities in (8-41), a task that is equivalent to deciding which part of equation (8-43) will be taken as $\Delta G°$. If we adopt the standard state of pure phases (including steam) at P and T, then when brucite and periclase are in equilibrium with steam, all phases will have unit activity, $\ln K = 0$, and $\Delta G_{r(P_e, T_e)} = \Delta G°_{(P_e, T_e)}$. We will refer to this case as standard state 1. This is the standard state commonly chosen for mineral–gas equilibria when variations in gas composition are important. However, in the case under discussion, the gas phase is assumed to be pure H_2O under all conditions. For pure gas reactions of this type, it is customary to adopt different standard states for minerals and steam—namely, pure solids at P and T, and pure ideal-gas H_2O at 1 bar and T (standard state 2).

We shall now investigate how these different standard-state choices affect the calculation of dehydration equilibria. The relationship between the two standard free energies and the total free energy of reaction at any pressure and temperature is

$$\Delta G_{r(P,T)} = \underbrace{\underbrace{\Delta G_{r(1,T)} + (P - 1)\Delta V_s}_{\Delta G°_{(2)}} + RT \ln f_{H_2O}}_{\Delta G°_{(1)}} \qquad (8\text{-}45)$$

where $\Delta G°_{(1)}$ and $\Delta G°_{(2)}$ represent the standard free energies for standard states 1 and 2, respectively. Standard state 1 requires that all phases be raised from reference conditions to P and T, and $\Delta G°_{(1)}$ must include the difference in the free energy of steam that results from raising steam from 1 bar and T to P and T—i.e., $RT \ln f_{H_2O}$. Thus, at equilibrium, $\Delta G°_{(1)} = \ln K_{(1)} = 0$ when all phases have standard-state compositions. In contrast, standard state 2 requires that the temperature of steam be raised from T_r to T_e, but the pressure on the steam phase remains at 1 bar. Solids are treated the same as in standard state 1. Thus, at equilibrium, the activities of pure solids remain unity, but the activity of H_2O is equal to the fugacity of H_2O ($f_{H_2O} = 1$), so the equilibrium constant for standard state 2 is equal to the fugacity of H_2O. The equilibrium constants must be correctly related to their appropriate standard free energies:

$$\ln K_{(1)} = 0 = -\frac{\Delta G°_{(1)}}{RT}$$

$$\ln K_{(2)} = \ln f_{H_2O} = -\frac{\Delta G°_{(2)}}{RT}$$

Let's see how this all fits together when we start substituting some real numbers.

Table 8-2 summarizes the relevant data for brucite dehydration at a total pressure of 1000 bars, and Figure 8-3 shows the variation of log K as a function of reciprocal temperature for both standard states 1 and 2. The first column in

TABLE 8-2

DATA FOR $Mg(OH)_2 \rightarrow MgO + H_2O$ AT $P = 1000$ bar

T(K)	Standard state 2				Standard state 1
	$\log K_{r(1,T)}$ (see note 1)	$\log K_{r(1,T)}$ (see note 2)	$\log K_{r(P,T)}$ (see note 3)	$\log f_{H_2O(P,T)}$ (see note 4)	$\log K_{r(P,T)}$ (see note 5)
500	-0.517	-0.577	-0.437	1.591	-2.028
600	0.898	0.769	0.885	2.149	-1.264
700	1.908	1.706	1.806	2.497	-0.691
800	2.666	2.388	2.475	2.711	-0.236
900	3.255	2.904	2.982	2.827	0.155

Note 1: Values of $\log K_{r(1,T)}$ calculated assuming $\Delta C_P = 0$, equivalent to $-\Delta H^\circ_{r(298)}/2.303RT + \Delta S^\circ_{r(298)}/2.303R$, where $\Delta H^\circ_{r(298)} = 81.24$ kJ and $\Delta S^\circ_{r(298)} = 152.59$ J·K^{-1}mol^{-1} (Robie et al., 1978).

Note 2: Values of $\log K_{r(1,T)}$ calculated for ΔH°_r and ΔS°_r both functions of temperature, evaluated from $\sum_i \nu_i \log K_{f_i}$ (Robie et al., 1978). These are the correct values for standard state 2 at 1 bar and T.

Note 3: Values of $\log K_{r(1000,T)}$ calculated as $\log K_{r(1,T)} - 999 \Delta V_s/2.303RT$, where $\Delta V_s = V_{periclase} - V_{brucite} = -1.3382$ J·bar^{-1} (Robie et al., 1978).

Note 4: Logarithm of the fugacity of pure steam at 1000 bar and T, interpolated from Burnham et al. (1969b).

Note 5: Values of $\log K_{r(1000,T)}$ for standard state 2 calculated by subtracting $\log f_{H_2O(1000,T)}$ from the value in column 4 (described in note 3); cf equation (8-45) for ΔG_r.

Table 8-2 shows $\log K_{r(1,T)}$ for standard state 2 calculated by neglecting heat-capacity data for the phases. Comparing these values with the true $\log K_r$ values in column 2, you can see that the $\Delta C_P = 0$ assumption is inappropriate here; the 10% error in $\log K$ at 900 K would result in a calculated equilibrium temperature that is about 60 K too low. Nonetheless, the true value of $\log K_{(2)}$ can easily be calculated at temperature intervals as small as desired. The same cannot be said for $\log K$ calculated at standard state 1, because this number includes $\log f_{H_2O}$ a complex function of temperature and pressure. This term must be obtained at every desired temperature, either by interpolation in a data table or by execution of a computer subroutine that calculates f_{H_2O} from some appropriate equation of state; both options are inconvenient compared to standard state 2, and this is one reason why standard state 2 is preferred. Another reason is that the numerical value of $\log K$ for standard state 2 has an important physical significance: it is equal to the fugacity of H_2O that must be present in the system in order for brucite and periclase to be in equilibrium. If the gas phase is pure H_2O, this condition can be achieved at only one temperature—i.e., the temperature at which $\log K$ for the dehydration equilibrium is exactly equal to the fugacity of H_2O in a pure steam phase. In the present example, this temperature is 859 K

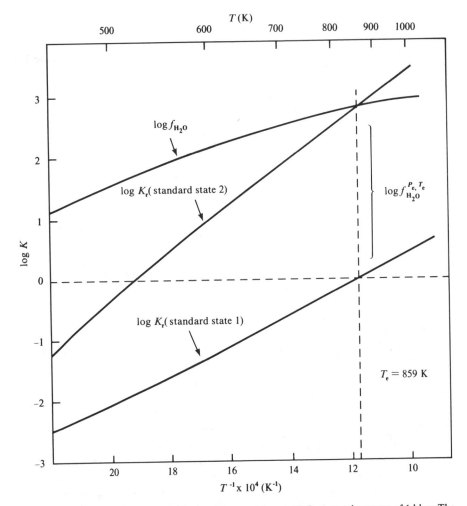

Figure 8-3. Plot of $\log K$ versus T^{-1} for brucite = periclase + H_2O at a total pressure of 1 kbar. The equilibrium temperature T_e is 859 K. See Table 8-1 for data.

when the total pressure is 1 kbar (see Figure 8-3). This is, of course, the same temperature at which $\log K$ for standard state 1 equals zero—the equilibrium condition of unit activity for all phases with reference to that state.

Determination of the equilibrium temperature at every isobar can be accomplished either by graphical means as shown in Figure 8-3 or by solving equation (8-43) for the $\Delta G_r = 0$ condition. The latter method is possible only when a computer program is available for calculating H_2O fugacities as a function of T and P. If the graphical method is chosen, then linear interpolation of $\log K$ (standard state 2) at 100 degree intervals of temperature is usually satisfactory. To demonstrate this point, $\log K$ for brucite dehydration at 1000 bars was calculated at 1 degree intervals using heat-capacity data, and the equilibrium f_{H_2O} was found to occur at 859 K. This compares favorably with the value of 858 K calculated by assuming a constant slope for $\log K_{(2)}$ between 800 and 900 K. Such a wide temperature range for interpolation is not generally recommended when using standard state 1; the slope of that line is not so well-behaved due to the inclusion of the $\log f_{H_2O}$ term. This point is emphasized in Figure 8-3 by the obvious curvature of $\log K_{(1)}$ versus reciprocal temperature at low temperature.

The equilibrium temperatures obtained for several isobars can be plotted as functions of pressure to give the dehydration curve for brucite = periclase + H_2O (Figure 8-4). Note the shape of the dehydration line—it is strongly curved, especially at low pressures. This is in sharp contrast to reactions involving only solids, which tend to be nearly linear over a wide P–T range. The difference is explained by recalling that ΔV for a solid–solid reaction is virtually independent of pressure and temperature, whereas ΔV for a reaction that

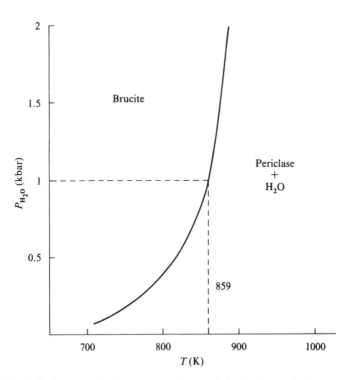

Figure 8-4. Dehydration curve for brucite = periclase + H_2O, showing equilibrium at 1 kbar and 859 K, as discussed in text. The rapidly changing slope of the curve at low pressures reflects the rapid changes in ΔV_r over this range.

liberates a gas phase decreases rapidly from low to high pressure, reflecting the much greater compressibility of the gas phase compared to the solids. Because ΔS for most reactions changes slowly with temperature, the change in ΔV causes a marked increase in the Clapeyron slope $(\Delta S/\Delta V)$ from low to high pressures for the solid–gas reaction.

Discussion: the meaning of pressure in natural systems

Although the thermodynamic pressure P is an exactly defined isotropic property of a system (a force per unit area), it may be difficult or even impossible to relate that P to the actual pressure that exists in a natural system. The *lithostatic pressure* is defined as a pressure that would result from a column of rock equal in height to the depth of the system; thus the lithostatic pressure depends on the assumed average density of the rock column. One practical problem that may arise is that the state of stress in a deeply buried body of rock may be far from isotropic, and there has been no general agreement in the literature on how to relate differential stresses to the thermodynamic pressure. Fortunately, most rocks are not very strong when they are subjected to deformational forces over very long periods of time. Under such conditions, rocks are probably incapable of maintaining very large differential stresses. Nevertheless, the problem still exists.

Another important consideration relates to fluid–rock interactions. The *hydrostatic pressure* is defined as a pressure that would be exerted on a system by a column of pure water at $25°C$ equal in height to the depth of the system. Such a condition is approximated only for shallow, water-filled fractures and pores. It is more common to refer to a *fluid pressure*. In most fluid-saturated systems at considerable depth, the situation is complicated. The fluid does not have free access to the earth's surface, so the fluid pressure is closer to lithostatic pressure. Under some conditions, the fluid pressure may even exceed the lithostatic pressure. Examples of overpressuring include rapid sedimentation, lateral tectonic compression, aquathermal pressuring, and dehydration of gypsum and clay minerals (Graf, 1982). One of the best examples is the exsolution of fluid during the cooling of a magma. The increase in pressure (above lithostatic) due to the large volume of released volatiles is sufficient to fracture the surrounding rock and is an important force both in volcanism and in the formation of some hydrothermal ore deposits. Thus, there are many significant problems that are difficult to assess in general terms. Many petrologists involved in thermodynamic calculations of rock–fluid equilibria in igneous or metamorphic environments have adopted the view that the fluid pressure on the system is equal to the lithostatic pressure, but Bruton and Helgeson (1983) have argued that fluid pressure may be substantially less than lithostatic pressure in the upper crust. Both mineral solubilities and dehydration equilibria are substantially dependent on the ratio of lithostatic pressure to fluid pressure.

Compositional dependence of K: stability of jadeite revisited

The purpose of this section is to show how mineral reactions involving solid solutions can be calculated as functions of temperature, pressure, and mineral composition, and how differences in the standard states chosen for the equilibrium constant affect the calculation. The reaction jadeite + quartz = analbite has been discussed in Chapter 4, but we can now write it in the form

$$\text{Na-pyroxene}_{(ss)} + \text{quartz} = \text{Na-plagioclase}_{(ss)}$$

to accommodate the possibility that the Na-pyroxene and plagioclase phases may vary in composition. The appropriate equilibrium constant is

$$K_{eq} = \frac{a_{ab}^{plag}}{a_{jd}^{px} a_{SiO_2}^{qz}} \tag{8-46}$$

which assumes that pure jadeite and pure albite are used as standard-state compositions. If we adopt the standard state of pure phases at P and T, then the activity of SiO_2 in quartz will always be unity. We may combine equations (8-30) and (8-37) to obtain

$$\ln K_{eq} = -\frac{\Delta H_{r(1,T)}^{\circ}}{RT} + \frac{\Delta S_{r(1,T)}^{\circ}}{R} - \frac{(P_e - 1)\Delta V_s}{RT} \tag{8-47}$$

which is equivalent to

$$\log K_{eq} = -\frac{\Delta H_{r(1,T)}^{\circ}}{2.303RT} + \frac{\Delta S_{r(1,T)}^{\circ}}{2.303R} - \frac{(P_e - 1)\Delta V_s}{2.303RT} \tag{8-48}$$

Now suppose that all phases are pure and have unit activities under all conditions. Because we have adopted a P and T standard state, $K = 1$ and $\log K = 0$ at any *equilibrium* pressure and temperature. Thus we can gather terms in (8-48) to obtain the equilibrium condition:

$$\log K_{(P_e, T_e)} = -\frac{\Delta H_{r(1,T_e)}^{\circ} + (P_e - 1)\Delta V_s}{2.303RT_e} + \frac{\Delta S_{r(1,T_e)}^{\circ}}{2.303R} = 0 \tag{8-49}$$

Equation (8-49) differs from (8-30) in the addition of a $P\Delta V_s$ pressure correction to the slope term.

Table 8-3 gives some relevant data needed to calculate $\log K$ for the jadeite + quartz = analbite reaction at 1 bar and 10,000 bars and at a number of temperatures. Figure 8-5 plots these data against reciprocal temperature. Note that both isobars have a common intercept at $1/T = 0$ but have different slopes. From equation (8-49), it is apparent that the common intercept is proportional to the entropy of reaction, whereas the slopes depend on the reaction enthalpy, on the volume change, and on the chosen pressure. The lines in Figure 8-5 appear straight, which implies that both ΔH and ΔS are independent of

TABLE 8-3
EQUILIBRIUM CONSTANTS FOR JADEITE + QUARTZ = ANALBITE

$T(K)$	$\log K_{f(1,T)}$ (see note 1)			$\log K_{r(1,T)}$ (see note 2)	$\log K_{r(10000,T)}$ (see note 3)
	Jadeite	Quartz	Analbite		
298.15	499.456	150.019	649.367	−0.108	−3.146
400	364.243	109.388	474.246	0.615	−1.649
500	285.029	85.602	371.653	1.022	−0.790
600	232.229	69.753	303.272	1.290	−0.220
700	194.533	58.442	254.455	1.480	0.186
800	166.278	49.968	217.866	1.620	0.488
900	144.316	43.388	189.428	1.724	0.718
1000	126.721	38.131	166.656	1.804	0.898

Note 1: Values of $\log K_{f(1,T)} = -\Delta G^{\circ}_{f(1,T)}/2.303RT$ for jadeite, quartz, and analbite from Robie et al. (1978).

Note 2: Values of $\log K_{r(1,T)}$ computed as $\sum_i v_i \log K_{f_i}$.

Note 3: Values of $\log K_{r(10000,T)}$ computed as $\log K_{f(1,T)} - 9999 \Delta V_s/2.303RT$, where $\Delta V_s = V_{ab} - (V_{qz} + V_{jd}) = 1.7342 \text{ J·bar}^{-1}$ (Robie et al., 1978).

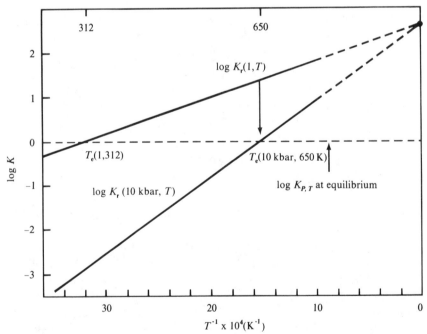

Figure 8-5. Plots of $\log K_r$ versus reciprocal temperature for jadeite + quartz = analbite at 1 bar and at 10 kbar.

temperature. As shown in Chapter 4, this implication is not true, but the slight curvature of the lines is not apparent at the scale of the figure. The common intercept for $1/T = 0$ is found at $\log K = 2.65$. When we multiply by $2.303R$, we obtain $\Delta S_r^\circ = 50.74 \text{ J} \cdot \text{K}^{-1}$, which is very close to the calorimetric value of $51.47 \text{ J} \cdot \text{K}^{-1}$ at 298.15 K.

Because jadeite, quartz, and analbite can be in equilibrium only when $\log K = 0$, the equilibrium temperature can be found at any pressure by noting the temperature of intersection of the desired isobar with the $\log K = 0$ line. The arrow in the figure shows the amount by which $\log K$ must be changed at 10 kbar from the 1 bar standard state to reach equilibrium. For instance, at 10 kbar the equilibrium temperature (T_e) has shifted from 312 K to 650 K. The length of the arrow at 10 kbar is equal to the pressure correction at 650 K. Alternatively, the equilibrium pressure may be found directly by choosing a temperature, calculating ΔH_r and ΔS_r at that temperature, and then solving equation (8-47) for the pressure at which $\log K = 0$. For example, at 800 K, we have

$$\frac{-\Delta H_{r(1,800)}^\circ}{RT} + \frac{\Delta S_{r(1,800)}^\circ}{R} - \frac{(P_e - 1)\,\Delta V_s}{RT} = 0$$

Using heat-capacity data from Chapter 5, we find $\Delta H_{r(1,800)}^\circ = 14.67 \text{ kJ} \cdot \text{mol}^{-1}$ and $\Delta S_{r(1,800)}^\circ = 49.36 \text{ J} \cdot \text{mol}^{-1} \cdot \text{K}^{-1}$. Thus, the equilibrium pressure is

$$P_e = \frac{-\Delta H_{r(1,800)} + T\,\Delta S_{r(1,800)}}{\Delta V_s} + 1 = 14.31 \text{ kbar}$$

This approach may be somewhat more cumbersome than the methods that solve for equilibrium directly in terms of reaction free energy, but the algebra is identical, and it makes little difference which point of view you take. However, when we allow for the possibility of solid solution, the equilibrium-constant method is definitely preferable. Before we tackle the problem rigorously, let's see how we can make qualitative predictions of the direction of shift in equilibrium due to solid solution in one of the phases.

For simplicity, let's assume dilution of the jadeite component in pyroxene solution while keeping the compositions of albite and quartz stoichiometric. If jadeite forms a stable solution with any other pyroxene component, then the free energy of that solution must be less than that of pure jadeite (remember that a stable solution must have a lower free energy than does a mechanical mixture of the same components). Thus the jadeite$_{(ss)}$ + quartz curves drawn as G–P or G–T cartoons must be shifted to lower G values relative to the pure jadeite + quartz curves. If the albite curve is unchanged, the solid solution results in a lower equilibrium pressure (in isothermal section) and a higher equilibrium temperature (in isobaric section) for the jadeite$_{(ss)}$ + quartz assemblage (Figure 8-6). This simple approach is a helpful crutch that provides significant qualitative information, as well as providing a check on calculated results. Now, on with the calculations.

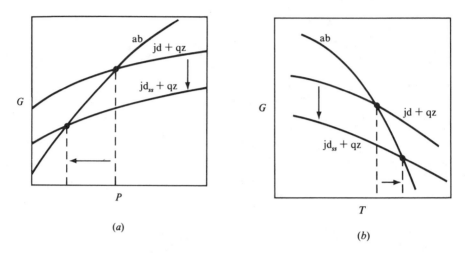

Figure 8-6. Cartoon plots of Gibbs free energy for the reaction jadeite + quartz = albite.   (a) Gibbs free energy versus pressure, showing that solid solution in jadeite lowers the equilibrium pressure relative to that of pure jadeite. The diagram is isothermal.   (b) Gibbs free energy versus temperature, showing that solid solution in jadeite raises the equilibrium temperature, relative to that of pure jadeite. The diagram is isobaric.

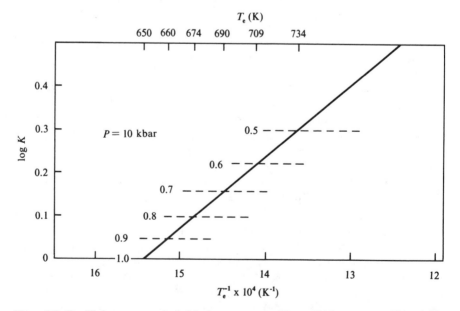

Figure 8-7. Equilibrium constant for jadeite + quartz = analbite at 10 kbar, contoured in activities of $NaAlSi_2O_6$ in Na-pyroxene. The intersection of each contour (dashed) with the log K curve (solid) is the equilibrium temperature for that activity.

If jadeite has nonunit activity, the $\log K_{eq}$ will be

$$\log K_{eq} = \log \frac{1}{a_{jd}^{px}} \qquad (8\text{-}50)$$

which cannot be zero so long as we continue to assume a pure-jadeite standard state. Figure 8-7 shows the 10-kbar isobar between 625 K and 830 K, contoured for activities of the jadeite component in pyroxene solution ranging from 1.0 to 0.5. The intersection of each constant-activity line with the isobar defines the equilibrium temperature for that composition, because each line was calculated by solving equation (8-50) for $\log K$ at the activity of interest. The activity range shown raises the equilibrium temperature at 10 kbar by more than 80 K. Similar calculations can be made at other pressures to yield a series of univariant lines for jadeite$_{(ss)}$ + quartz = analbite, each contoured for a specific activity of the jadeite component (Figure 8-8). Note that each phase boundary in Figure 8-8

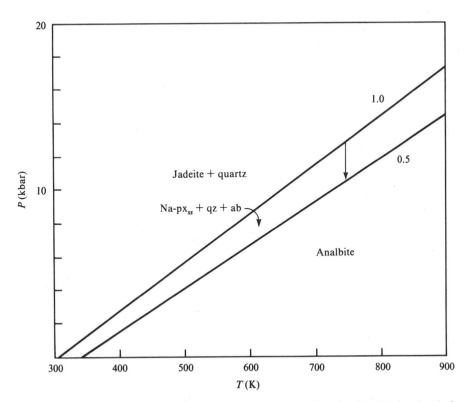

Figure 8-8. Calculated phase diagram for jadeite + quartz = analbite, showing divariant band of Na-pyroxene solid solution + quartz + analbite, with contour lines representing the activity of the jadeite component.

is univariant so long as the jadeite activity is specified, but the area labeled Na-px$_{(ss)}$ + quartz + albite is divariant (at least) because the system can no longer be described in terms of two components.

Throughout this discussion, we have illustrated the effect of varying jadeite activity on the pressure and temperature of the reaction, but we have not recommended a method for calculating jadeite activity from a chemical analysis of pyroxene. We could resort to an ideal-mixing model and substitute the mole fraction of jadeite in pyroxene for the activity of jadeite. As discussed in Chapter 6, these mole fractions are model-dependent, and the assumption of ideality is not always appropriate. Real-solution models are more realistic, but the data required to generate such models are often unavailable. The problem is significantly magnified for cases where a mineral solid solution may equilibrate with an aqueous phase because in such cases we must account for the distribution of ions between a nonideal crystalline solution and a nonideal aqueous phase. As we shall see in the next section, these complications introduce severe theoretical difficulties into the evaluation of an equilibrium constant involving solid solutions.

Compositional dependence of K: magnesian calcites

The effect of composition on the equilibrium constant is a complex and poorly understood subject in which thermodynamic and kinetic ambiguities abound. The compositional dependence refers to both the compositions of mineral solid solutions and, where appropriate, the variation in any coexisting fluid composition. The ambiguities are both conceptual and empirical. Definitions of multicomponent distributions among phases at equilibrium have been represented in different and sometimes contradictory ways (Thorstenson and Plummer, 1977; Lafon, 1978; Garrels and Wollast, 1978; Berner, 1978).

Furthermore, direct measurement of activity coefficients for a variable component in a solid phase is not possible as it is for aqueous solutions. It is usually accomplished only by indirectly measuring a coexisting vapor or liquid phase presumed to be in equilibrium. Even so, the number of such measurements for geochemical applications is typically limited. Most thermodynamic measurements have been made on pure, ideal, end-member compositions rather than solid solutions. At low temperatures, these data are extremely difficult to obtain because the kinetic problems are abundant.

As an introduction to the thermodynamics of multicomponent solids in equilibrium with a fluid phase, we present the derivations and results of Thorstenson and Plummer (1977) on magnesian calcites. In shallow, marine environments a high-magnesium calcite commonly precipitates, containing 10 to 20 mole-percent $MgCO_3$. During diagenesis, high-magnesium calcite converts to low-magnesium calcite (<5 mole-percent $MgCO_3$), suggesting that low-magnesium calcites are thermodynamically more stable.

If a low-magnesium calcite (lmc) has achieved stable thermodynamic equilibrium with an aqueous phase (i.e., saturation), then the free energy for the reaction

$$Ca_{(1-x)}Mg_xCO_3 = (1-x)\,Ca^{2+}_{(aq)} + x\,Mg^{2+}_{(aq)} + CO^{2-}_{3(aq)} \qquad (8\text{-}51)$$

will be at a minimum. The equilibrium constant for reaction (8-51) is

$$K_{lmc}(x) = a^{(1-x)}_{Ca^{2+}_{(aq)}} a^x_{Mg^{2+}_{(aq)}} a_{CO^{2-}_{3(aq)}} \qquad (8\text{-}52)$$

The two components of calcite can be written in different ways, but we prefer to follow the original discussion of Thorstenson and Plummer (1977) and use $CaCO_3$ and $MgCO_3$. Thus, another consequence of equilibrium is that

$$\mu^{lmc}_{CaCO_3} = \mu^{aq}_{CaCO_3} \qquad (8\text{-}53)$$

and

$$\mu^{lmc}_{MgCO_3} = \mu^{aq}_{MgCO_3} \qquad (8\text{-}54)$$

which implies that

$$a^{lmc}_{CaCO_3} K_{cc} = a_{Ca^{2+}_{(aq)}} a_{CO^{2-}_{3(aq)}} \qquad (8\text{-}55)$$

and

$$a^{lmc}_{MgCO_3} K_{mg} = a_{Mg^{2+}_{(aq)}} a_{CO^{2-}_{3(aq)}} \qquad (8\text{-}56)$$

where cc is calcite and mg is magnesite. Standard-state conditions including 298 K and 1 atm are assumed for the low-magnesium calcite and the associated components.

If we now assume that the solid lmc reacts with a fixed composition, then, in effect, it behaves as a one-component solid. This condition is defined as stoichiometric saturation, and it refers to "equilibrium between an aqueous phase and a multicomponent solid in situations where, owing to kinetic restrictions, the composition of the solid remains invariant, *even though the solid phase may be part of continuous compositional series*" (Thorstenson and Plummer, 1977, p. 1209). In this situation, the equilibrium restraints of equations (8-53) and (8-54) no longer apply. Equation (8-52), however, is still valid and, for any given value of x, there is a fixed calcite composition and a unique solution composition. Thorstenson and Plummer (1977) assumed that their estimates of $K(x)$ from dissolution experiments by Plummer and Mackenzie (1974) over a wide range of Mg-calcite compositions represented thermodynamic equilibrium. Their results, plotted in Figure 8-9, show three interesting trends.

1. The solubility product goes through a minimum around 4 mole-percent $MgCO_3$. A Mg-calcite in the range of 1 to 5 mole-percent will be less soluble than pure, well-crystallized calcite. Thus, a small amount of solid substitution decreases the solubility compared to the pure host mineral.

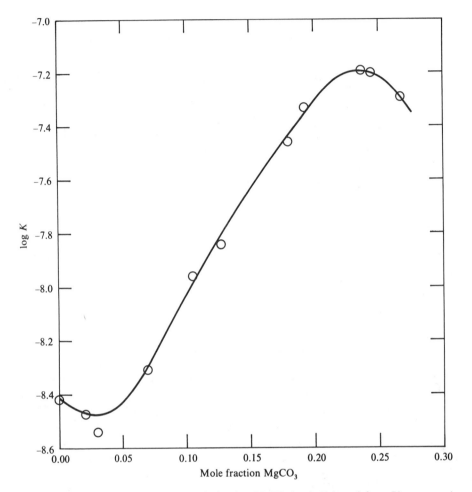

Figure 8-9. Plot of $\log K(x)$ against mole fraction $MgCO_3(=x)$. (Adapted from Plummer and Mackenzie, 1974)

2. From about 5 to 20 mole-percent, the solubility product increases almost linearly to more than an order of magnitude greater than that for pure calcite. Thus, extensive solid substitution can increase the solubility of a multicomponent phase.

3. A maximum in the solubility product is reached around 23 mole-percent, after which the value decreases. Continually decreasing $K(x)$ would be expected until 50 mole-percent is reached, where dolomite would be the stable phase and the $K(x)$ would be somewhere in the vicinity of K_{cc} or lower.

The compositional dependence of $K(x)$ can be derived from reaction (8-51). Equation (8-52) can be rearranged to

$$K(x) = a_{Ca^{2+}} \, a_{CO_3^{2-}} \left(\frac{a_{Mg^{2+}}}{a_{Ca^{2+}}} \right)^x = IAP \left(\frac{a_{Mg^{2+}}}{a_{Ca^{2+}}} \right)^x \qquad (8\text{-}57)$$

where $IAP = a_{Ca^{2+}} a_{CO_3^{2-}}$. In logarithmic form, we have

$$\log K(x) = \log IAP + x \log \frac{a_{Mg^{2+}}}{a_{Ca^{2+}}} \qquad (8\text{-}58)$$

and

$$\frac{\partial [\log K(x)]}{\partial x} = \log \frac{a_{Mg^{2+}}}{a_{Ca^{2+}}} \qquad (8\text{-}59)$$

Equation (8-59) can be used to calculate the activity ratios for $Mg^{2+}_{(aq)}/Ca^{2+}_{(aq)}$ that are shown in Figure 8-10. From this plot, two important points should be apparent. First, the activity ratios for the aqueous species are extremely high and outside the range of natural-water compositions for the range of compositions of natural high-Mg calcites (10 to 20 mole-percent). Therefore, the chemistry of high-Mg calcites is not determined by thermodynamic equilibrium. This example

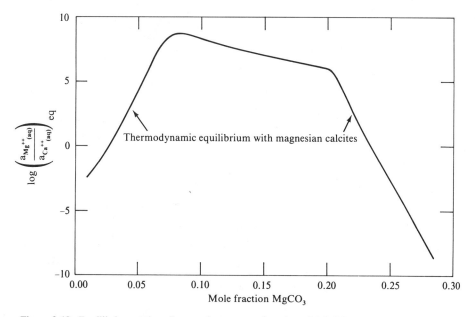

Figure 8-10. Equilibrium ratios of $a_{Mg^{2+}}/a_{Ca^{2+}}$ as a function of $MgCO_3$ content in magnesian calcites. The values plotted here for the activity ratio are those necessary to maintain thermodynamic equilibrium between an aqueous phase and magnesian calcites.

might be typical for highly nonideal behavior, which can be further demonstrated by calculating the activity and activity coefficient for the $MgCO_3$ component in solid Mg-calcite. From the criteria for thermodynamic equilibrium and the Gibbs–Duhem equation, Thorstenson and Plummer (1977) have shown that

$$\log a_{CaCO_{3(s)}} = -x\frac{\partial[\log K(x)]}{\partial x} + \log K(x) - \log K_{cc} \qquad (8\text{-}60)$$

$$\log a_{MgCO_{3(s)}} = (1 - x)\frac{\partial[\log K(x)]}{\partial x} + \log K(x) - \log K_{mg} \qquad (8\text{-}61)$$

which permits the calculation of the appropriate activities and activity coefficients shown in Figure 8-11. As can be seen, the solvent activity coefficient shows little variation and can be approximated by regular solution theory up to about 8 mole-percent. The solute activity coefficient, however, is strongly nonideal and

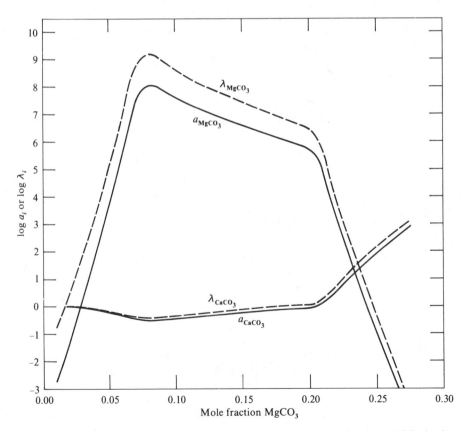

Figure 8-11. Activities and activity coefficients of the components $CaCO_3$ and $MgCO_3$ in the magnesian calcites.

increases by more than 10^9 at 8 mole-percent. Likewise, at >20 mole-percent, both λ_{CaCO_3} and λ_{MgCO_3} are strongly nonideal. Similarly, nonideal behavior has been indicated by work on $SrCO_3$ in calcite (Kinsman and Holland, 1969) and $ZnCO_3$ in calcite (Crockett and Winchester, 1966), although Mg-calcite may be one of the most extreme examples. The data on Mg-calcites are rather extreme compared to other systems probably because biogenic calcites were used in the original experiments of Plummer and Mackenzie (1974). Synthetic, inorganically prepared samples should show considerably less deviation from ideality.

SUMMARY

- The equilibrium constant K_{eq} for a chemical reaction is defined as the continuous product of the activities of all reacting species, each raised to an exponent equal to the stoichiometric coefficient for that species: $K_{eq} = \prod_i a_i^{\gamma_i}$.

- The equilibrium constant is related to the standard free energy of the reaction by the fundamental equation $\Delta G^\circ = -RT \ln K$. The standard state chosen for each of the activities in the equilibrium constant must be reflected in the calculation of ΔG°; thus for a given closed system at any P, T, or composition, there is no unique value for either K or ΔG°.

- Different constants K have been defined for different situations. Some of the most important are the following.
  1. The distribution coefficient K_D describes the equilibrium of mass between coexisting solids of variable composition (using mole fractions) and/or liquids (using molalities).
  2. The stoichiometric equilibrium constant K^{stoic} is written for aqueous species in terms of molalities rather than activities.
  3. The apparent equilibrium constant K^{app} is written for aqueous species in terms of molalities except for H^+, which is written as a_{H^+}.
  4. The solubility-product constant K_{sp} is written for the specific case of the equilibrium saturation of a single pure solid with respect to an aqueous phase. The K_{sp} is a true equilibrium constant because all species are written as activities.
  5. The stoichiometric solubility-product constant K_{sp}^{stoic} takes the form of a K_{sp}, but all activities are replaced by molalities.

- The combined-temperature-and-pressure effect on the equilibrium constant is given by

$$\ln K = -\frac{\Delta H_r^\circ}{RT} + \frac{\Delta S_r^\circ}{R} - \int_{P_1}^{P_2} \frac{\Delta V_r^\circ}{RT} \tag{8-62}$$

where ΔH_r°, ΔS_r°, and ΔV_r° must agree with the standard state chosen for all activities in the equilibrium constant. This equation can be used to solve for the equilibrium P and T of a given reaction.

- If $\Delta C_{P_r} = 0$, then the enthalpies and entropies in (8-62) may be replaced by constant values, and $\ln K$ is a linear function of reciprocal temperature if pressure is held constant. This assumption may be reasonable if a small-enough temperature range is considered, but it is generally invalid when aqueous species are involved. We recommend using heat-capacity data whenever possible.

- If ΔV_r° is independent of pressure, then the last term is (8-62) is replaced by $-\dfrac{\Delta V_r^\circ}{RT}(P_2 - P_1)$. This is an excellent approximation for reactions involving solids only, and it is adequate for some reactions involving aqueous species. If a gas phase is involved, however, the simplification is invalid, and special provisions must be taken to account for changes in the volume of the gas as a function of pressure. This correction may be conveniently expressed by the fugacity of the pure gas, which implies a standard-state choice of 1 bar and T for the gas species.

- The effect of variable composition on the equilibrium constant is problematical and not well understood; the problems involved relate primarily to the most-correct way to calculate the activities of components in solid solution from compositional data. Once the activities are known, calculation of equilibrium pressure or temperatures as a function of composition can be straightforward.

PROBLEMS

1. Write the equilibrium-constant expression for the reaction
$$CaCO_{3(s)} + 2H^+_{(aq)} + SO^{2-}_{4(aq)} + H_2O_{(l)} \rightleftharpoons CaSO_4 \cdot 2H_2O_{(s)} + CO_{2(g)}$$
assuming that solids are pure crystalline phases and that the gas is ideal.

2. What is the stoichiometric solubility product constant for kaolinite dissolution in acid? Use $H_4SiO_{4(aq)}$ for aqueous silica.

3. a) Calculate a revised value for $C^\circ_{P(CO^{2-}_{3(aq)})}$ at 25°C based on $C^\circ_{P(CaCO_{3(aq)})}$. (*ans:* $-184.3 \text{ J} \cdot \text{mol}^{-1}$)

   b) Using this revised value, calculate a new temperature-dependent function to replace equation (8-34).

   c) Compare the calcite solubility curve determined by the equation from part b with the curve from Plummer and Busenberg (1982). What does this comparison indicate about the value of $C^\circ_{P(CO^{2-}_{3(aq)})}$?

4. Using the heat-capacity, enthalpy, and entropy data for aragonite that follow, calculate the solubility-product constant and plot $\log K_{sp}$ versus $1/T$ for 0°C to 100°C. Compare with the curve measure by Plummer and Busenberg (1982), who report the equation

$$\log K(\text{aragonite}) = -171.9773 - 0.077993T + \frac{2903.293}{T} + 71.595 \log T$$

For aragonite: $\Delta H_f^\circ = -1207.43$ J·mol^{-1}; $S^\circ = 87.99$ J·mol^{-1}·K^{-1}; $C_P^\circ = 51.196 + 0.1054T$.

5. The distribution of F and OH between (F,OH)-apatite and (F,OH)-phlogophite can be written

$$Ca_5(PO_4)_3OH + \tfrac{1}{2}KMg_3Si_3AlO_{10}F_2 = Ca_5(PO_4)_3F + \tfrac{1}{2}KMg_3Si_3AlO_{10}(OH)_2$$
$$\text{OH-ap} \qquad\quad \text{F-ph} \qquad\qquad\quad \text{F-ap} \qquad\qquad \text{OH-ph}$$

Write an expression for the distribution coefficient K_D for F = OH exchange between apatite and phlogopite. If the standard free energy for the reaction is $\Delta G_r^\circ = -90000 + 34.08T$ (pure OH and F phases at P and T), and if both F and OH mix ideally in phlogopite and apatite, then calculate the temperature at which a phlogopite of composition F/(F + OH) = 0.12 was in exchange equilibrium with an apatite in which F/(F + OH) = 0.95. Assume that no other anions occupy the exchange site and that ΔG_r° is independent of total pressure. (*Ans:* 1200 K)

6. Use the equation given in Problem 5 to construct a generalized graph of X_F^{apatite} versus $X_F^{\text{phlogopite}}$ contoured in isotherms ranging from 1500 K to 1000 K. If the best one can hope for in analyzing F in these phases is $\pm 10\%$ relative, look at the graph and decide whether apatite–biotite pairs can be used as a realistic igneous geothermometer. What other complications could arise in "real-world" applications? (See Ludington, 1978).

7. From the following data, construct a decarbonation curve (P_{CO_2} versus T; cf Figure 8-4 for dehydration) for the reaction

$$\text{magnesite} = \text{periclase} + CO_2$$
$$MgCO_3 = MgO + CO_2$$

for total pressures of 250, 500, 1000, and 2000 bars.
Data:

$$\Delta H_{r(1,298)}^\circ = 118.28 \text{ kJ}$$
$$\Delta S_{r(1,298)}^\circ = 175.64 \text{ J·K}^{-1}$$
$$\Delta V_{s(1,298)}^\circ = -16.77 \text{ cm}^3$$

The heat capacity change for the reaction is

$$\Delta C_P = 71.912 - 0.05616T - 1386.1T^{-1/2} + 2.0766 \times 10^6 T^{-2}$$

Use the following fugacity coefficients for CO_2:

T(K)	250 bars	500 bars	1000 bars	2000 bars
700	0.996	1.035	1.188	1.777
800	1.020	1.075	1.241	1.805
900	1.033	1.096	1.267	1.800
1000	1.039	1.106	1.278	1.780
1100	1.043	1.111	1.280	1.753
1200	1.044	1.112	1.276	1.723
1300	1.044	1.111	1.270	1.691

8. Consider the reaction

$$\underset{\text{grossular}}{Ca_3Al_2Si_3O_{12}} + \underset{\alpha\text{-quartz}}{SiO_2} = \underset{\text{anorthite}}{CaAl_2Si_2O_8} + \underset{\text{wollastonite}}{2\,CaSiO_3}$$

Suppose that the log of the equilibrium constant for this reaction is given by

$$\log K_{(P,T)} = -\frac{2597}{T} + 3.421 - 0.171\frac{P}{T}$$

where the standard states are pure phases at P and T, for P in bars and T in kelvins. Calculate the pseudounivariant line (up to 5 kbar) that applies when pure α-quartz, anorthite, and wollastonite are in equilibrium with a garnet in which $X_{Ca} = 0.7$ in the eightfold site and $X_{Al} = 0.85$ in the sixfold site. Use the ideal-mixing-on-sites model to determine the activity of the grossular component in garnet. Compare your curve with the true univariant curve that applies when all phases are pure.

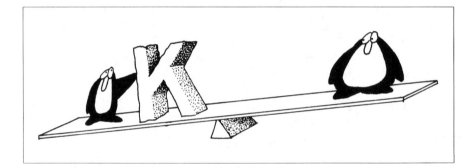

9

MINERAL EQUILIBRIA
II. CHEMICAL POTENTIAL
DIAGRAMS

Activity diagrams depicting chemical equilibrium among minerals and aqueous solutions are important references for the modern geologist. Such diagrams facilitate prediction and interpretation of the chemical environments in which mineral assemblages form in geochemical processes.

H. C. HELGESON, T. H. BROWN, and R. H. LEEPER (1969)

You now have a sense of what equilibrium constants are and how they can be used to define virtually any kind of geochemical equilibrium. In this chapter, we examine the very important problem of selecting those thermodynamic variables that are most important in analyzing each such equilibrium. The problem stems from the fact that systems are very complex chemically, so the challenge is to isolate those chemical parameters (represented by the chemical potentials or activities of components in the system) that are most effective in controlling or describing mineral reactions. For the purposes of this chapter, we shall hold temperature and pressure constant, thus concentrating only on chemical potential variations.

This approach is most important when mineral reactions involve either gases or ionized species. Examples are

$$\underset{\text{magnetite}}{\text{Fe}_3\text{O}_4} + 3\,\text{S}_{2(g)} = 3\,\underset{\text{pyrite}}{\text{FeS}_2} + 2\,\text{O}_{2(g)}$$

and

$$\underset{\text{albite}}{\text{NaAlSi}_3\text{O}_8} + 4\,\text{H}^+_{(aq)} + 4\,\text{H}_2\text{O}_{(l)} = \text{Na}^+_{(aq)} + \text{Al}^{3+}_{(aq)} + 3\text{H}_4\text{SiO}_{4(aq)}$$

In the first example, it is reasonable to choose the chemical potentials of O_2 and S_2 gas as independent variables in the system because their relative values obviously will determine whether magnetite or pyrite is the more stable mineral. In the second example, the solubility of albite in liquid water clearly is a function of the activities of Na^+, Al^{3+}, silicic acid, and pH. However, the situation is more complicated because aluminum forms a number of aqueous complexes as a function of pH. How can these interrelations best be shown on a two-dimensional phase diagram? Another problem is that the choice of variables may depend on the temperature of the system. The reaction between muscovite and K-feldspar may be written either in terms of ions and aqueous silicic acid as

$$3\, KAlSi_3O_8 + 2\, H^+ + 12\, H_2O_{(l)} = KAl_2Si_3AlO_{10}(OH)_2 + 2\, K^+ + 6\, H_4SiO_{4(aq)}$$

or in terms of molecular species and quartz as

$$3\, KAlSi_3O_8 + 2\, HCl_{(aq)} = KAl_2Si_3AlO_{10}(OH)_2 + 2\, KCl_{(aq)} + 6\, SiO_2$$

The first representation is more appropriate for surface conditions in the presence of liquid water, whereas the second equation may better describe high temperatures, at which the chloride species are predominantly associated and the H_2O is a supercritical steam phase.

We begin the chapter by considering mineral–gas reactions at $25°C$, and we then move on to "mixed-volatile" mineral reactions involving H_2O and CO_2 at elevated temperatures and pressures. The chapter concludes with a long look at mineral solubilities and stability diagrams in the presence of dissolved aqueous species. The reader should be aware of the fact that the important special case of oxidation–reduction reactions is postponed until Chapter 10.

9-1 THE CHEMICAL POTENTIAL DIAGRAM: DERIVATION OF SLOPE CONDITIONS

Consider first the possible equilibrium between calcite and fluorite in the presence of liquid water:

$$\underset{\text{calcite}}{CaCO_3} + 2\, HF_{(g)} = \underset{\text{fluorite}}{CaF_2} + CO_{2(g)} + H_2O_{(l)} \tag{9-1}$$

The equilibrium constant may be written as

$$K_{eq} = \frac{f_{CO_2}}{f_{HF}^2} \tag{9-2}$$

If we take the standard states of both CO_2 and HF to be the pure gases at 1 bar and $25°C$, then the fugacities of both species will be numerically equal to their activities—and will be virtually identical to their partial pressures as well. The activities of both mineral phases and that of liquid water do not appear in the equilibrium constant because they are assumed to have unit activity (because they are physically equivalent to the standard-state conditions) at the pressure and temperature of interest. The gaseous components, however, are free to vary their fugacities so long as their combined total pressures do not exceed 1 bar, the pressure on the system.

Taking logarithms of equation (9-2), we have

$$\log K = \log f_{CO_2} - 2\log f_{HF} \tag{9-3}$$

which may be rearranged in the form

$$\log f_{CO_2} = 2\log f_{HF} + \log K \tag{9-4}$$

Because $\log K$ is a constant at fixed P and T, equation (9-4) must describe a straight line with a slope of $+2$ in coordinates $\log f_{CO_2} - \log f_{HF}$. This line represents equilibrium between calcite and fluorite as a function of the changing activities of gaseous HF and CO_2. By consulting appropriate tables of thermodynamic data and by taking a value of 22.564 for $\log K$ at $25°C$, one can easily draw the line (Figure 9-1). Starting at any point on the line, increasing the fugacity of CO_2 will drive the equilibrium to the left, converting fluorite to calcite. Increasing the fugacity of HF must have the opposite effect, as shown by the labeling of the isobaric-isothermal divariant fields.

The coordinate axes could just as correctly have been labeled in terms of $\log a_{CO_2}$ or even $\log P_{CO_2}$. As a result of both the standard-state choice and the very low pressures involved (i.e., less than 1 bar), the activities, fugacities, and partial pressures of the gaseous species are numerically equivalent. We could also label the figure in terms of chemical potentials. To see how this could be done, multiply both sides of (9-4) by $2.303RT$:

$$2.303RT\log f_{CO_2} = 2(2.303RT\log f_{HF}) + 2.303RT\log K \tag{9-5}$$

By comparison with the familiar equations (5-62) and (8-9), we make appropriate substitutions for the free energy of the reaction and the chemical potentials of the reactant species:

$$\mu_{CO_2} - \mu_{CO_2}^\circ = 2(\mu_{HF} - \mu_{HF}^\circ) - \Delta G_r^\circ \tag{9-6}$$

By rearrangement, we have

$$\mu_{CO_2} = 2\mu_{HF} - \Delta G_r^\circ + \mu_{CO_2}^\circ - 2\mu_{HF}^\circ \tag{9-7}$$

The standard-state chemical potentials of CO_2 and HF are equal to the free energies of 1 mol of pure gas at 1 bar and T. Taking these values to be -394.38 and $-275.40 \text{ kJ} \cdot \text{mol}^{-1}$, respectively, we have

$$-\Delta G_r^\circ + \mu_{CO_2}^\circ - 2\mu_{HF}^\circ = 22.564 \times 298.15 \times 0.008314 \times 2.303$$

$$- 394.38 + 2(275.40)$$

$$= 285.23 \text{ kJ}$$

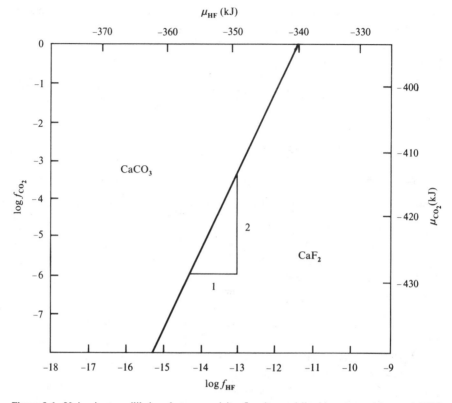

Figure 9-1. Univariant equilibrium between calcite, fluorite, and liquid water at 1 bar and 25°C, showing alternative log fugacity and chemical potential scales. The line has a slope of $+2$, which is determined by the stoichiometric coefficient for HF.

which leads to

$$\mu_{CO_2} = 2\mu_{HF} + 285.23 \qquad (9\text{-}8)$$

Using this equation, we can plot the calcite–fluorite equilibrium line in terms of chemical potentials, as shown on the alternative axes in Figure 9-1.

Suppose that no thermodynamic data were available for the reaction. It would still be possible to plot Figure 9-1 in schematic terms. Regardless of the value for $\log K$, it is clear from (9-4) (1) that the equilibrium line must be linear providing $\log K$ does not change, (2) that the slope of the line must be $+2$, and (3) that calcite must be stable on the high f_{CO_2} side and fluorite must be stable on the high f_{HF} side. It is also apparent that the slope of the line is determined solely by the stoichiometric coefficients of the gaseous reactants, because it is derived directly from the exponents of the fugacities in the expression for the equilibrium constant (cf (9-2) and (9-3)).

A more general derivation of the slope condition proceeds from the Gibbs–Duhem equation. By considering a system at constant T and P in which equilibrium is governed by the chemical potentials of two independent components 1 and 2 (analogous to CO_2 and HF in the case under consideration), we may write at equilibrium

$$n_1\,d\mu_1 + n_2\,d\mu_2 = 0 \qquad \text{(cf 5-26)}$$

Because the number of moles at equilibrium is determined by the stoichiometric coefficients of reaction, we may also write

$$v_1\,d\mu_1 + v_2\,d\mu_2 = 0 \qquad (9\text{-}9)$$

By rearrangement and indefinite integration, we obtain

$$\mu_1 = -\left(\frac{v_2}{v_1}\right)\mu_2 + \text{constant} \qquad (9\text{-}10)$$

which is equivalent, through substitution of (5-62) to

$$\ln a_1 = -\left(\frac{v_2}{v_1}\right)\ln a_2 + \text{constant} \qquad (9\text{-}11)$$

The integration constants in equations (9-10) and (9-11) are not the same; each is equal to the appropriate free-energy terms and equilibrium constants, respectively, in equations (9-8) and (9-4). For the calcite–fluorite example, $v_1 = v_{CO_2} = +1$, and $v_2 = v_{HF} = -2$, giving a *positive* slope $-(-2/1) = 2$. In this case, the gaseous species appear on *opposite sides* of equation (9-1). In contrast, it should be apparent that a reaction written with the two reactants that control the equilibrium on the *same side* of the equation must have a *negative* slope in chemical potential or log activity coordinates, because the ratio of the stoichiometric coefficients will always be positive, regardless of whether the reactants in question appear on the lefthand or the righthand side of the equal sign. Finally, if the reaction is controlled by a *single* activity or chemical potential, then the equilibrium line must be normal to the coordinate axis of that component, because equilibrium in such a case will occur at a unique log activity or chemical potential value. These cases are summarized in Figure 9-2, along with examples of each reaction type. Diagrams such as Figure 9-1 and those shown schematically in Figure 9-2 are variously called *activity diagrams*, *fugacity diagrams*, *partial pressure diagrams*, or *chemical potential diagrams*, depending on conditions of equilibrium, choice of standard state, and convenience of presentation. It is important to point out that the partial pressure diagram is valid only for total pressures of 1 bar and under conditions where the partial pressures of the reactant species are much less than 1 bar and thus essentially indistinguishable from their fugacities. The other diagrams are, of course, valid for any pressure and temperature, so long as both conditions are held constant.

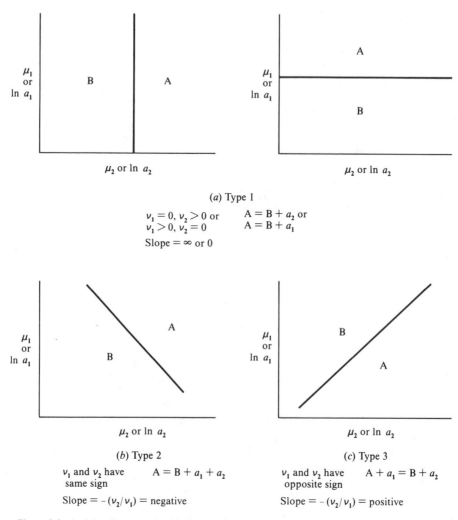

(a) Type 1

$$\nu_1 = 0, \nu_2 > 0 \text{ or} \qquad A = B + a_2 \text{ or}$$
$$\nu_1 > 0, \nu_2 = 0 \qquad A = B + a_1$$
$$\text{Slope} = \infty \text{ or } 0$$

(b) Type 2

ν_1 and ν_2 have $A = B + a_1 + a_2$
same sign

Slope $= -(\nu_2/\nu_1)$ = negative

(c) Type 3

ν_1 and ν_2 have $A + a_1 = B + a_2$
opposite sign

Slope $= -(\nu_2/\nu_1)$ = positive

Figure 9-2. Activity diagrams classified according to reaction types. Here A or B represents any mineral or mineral assemblage, and a_1 and a_2 are the activities of components 1 and 2 with stoichiometric coefficients ν_1 and ν_2, respectively.

9-2 MIXED–VOLATILE REACTIONS AT ELEVATED TEMPERATURE AND PRESSURE

When more than one reaction appears on an activity diagram, intersections may occur. Such intersections may result in isothermal-isobaric invariant points that can be analyzed using Schreinemakers techniques (cf Chapter 4). Consider, for example, the ternary system $MgO–CO_2–H_2O$ with mineral phases periclase (MgO, per), brucite ($Mg(OH)_2$, bru), and magnesite ($MgCO_3$, mag). Let these phases coexist with a high-temperature fluid phase in which the activities of CO_2 and H_2O are assumed to be independently variable. Thus, if total pressure and temperature are constant, the only system variables are a_{H_2O} and a_{CO_2}. The possible reactions are

(mag)	bru $= $ per $+ H_2O$
(bru)	mag $= $ per $+ CO_2$
(per)	mag $+ H_2O = $ bru $+ CO_2$

If two of these reactions intersect, an invariant point must result. The reason for this invariant point is that each reaction represents equilibrium between products and reactants and defines a restriction on the system that can be expressed as $\Delta G_{r(T,P)} = 0$. A system with two variables (a_{H_2O} and a_{CO_2}) and two restrictions (the equilibrium conditions for two reactions) is invariant by definition. Furthermore, because the third reaction is not independent (it can be obtained by algebraic combination of the other two), it also must intersect at the invariant point. An "instant" Schreinemakers analysis shows that the sequence of lines around the invariant point must be (bru)–(mag)–(per) in order to ensure that no field occupies an angular sector greater than 180° and that the rules regarding the proper placement of metastable extensions are obeyed. Be sure to stop at this point and convince yourself that the above sequence is correct. Moreover, because we have chosen to construct an activity diagram, the slopes of the lines are determined by the stoichiometric coefficients: reaction (mag) must be normal to the a_{H_2O} axis, reaction (bru) must be normal to the a_{CO_2} axis, and reaction (per) has a slope of $+1$. Furthermore, the stability field of periclase must lie in regions of lowest a_{H_2O} and a_{CO_2}, because increasing either activity must favor the formation of brucite or magnesite. The only possible solution is shown schematically in Figure 9-3.

Because thermodynamic data are available for all phases in the system, it is possible to calculate the activity diagram for any temperature and pressure of interest. For standard states of pure ideal-gas CO_2 and pure ideal-gas H_2O at 1 bar and T, the activities are replaced by fugacities, and the equilibrium constants become

$$K_{(mag)} = f_{H_2O} \qquad (9\text{-}12)$$

$$K_{(bru)} = f_{CO_2} \qquad (9\text{-}13)$$

$$K_{(per)} = \frac{f_{CO_2}}{f_{H_2O}} \qquad (9\text{-}14)$$

assuming that the compositions of magnesite, brucite, and periclase remain stoichiometric under all conditions and thus may assume unit activities. Using the techniques outlined in the previous chapter, we can calculate these equilibrium constants from standard free energies, entropies, and heat capacities

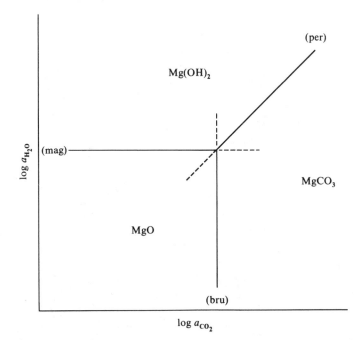

Figure 9-3. Schematic analysis of the invariant point for periclase + brucite + magnesite + fluid, plotted as on an activity diagram.

of all the species. The calculation of reaction (9-12) has been discussed in Chapter 8. Results for a specific set of data are shown in Figure 9-4, plotted in terms of $\log K$ versus reciprocal temperature, assuming a total fluid pressure on the system of 2 kbar. Note that the intersection of reaction (mag) with the 700-K isotherm ($\log K = 1.90$) gives the intersection of (mag) with the $\log f_{H_2O}$ axis on the fugacity diagram, because the equilibrium constant for that reaction is equal to the fugacity of H_2O. Likewise, the intersection of reaction (bru) at 700 K ($\log K = 0.50$) provides the intersection with $\log f_{CO_2}$ axis and determines the coordinates of the invariant point for that isotherm (Figure 9-5).

We must pause at this point to consider an interesting complication. Any activity diagram that involves volatile reactants such as CO_2, O_2, S_2, H_2O, or HF can be calculated in this striaghtforward manner, but the resulting diagram may not be physically meaningful. Return briefly to Figure 9-1. So long as the fugacity of CO_2 stays below 1 bar, the entire range of the diagram is physically accessible according to a model of a 1-atm system in equilibrium with very small partial pressures of HF and CO_2. If calcite and fluorite were not in equilibrium, then the fugacities of CO_2 and HF could—in a thought experiment, at least—be *independently* varied throughout the calcite and fluorite fields. The same case

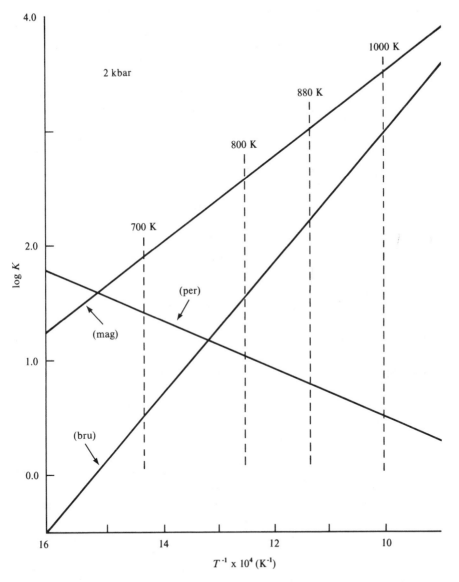

Figure 9-4. Plot of log K versus reciprocal temperature for the three reactions in the system MgO–H_2O–CO_2, calculated for a total fluid pressure of 2 kbar. The reactions represented as (bru), (mag), and (per) appear in the text near the beginning of Section 9-2. Four isotherms used in the plotting of Figure 9-6 are emphasized by dashed lines.

applies for volatiles such as O_2 and S_2 because the magnitudes of their fugacities in mineral reactions are very small—i.e., much less than 1 bar.

The only way to treat the $MgO-CO_2-H_2O$ system similarly is to assume that a very significant proportion of the 2-kbar fluid phase is composed of an inert gas species such as argon. In this case, the total pressure would be given by $P_{total} = P_{Ar} + P_{H_2O} + P_{CO_2}$, and the partial pressures (and thus the fugacities) of H_2O and CO_2 could be independently varied.

It is clear that this model has little if any geological significance. A more likely case has the total pressure made up entirely of CO_2 and H_2O, so that

$$P_{total} = P_{H_2O} + P_{CO_2} \qquad (9\text{-}15)$$

Under these more realistic circumstances, only one of the two volatile activities can be varied independently, because of the extra restriction introduced by equation (9-15). By adopting the simplifying assumption of ideal mixing of H_2O and CO_2, we have

$$P_{total} = X_{H_2O}P_{total} + X_{CO_2}P_{total} \qquad (9\text{-}16)$$

By adopting the Lewis–Randall rule (cf Section 6-1), we may introduce fugacities by the relation

$$P_{total} = \frac{f_{H_2O}}{f^{\circ}_{H_2O}} P_{total} + \frac{f_{CO_2}}{f^{\circ}_{CO_2}} P_{total} \qquad (9\text{-}17)$$

where the superscripted fugacities refer to the *pure* gas at P and T. Because, by definition, $f^{\circ}_{H_2O} = \Gamma_{H_2O}P_{total}$ and $f^{\circ}_{CO_2} = \Gamma_{CO_2}P_{total}$, we have the simple condition for the ideal mixing of pure (real) CO_2 and H_2O:

$$P_{total} = \frac{f_{H_2O}}{\Gamma_{H_2O}} + \frac{f_{CO_2}}{\Gamma_{CO_2}} \qquad (9\text{-}18)$$

Because the fugacity coefficients are functions of pressure and temperature only, and because $X_{CO_2} + X_{H_2O} = 1$, there is a *unique* path of fluid compositions in activity space that is accessible to the system. For instance, taking $\Gamma_{H_2O} = 0.245$ and $\Gamma_{CO_2} = 1.76$ at 2 kbar and 700 K, we obtain for $X_{CO_2} = 0.5$ the value $f_{CO_2} = X_{CO_2}f^{\circ}_{CO_2} = X_{CO_2}P_{total}\Gamma_{CO_2} = 1760$ bar. Similarly, f_{H_2O} in the pure ideal CO_2-H_2O system has a value of 245 bar when $X_{CO_2} = 0.5$. Solving the same equations for a range of X_{CO_2} values results in the total-pressure line shown in Figure 9-5. The arrow points to the composition of $X_{CO_2} = 0.5$ used in the preceeding example calculation. Note that the total-pressure line never intersects the periclase field but does cross the brucite–magnesite reaction at a CO_2 fugacity of about 16 bars, equivalent to a very low X_{CO_2} of 0.004.

Figure 9-6 is a polythermal fugacity diagram for the $MgO-CO_2-H_2O$ system drawn at 700, 800, 880, and 1000 K. The enlargement of the periclase field is governed by the increasing equilibrium constants for reactions (bru) and (mag),

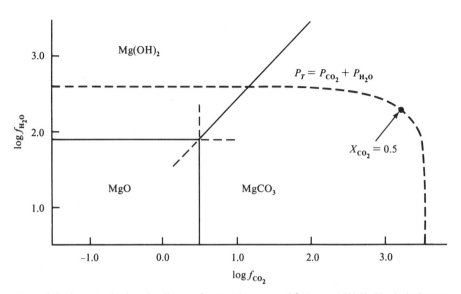

Figure 9-5. Quantitative fugacity diagram for a total pressure of 2 kbar and 700 K. The dashed curve represents the unique locus of possible fugacities in the system when the total pressure is equal to the sum of the partial pressures of H_2O and CO_2. Under these conditions, all other areas of the diagram are inaccessible.

as shown in Figure 9-4. The isotherms of the P_{total} line also shift with temperature, mostly as a result of the significant increase in the fugacity coefficient of H_2O with temperature at 2 kbars. Note that, as a result of the enlargement of the periclase field, all three reactions eventually become stable. A special situation arises at about 880 K when the total-pressure line actually intersects the invariant point at a log f_{CO_2} value of 2.24 (174 bars). Because the fugacity coefficient of pure CO_2 is about 1.80 under these conditions, this intersection corresponds to $X_{CO_2} = 174/(1.80 \times 2000) = 0.048$ for the ideal-mixing case.

The phase relations for the H_2O–CO_2 fluid can be more conveniently represented on a graph that plots temperature against mole fraction of CO_2 (Figure 9-7). This type of plot has the advantage of directly relating the equilibrium boundaries to fluid *composition*, which means that the boundaries can be located experimentally if the mole fraction of CO_2 in the experimental fluid can be measured. These diagrams are used extensively in metamorphic petrology, and many excellent summaries of their construction and use are available in the literature. Greenwood (1967) derived the slope conditions for mixed volatile reactions in T–X_{CO_2} coordinates and showed how reaction boundaries in the system MgO–SiO_2–H_2O–CO_2 can be determined experimentally. Both Kerrick (1974) and Ferry and Burt (1982) have reviewed the alternative representations of mixed-volatile equilibria with special attention to

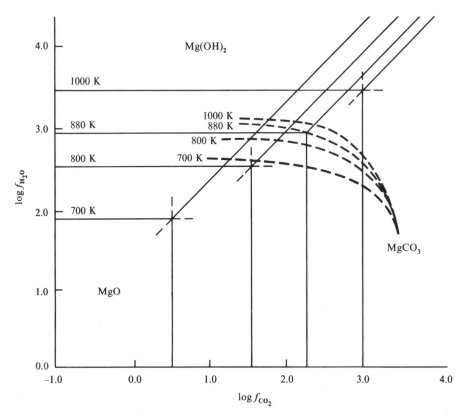

Figure 9-6. Polythermal fugacity diagram for phases in the system $MgO-H_2O-CO_2$ at a total pressure of 2 kbar. Phase boundaries at 700 K, 800 K, 880 K, and 1000 K are read from appropriate intersections in Figure 9-4. Dashed lines are partial traces of the $P_{total} = P_{H_2O} + P_{CO_2}$ line at each of the four temperatures.

metamorphic applications, and Kerrick and Slaughter (1976) discussed the calculation of $T-X_{CO_2}$ diagrams from fundamental thermodynamic data, emphasizing the problems that arise in transferring the same equilibria represented by the plot of $\log K$ versus $1/T$ (such as Figure 9-4) to the $T-X_{CO_2}$ diagram (such as Figure 9-7). Because of the recent availability of activity coefficients for the H_2O-CO_2 system at elevated temperature and pressure, rigorous calculations of $T-X_{CO_2}$ diagrams that do not rely on assumptions of ideality are now possible. It should be emphasized, however, that specific mixing data for fluid-phase components are *not* required for the calculation of activity or fugacity diagrams, which is one of the benefits of these diagrams. Mixing models are required only when the activities or fugacities must be related to the composition of the fluid phase.

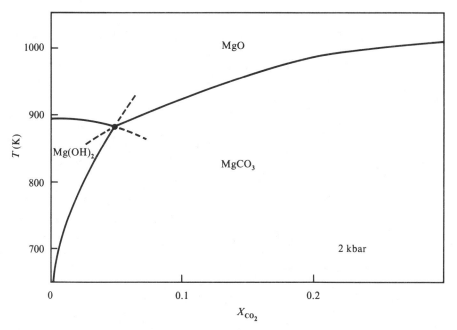

Figure 9-7. A T–X_{CO_2} diagram for phases in the system MgO–H$_2$O–CO$_2$ constructed from data in Figure 9-6, assuming ideal mixing in a H$_2$O–CO$_2$ fluid.

9-3 AQUEOUS ACTIVITY DIAGRAMS

Thus far, we have limited our discussion to equilibria involving the activities of gaseous species. Activity calculations of mineral–gas equilibria are straight-forward and can easily be corrected for temperatures above 25°C because the entropies and heat capacities of most common gases are fairly well known. However, the utility of such diagrams at low temperatures, although they are theoretically correct, is open to some question. Consider the partial-pressure diagram for the system Cu–O–S at 25°C and 1 atm (Figure 9-8). This diagram can easily be calculated from available thermodynamic data (see Problem 2), and it shows the relative stability of metallic Cu, the two copper oxides, and the two most-common copper sulfides in terms of the partial pressures of O$_2$ and S$_2$ gases. Although the diagram is completely rigorous and correct from a thermodynamic viewpoint, it cannot have much meaning in a physical sense. For instance, the boundary between Cu and Cu$_2$O shows an equilibrium partial pressure of oxygen of $10^{-51.3}$. A system at atmospheric pressure must have a volume of about 4×10^{16} km^3 (more than 10,000 times the volume of the earth) in order to find *one* oxygen molecule that exerts a partial pressure of $10^{-51.3}$ bar. What thermodynamics must be telling us from this exercise is that alternative

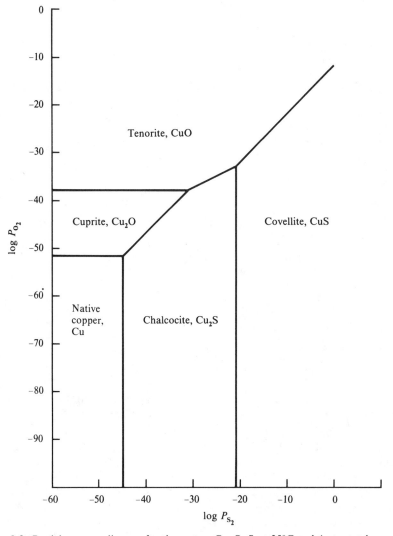

Figure 9-8. Partial-pressure diagram for the system Cu–O–S at 25°C and 1 atm total pressure. (Garrels and Christ, 1965)

ways to write the reaction will be more meaningful. One way around the problem is to express these reactions in terms of chemical potentials rather than partial pressures. However, it may be more desirable to formulate the reactions in terms of species that are actually present. For near-surface conditions where liquid water is present, dissolved aqueous species may be abundant and are likely to directly affect such reactions. The remainder of this chapter will show how activity diagrams are used to represent the stabilities of minerals in equilibrium with dissolved aqueous species.

Aqueous activity diagrams can show clearly the effect of pH, ligand concentration, temperature, and pressure on solution–mineral equilibria. Suppose we wish to study the weathering of feldspars, for example, and we want to know what happens to the aluminum when albite is dissolved in water. If the reaction were congruent, we might expect the stoichiometric reaction:

$$NaAlSi_3O_{8(s)} + 4H^+_{(aq)} + 4H_2O_{(l)} \rightarrow Na^+_{(aq)} + Al^{3+}_{(aq)} + 3H_4SiO_{4(aq)}$$

However, this reaction is a function of the pH of the solution, and both aluminum and dissolved silica can hydrolyze to different degrees, depending on the pH. Aluminum and silica can also react to form insoluble phases such as gibbsite and kaolinite.

First, we shall explore the effect of pH on aluminum hydrolysis and develop an activity diagram that describes this effect. The first hydrolysis constant for aluminum is based on the reaction

$$Al^{3+}_{(aq)} + H_2O_{(l)} \rightleftharpoons AlOH^{2+}_{(aq)} + H^+_{(aq)} \tag{9-19}$$

with an equilibrium constant of

$$K_1 = \frac{a_{AlOH^{2+}} \, a_{H^+}}{a_{Al^{3+}} \, a_{H_2O}} = 10^{-5.0} \quad \text{(at 25°C)} \tag{9-20}$$

By assuming $a_{H_2O} = 1$ and taking the logarithm, we obtain

$$\log K_1 = \log a_{AlOH^{2+}} - pH - \log a_{Al^{3+}} = -5.0$$

or

$$\log a_{Al^{3+}} = \log a_{AlOH^{2+}} + 5 - pH \tag{9-21}$$

Equation (9-21) tells us several things. At a pH of 5, the activities of $Al^{3+}_{(aq)}$ and $AlOH^{2+}_{(aq)}$ are equal. At a pH greater than 5, the activity of $AlOH^{2+}_{(aq)}$ becomes dominant; at a pH less than 5, the activity of $Al^{3+}_{(aq)}$ becomes dominant. Furthermore, when the activity of one ion becomes dominant and therefore nearly constant, the activity of the other ion is a function of pH with a slope of $+1$ or -1. For example, when $AlOH^{2+}_{(aq)}$ is dominant, we can represent its activity as a constant K', and $\log K' = K$:

$$\log a_{Al^{3+}_{(aq)}} = K + 5 - pH \tag{9-22}$$

where $\log a_{Al^{3+}_{(aq)}}$ is a function of pH with a slope of -1. These characteristics allow us to draw a diagram for the first hydrolysis of aluminum as shown in Figure 9-9. The total aluminum activity Al_T is assumed to be 10^{-7} mol·L^{-1}. Thus, when Al^{3+} or $AlOH^{2+}$ is dominant, its activity can be represented by a horizontal line at $\log a_{Al_i} = -7$, independent of pH. This condition is a mass-balance constraint:

$$a_{Al^{3+}_{(aq)}} + a_{AlOH^{2+}_{(aq)}} = 10^{-7} m \qquad (9\text{-}23)$$

At pH $= 5$, these activities must be equal to each other and therefore equal to 0.5×10^{-7}. This condition is plotted as point A in Figure 9-9. The line representing $\log a_{Al^{3+}}$ at pH > 5 is given by

$$\log a_{Al^{3+}} = -2 - pH$$

and that representing $\log a_{AlOH^{2+}}$ is given by

$$\log a_{AlOH^{2+}} = -12 + pH$$

Now we have four straight lines that must connect to form two continuous lines that cross at point A. In the immediate vicinity of point A, the lines must become curved because the two ions are present in comparable activities. The analytical equation that represents $\log a_{Al^{3+}}$ for pH values of 4 through 6 is found to be

$$a_{Al^{3+}} = \frac{10^{-pH-7}}{10^{-pH} + 10^{-5}} \qquad (9\text{-}24)$$

by simultaneously solving the two preceding equations.

Next we shall see what happens when all four hydrolysis products $AlOH^{2+}$, $Al(OH)_2^+$, $Al(OH)_3^0$, and $Al(OH)_4^-$ are taken into consideration. Using the four

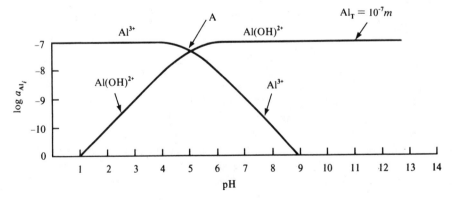

Figure 9-9. Activities of $Al^{3+}_{(aq)}$ and $AlOH^{2+}_{(aq)}$ as a function of pH.

stability constants at 25°C,

$$K_1 = 10^{-5.0}$$

$$K_2 = 10^{-10}$$

$$K_3 = 10^{-16}$$

$$K_4 = 10^{-22}$$

we can derive the basic equations:

$$\log a_{Al^{3+}} = \log a_{AlOH^{2+}} - pH + 5$$

$$\log a_{AlOH^{2+}} = \log a_{Al(OH)_2^+} - pH + 5$$

$$\log a_{Al(OH)_2^+} = \log a_{Al(OH)_3^\circ} - pH + 6$$

$$\log a_{Al(OH)_3^\circ} = \log a_{Al(OH)_4^-} - pH + 6$$

From these equations, we can tabulate the crossover points:

A $a_{Al^{3+}} = a_{AlOH^{2+}} = a_{AlOH_2^+}$ (pH = 5)

B $a_{Al(OH)_2^+} = a_{Al(OH)_3^\circ} = a_{Al(OH)_4^-}$ (pH = 6)

C $a_{Al(OH)_4^-} = a_{Al^{3+}}$ (pH = 5.5)

D $a_{Al(OH)_3^\circ} = a_{AlOH^{2+}}$ (pH = 5.5)

E $a_{Al(OH)_4^-} = a_{AlOH^{2+}}$ (pH = 5.7)

F $a_{Al(OH)_3^\circ} = a_{Al^{3+}}$ (pH = 5.3)

By estimating the activities at crossover points and solving all of the straight-line segments, we can draw the activity diagram for aluminum species as a function of pH (Figure 9-10). This diagram illustrates that the predominant species for pH

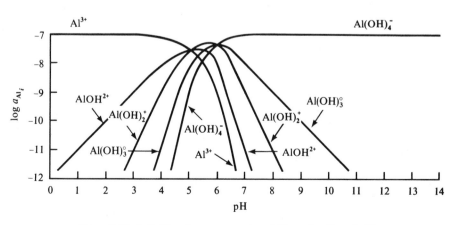

Figure 9-10. Activities of aqueous species of Al as a function of pH.

< 5 is $Al^{3+}_{(aq)}$, that for $pH = 5$ to 6 is $Al(OH)^+_{2(aq)}$, and that for $pH > 6$ is $Al(OH)^-_{4(aq)}$.

These logarithmic-activity diagrams were advocated by Lars Gunnar Sillén (1967), and they are commonly called master-variable diagrams because they show the functional dependence of some important variable (such as an aluminum hydrolysis specie) on pH, the independent (or master) variable. Numerous examples of these diagrams are presented by Butler (1964*a,b*; 1982) and Stumm and Morgan (1981).

Another common use of activity diagrams is to show how the solubility of a mineral depends on pH. The solubility-product constant of a mineral is always a fixed value at constant temperature and pressure, but the measured solubility varies with pH, ionic strength, and composition. To continue the story of the weathering of albite, suppose we want to know whether gibbsite precipitates during albite dissolution. The answer to this question requires a knowledge of gibbsite solubility as a function of pH. A solubility diagram for gibbsite can easily be derived from the known solubility-product constant and the hydrolysis constants K_1 through K_4. We begin with the simple dissolution of gibbsite,

$$Al(OH)_{3(c)} + 3\,H^+ = Al^{3+} + 3\,H_2O \tag{9-25}$$

for which

$$K_{sp} = \frac{a_{Al^{3+}}\, a^3_{H_2O}}{a^3_{H^+}} = 10^{8.11} \tag{9-26}$$

or

$$\log a_{Al^{3+}} = 8.11 - 3pH \tag{9-27}$$

Equation (9-27) describes a straight-line relationship between the free-aluminum activity and the pH. At pH values near and above 5, the other hydrolysis products will affect the solubility. By rearrangement of (9-21), we have

$$\log a_{AlOH^{2+}} = \log a_{Al^{3+}} - 5 + pH \tag{9-28}$$

Then, by substitution, we obtain

$$\log a_{AlOH^{2+}} = 3.11 - 2pH \tag{9-29}$$

Similarly, we can derive

$$\log a_{Al(OH)^+_2} = -1.89 - pH \tag{9-30}$$

$$\log a_{Al(OH)^\circ_3} = -7.89 \tag{9-31}$$

$$\log a_{Al(OH)^-_4} = -13.89 + pH \tag{9-32}$$

These linear relationships are shown in Figure 9-11. Rather than arbitrarily setting the total aluminum activity as constant at $10^{-7}\, m$, we allow the activity to be set by the solubility-product constant and the hydrolysis constants, which act together as a mass-balance constraint. The solubility is the sum total of all the

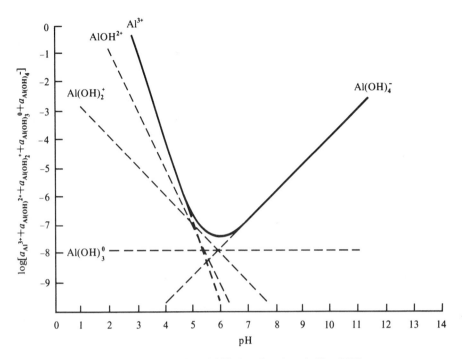

Figure 9-11. Gibbsite solubility as a function of pH at 25°C.

aluminum specie activities, as seen in Figure 9-11. For example, at the crossover point between Al^{3+} and $AlOH^{2+}$ the activities are calculated as

$$a_{Al^{3+}} = 1.29 \times 10^{-7} = a_{AlOH^{2+}} = a_{Al(OH)_2^+}$$

$$a_{Al(OH)_3^0} = 1.29 \times 10^{-8}$$

$$a_{Al(OH)_4^-} = 1.29 \times 10^{-9}$$

The total activity is then 1.43×10^{-7} m. A single general expression can be derived for the entire solubility curve as follows. We begin with the mass-balance expression

$$a_{Al_T} = a_{Al^{3+}} + a_{AlOH^{2+}} + a_{Al(OH)_2^+} + a_{Al(OH)_3^0} + a_{Al(OH)_4^-} \tag{9-33}$$

Next we substitute the mass-action expressions (see equations (9-27) through (9-32)):

$$a_{Al_T} = 10^{-3pH+8.11} + 10^{-2pH+3.11} + 10^{-pH-1.89}$$

$$+ 10^{-7.89} + 10^{pH-13.89} \tag{9-34}$$

Equation (9-34) can be programmed on any programmable hand calculator, and then the minimum in the solubility curve can easily be plotted.

Solubility diagrams such as Figure 9-11 can be derived for master variables other than pH. Carbonate-mineral solubilities will depend on P_{CO_2} and pH; sulfate-mineral solubilities will depend on chloride and sulfate activities; and silicate-mineral solubilities will depend on cation activities and dissolved-silica activity. Occasionally, the actual data that you need to produce a solubility diagram are derived by a procedure just the reverse of the one shown here. That is, the solubility of a solid is measured over a range of pH (or ionic strength, or ligand concentration, or temperature), the data are corrected for nonideality, and the results are fitted according to an assumed number of dissolved species with specified stoichiometrics. A good example is the gibbsite-solubility investigation of May et al. (1979).

Our understanding of how mineral solubilities are affected by different chemical constituents is certainly improved with solubility diagrams, but they are merely qualitative tools in the interpretation of natural-water geochemistry. These diagrams can show the effect of only a limited number of factors—e.g., Figure 9-11 applies only to the infinite-dilution standard state at a fixed temperature. Natural waters can vary enormously with respect to ionic strength, composition, temperature, and pressure. How then can the geochemistry of natural waters be related in a quantitative manner to mineral solubilities? That is the subject matter of the remainder of this chapter.

9-4   THE SATURATION INDEX

When an aqueous solution has reached saturation equilibrium with respect to a mineral such as calcite, then the reaction

$$CaCO_{3(s)} = Ca^{2+}_{(aq)} + CO^{2-}_{3(aq)} \tag{9-35}$$

is at reversible equilibrium, and the mass-action expression

$$K_{sp} = \frac{a_{Ca^{2+}} a_{CO_3^{2-}}}{a_{CaCO_3}} = 10^{-8.48} \tag{9-36}$$

applies to the equilibrium. A natural water may or may not be at saturation equilibrium, depending on the presence or absence of calcite, the available surface area, the residence time of the water, the presence of inhibiting agents, and the temperature. If it is at equilibrium, then the ion activity product (IAP) calculated from the water sample should be the same as the solubility product constant; i.e.,

$$(a_{Ca^{2+}} a_{CO_3^{2-}})_{water} = (a_{Ca^{2+}} a_{CO_3^{2-}})_{equilibrium} = 10^{-8.48}$$

or

$$\frac{(a_{Ca^{2+}} a_{CO_3^{2-}})_{water}}{(a_{Ca^{2+}} a_{CO_3^{2-}})_{equilibrium}} = \frac{IAP}{K_{sp}} = 1 \tag{9-37}$$

Because IAP values may vary by orders of magnitude, it is more convenient to take the logarithm of the ratio, commonly called the saturation index (SI):

$$SI = \log \frac{IAP}{K_{sp}} = 0 \quad \text{at equilibrium} \qquad (9\text{-}38)$$

If a water is supersaturated, then the SI is positive, and the mineral has a tendency to precipitate; if the water is undersaturated, then the SI is negative, and the mineral has a tendency to dissolve:

$$\log \frac{IAP}{K_{sp}} > 0 \quad \text{if supersaturated}$$

and

$$\log \frac{IAP}{K_{sp}} < 0 \quad \text{if undersaturated}$$

Obviously the SI has the form of a free energy, and some authors have chosen to describe the degree of saturation equilibrium in terms of free energies, or the "disequilibrium index" (DI):

$$DI = RT \ln \frac{IAP}{K_{sp}} \qquad (9\text{-}39)$$

Recalling equations (5-62) and (8-9), we have

$$DI = RT \ln \frac{IAP}{K_{sp}} = -RT \ln K + RT \ln IAP$$

$$= \Delta G_r^\circ + RT \ln IAP = \Delta G_r$$

The disequilibrium index is not used as much as equation (9-38) for the simple reason that people tend to remember solubility-product constants more easily than free energies of dissolution.

Now, if it is possible to calculate activities of free ions from a given water analysis, then it is a relatively simple matter to calculate quantitatively the degree of saturation. Ionic strength and compositional influences on the solubility are corrected for in the activity calculations, and the temperature correction can be applied to both the IAP and K_{sp}. The trick is to calculate reliable aqueous activities. These calculations are largely covered in Chapter 7 and Appendix E. They involve corrections for ion pairing and complexing as well as activity-coefficient calculations. These corrections can be very difficult, requiring the use of a moderate-to-large computer system. The method of computation and computer programs that may be used for this purpose are discussed in Appendix E.

Diagrams of SI are presented in several different ways, and the objectives and the field situation usually dictate the preferred approach. The following examples cover some of the range of possibilities.

Water–mineral equilibrium relations in the oceans, in lakes, in marine sediments, and in lake sediments are readily visualized as a function of depth. Emerson (1976) and Emerson and Widmer (1978) investigated the degree of saturation for several iron minerals in the Greifensee, a Swiss lake, by calculating the IAP from pore-water analyses. Their results for vivianite are shown in Figure 9-12, where near-equilibrium saturation is demonstrated. In this diagram, $-\log \text{IAP}$ is plotted, and $-\log K_{sp}$ (or pK_{sp}) is shown by the solid lines.

Pearson and Rettman (1976) studied the geochemistry and isotope hydrology of the Edwards limestone aquifer in central Texas and examined the

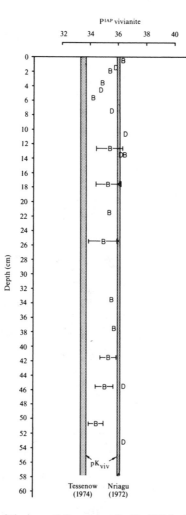

Figure 9-12. Negative log of the ion activity product for $Fe_3(PO_4)_2$, based on data from two cores. Error bars result from different methods of preparation of phosphate samples from one of the cores. (Emerson, 1976)

saturation state with respect to calcite, dolomite, gypsum, celestite, strontianite, and fluorite. Several distinct water groups were characterized, and the saturation indices for three of these groups (recharge, fresh water, and saline) are shown in Figures 9-13 and 9-14. These plots are shown as frequency-distribution diagrams, which show the dispersion in the data. All water types are tightly grouped around saturation equilibrium for calcite. Dolomite SI values, however, show progressively greater degree of saturation going from the recharge to fresh water to saline water. In the saline waters, the SI values for gypsum and celestite are

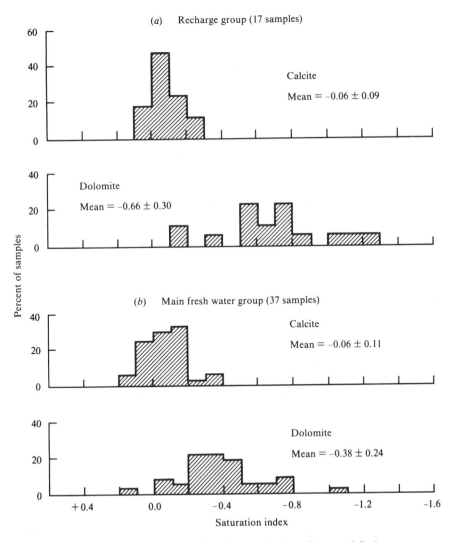

Figure 9-13. Distribution of saturation indices for samples in recharge and fresh-water groups. (Pearson and Rettman, 1976)

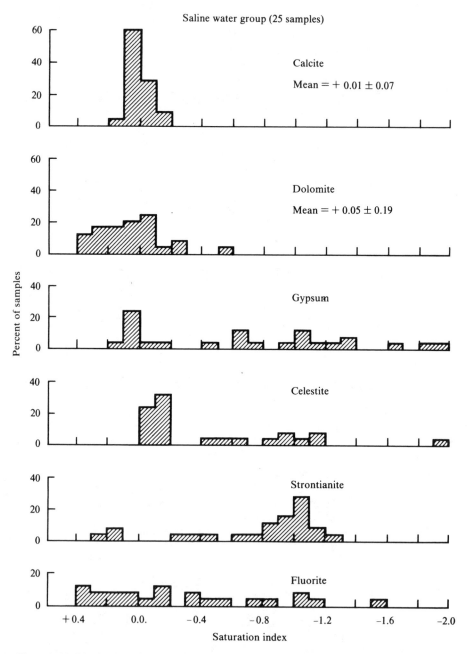

Figure 9-14. Distribution of saturation indices for samples in saline water group. (Pearson and Rettman, 1976)

rather dispersed but clearly limited by their solubilities, whereas the SI values for strontianite and fluorite do not show a clear solubility limit.

Van Breemen (1976) examined water–mineral equilibria in acid sulfate soil waters and found some interesting trends in the IAP values for gypsum and jarosite. Over 100 water samples were analyzed, and IAP values were plotted as a function of ionic strength, redox potential (E_H), and pH as shown in Figures 9-15 and 9-16 for gypsum and jarosite. Gypsum provides an excellent solubility limit, as shown by the leveling-off of IAP values at K_{sp} for gypsum. Gypsum consistently shows a rapid approach to equilibrium. Jarosite, however, shows several orders of magnitude of supersaturation, even though the mineral is found to be present and presumably precipitating in the soil water. In contrast to gypsum, jarosite must have a strong kinetic barrier to saturation equilibrium. Nordstrom, Jenne, and Ball (1979) observed the same phenomenon in acid mine drainage from northern California.

These examples show the variety of representations that have been used to interpret water–mineral equilibria with the saturation index at low temperatures. Several problems make it difficult to use the SI approach at high temperatures— for example, the inherent errors in calculating ion pairs and complexes at high temperature make the SI values rather imprecise. There has been much progress on high-T-and-P electrolyte solutions, but several alternative theories exist and these have not yet been resolved (see Rickard and Wickman, 1981).

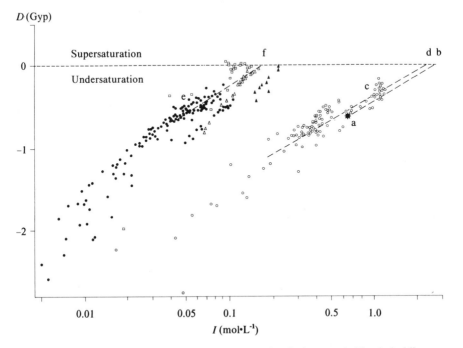

Figure 9-15. Disequilibrium index of gypsum plotted against ionic strength. The dashed lines are hypothetical evaporation paths. (van Breeman, 1976)

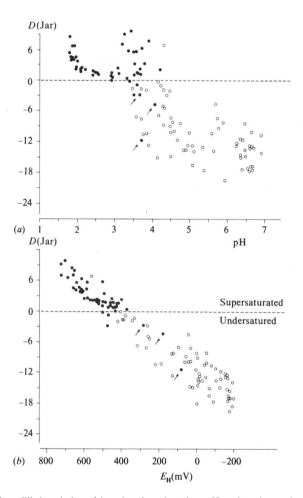

Figure 9-16. Disequilibrium index of jarosite plotted against pH and against redox potential E_H. Arrows indicate samples from the field containing both jarosite and pyrite; open circles indicate samples containing no jarosite; dark circles indicate samples containing jarosite. (van Breeman, 1976)

9-5 MINERAL STABILITY DIAGRAMS

For the graphical representation of several mineral phases and the effect of the aqueous fluid composition and temperature on their stabilities, the most attractive approach is a mineral stability diagram. These diagrams, similar in kind to the partial pressure diagrams discussed earlier, are orthogonal plots in which the most important aqueous ion activities are the coordinates. We shall consider the two systems $MgO-SiO_2-H_2O$ and $Na_2O-K_2O-Al_2O_3-SiO_2-H_2O-HCl$ and shall derive the stability diagram for each.

MgO–SiO$_2$–H$_2$O

In a three-component system at constant T and P, the description of a single phase requires two independent variables because there are two degrees of freedom by the phase rule. If we take these variables to be the activities or chemical potentials of appropriate species, then in this coordinate system any reactions between two phases will be univariant, and three phases will produce an invariant point. The specific choice of variables is dictated by how we wish to represent the formulas of the reacting species. In the MgO–SiO$_2$–H$_2$O system, $a_{Mg^{2+}_{(aq)}}$ and $a_{H_4SiO_{4(aq)}}$ might seem like obvious choices. Now let's consider the phases that occur in this system:

brucite	Mg(OH)$_2$
chrysotile	Mg$_3$Si$_2$O$_5$(OH)$_4$
sepiolite	Mg$_4$Si$_6$O$_{15}$(OH)$_2$ · 6 H$_2$O
talc	Mg$_3$Si$_4$O$_{10}$(OH)$_2$
anthophyllite	Mg$_7$Si$_8$O$_{22}$(OH)$_2$
enstatite	MgSiO$_3$
forsterite	Mg$_2$SiO$_4$
quartz	SiO$_2$

By writing the stoichiometric aqueous solubilities of each phase, we find that the aqueous species Mg^{2+}, H$^+$, and H$_4$SiO$_4$ occur throughout. For example, the dissolution of chrysotile can be written as

$$\text{Mg}_3\text{Si}_2\text{O}_5(\text{OH})_4 + 6\,\text{H}^+ \rightleftharpoons 3\,\text{Mg}^{2+} + 2\,\text{H}_4\text{SiO}_4 + \text{H}_2\text{O}$$

$$K_{\text{chrysotile}} = \frac{a^3_{\text{Mg}^{2+}} \cdot a^2_{\text{H}_4\text{SiO}_4}}{a^6_{\text{H}^+}} = 10^{34.1} \tag{9-40}$$

and

$$\log K_{\text{chrysotile}} = 3\log \frac{a_{\text{Mg}^{2+}}}{a^2_{\text{H}^+}} + 2\log a_{\text{H}_4\text{SiO}_4} = 34.1 \tag{9-41}$$

In every dissolution reaction, $a_{H_4SiO_4}$ and the ratio $a_{Mg^{2+}}/a^2_{H^+}$ occur as compositional variables in the aqueous solution that determine the solubility. Therefore, every solubility can be represented by a straight line—such as equation (9-41) for chrysotile—on a diagram with $\log(a_{Mg^{2+}}/a^2_{H^+})$ and $\log a_{H_4SiO_4}$ as coordinates. Field observations tell us that anthophyllite, enstatite, and forsterite are unstable in a weathering environment (25°C in the presence of water) and should be unstable on a 25°C mineral stability diagram. We shall go through the calculations to see whether the thermodynamic data are consistent with this observation. For brucite, we have

$$Mg(OH)_2 + 2H^+ = Mg^{2+} + 2H_2O$$

$$\log \frac{a_{Mg^{2+}}}{a_{H^+}^2} = 16.7 \tag{9-42}$$

For sepiolite,

$$Mg_4Si_6O_{15}(OH)_2 \cdot 6H_2O + 8H^+ + H_2O = 4Mg^{2+} + 6H_4SiO_4$$

$$4\log \frac{a_{Mg^{2+}}}{a_{H^+}^2} = 31.25 - 6\log a_{H_4SiO_4} \tag{9-43}$$

For talc,

$$Mg_3Si_4O_{10}(OH)_2 + 6H^+ + 4H_2O = 3Mg^{2+} + 4H_4SiO_4$$

$$3\log \frac{a_{Mg^{2+}}}{a_{H^+}^2} = 21.2 - 4\log a_{H_4SiO_4} \tag{9-44}$$

For anthophyllite,

$$Mg_7Si_8O_{22}(OH)_2 + 14H^+ + 8H_2O = 7Mg^{2+} + 8H_4SiO_4$$

$$7\log \frac{a_{Mg^{2+}}}{a_{H^+}^2} = 73.3 - 8\log a_{H_4SiO_4} \tag{9-45}$$

For enstatite,

$$MgSiO_3 + 2H^+ + H_2O = Mg^{2+} + H_4SiO_4$$

$$\log \frac{a_{Mg^{2+}}}{a_{H^+}^2} = 11.4 - \log a_{H_4SiO_4} \tag{9-46}$$

For forsterite,

$$Mg_2SiO_4 + 4H^+ = 2Mg^{2+} + H_4SiO_4$$

$$2\log \frac{a_{Mg^{2+}}}{a_{H^+}^2} = 29.1 - \log a_{H_4SiO_4}$$

For quartz,

$$SiO_{2(qtz)} + 2H_2O = H_4SiO_{4(aq)}$$

$$\log a_{H_4SiO_4} = -4.0$$

As we have mentioned previously, the slopes of the lines are equal to the ratios of the respective stoichiometric coefficients, and the positions of the axes are equal to the values for the respective equilibrium constants, as shown in Figure 9-17. These solubility lines have both stable and metastable sections. Recognizing that each line represents equilibrium saturation and that below each line only the solution is stable, we can see that brucite is the stable phase up to

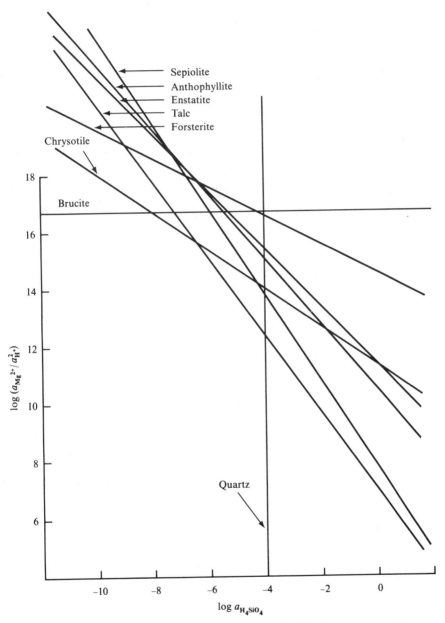

Figure 9-17. Solubility diagram for minerals in the $MgO–SiO_2–H_2O$ system at 25°C.

$a_{H_4SiO_4} = 10^{-8}$; then chrysotile is stable up to about $a_{H_4SiO_4} = 10^{-6.9}$; and finally, talc is stable at higher activities of silica. Beyond these limits, the lines become metastable, and thus the entire line segments for sepiolite, anthophyllite, enstatite, and forsterite are metastable. The latter three minerals show metastability limits that indeed are consistent with field observations. Sepiolite, however, is found in several low-temperature aquatic environments (Hathaway and Sachs, 1965; Wollast et al., 1968), suggesting that either (1) it is a metastable precursor to talc with a lower kinetic barrier to precipitation, or (2) there is some error in the thermodynamic data large enough to reverse the relative stabilities of these two minerals. Both are good possibilities, and further research is warranted to resolve this question.

To complete the stability diagram, we should draw the phase boundaries between the various minerals. For example, from the crossover point between brucite and chrysotile, there should be another line intersecting at the same point and sloping to the right of the chrysotile line and to the left of the brucite line. The analytic expression for the brucite–chrysotile equilibrium can be derived simply by combining equation (9-40) with equation (9-42):

$$Mg_3Si_2O_5(OH)_4 + 4H^+ + H_2O = 2Mg^{2+} + 2H_4SiO_4 + Mg(OH)_2$$

$$2\log\frac{a_{Mg^{2+}}}{a_{H^+}^2} = 17.4 - 2\log a_{H_4SiO_4} \tag{9-47}$$

Likewise, for the chrysotile–talc equilibrium, we have

$$Mg_3Si_4O_{10}(OH)_2 + 5H_2O = Mg_3Si_2O_5(OH)_4 + 2H_4SiO_4$$

$$2\log a_{H_4SiO_4} = -12.9$$

$$\log a_{H_4SiO_4} = -6.45 \tag{9-48}$$

For talc–quartz equilibrium,

$$Mg_3Si_4O_{10}(OH)_2 + 6H^+ + 2H_2O = 3Mg^{2+} + SiO_2 + 3H_4SiO_4$$

$$3\log\frac{a_{Mg^{2+}}}{a_{H^+}^2} = 17.2 - 3\log a_{H_4SiO_4} \tag{9-49}$$

From equations (9-47) through (9-49), we can draw the stability boundaries for talc, quartz, brucite, and chrysotile. Similar equations for sepiolite complete the phase diagram as shown in Figure 9-18. We can see what phases may be stable in hydrogeologic environments from the limits of $a_{Mg^{2+}}/a_{H^+}^2$ and $a_{H_4SiO_4}$ known to occur in natural waters. These limits are shown as the box bounded by a dashed line in Figure 9-18. As you can see, every phase is included within the natural-water composition limit, and their relative stabilities depend on the magnesium and silica concentrations and the pH. Experimental measurements of this system at high temperatures show the disappearance of chrysotile and the appearance of forsterite in its place (Figure 9-19). At these temperatures (300°C to 600°C),

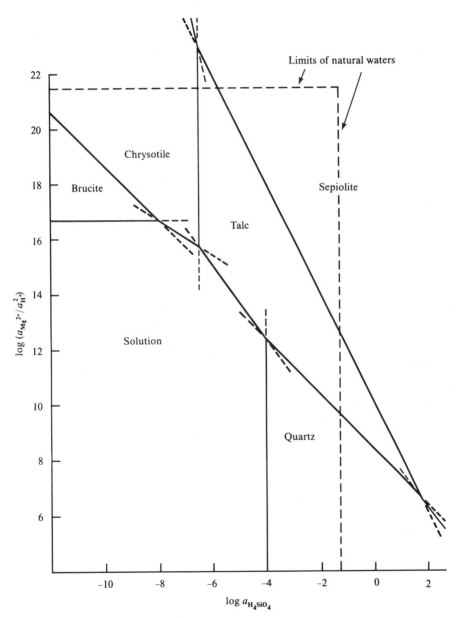

Figure 9-18. Fluid-mineral stability diagram for the MgO–SiO$_2$–H$_2$O system at 25°C.

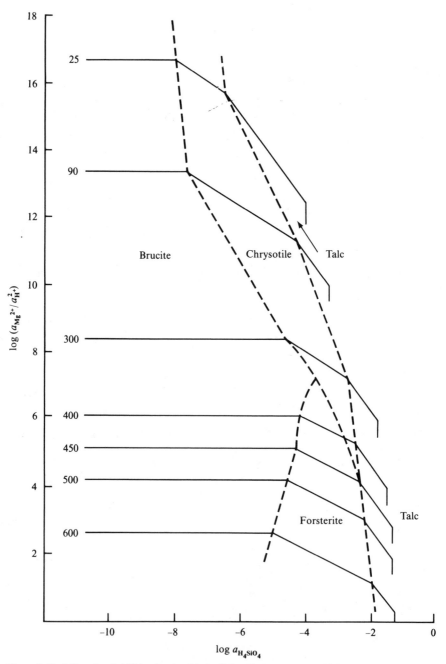

Figure 9-19. Mineral solubilities in the $MgO–SiO_2–H_2O$ system at high temperatures. (After Eugster, 1981, and Hemley et al., 1977a,b)

enstatite and anthophyllite also appear, and their phase relations have been worked out by Hemley et al. (1977*a*,*b*) and Frantz and Popp (1979).

$Na_2O-K_2O-Al_2O_3-SiO_2-HCl-H_2O$

The $Na_2O-K_2O-Al_2O_3-SiO_2-HCl-H_2O$ system is quite important because it contains the common rock-forming minerals albite $(NaAlSi_3O_8)$, microcline $(KAlSi_3O_8)$, muscovite $(KAl_3Si_3O_{10}(OH)_2)$, paragonite $(NaAl_3Si_3O_{10}(OH)_2)$, pyrophyllite $(Al_2Si_4O_{10}(OH)_2)$, kaolinite $(Al_2Si_2O_5(OH)_4)$, and gibbsite $(Al(OH)_3)$. These minerals and their various possible assemblages in geological environments give us clues to the conditions of formation of several different rock types—including information about such processes as diagenesis, weathering, hydrothermal alteration, and metamorphism. Let us assume that solution is always present, alumina is conserved in the solid phases, silica is fixed at quartz saturation, and P and T are constant. Then the number of independent variables is reduced to only two—such as μ_{Na_2O} and μ_{K_2O}, or $\ln(a_{Na^+}/a_{H^+})$ and $\ln(a_{K^+}/a_{H^+})$. Because pyrophyllite, kaolinite, and gibbsite do not contain any alkali metals, their stabilities are governed by silica activities and temperature. For the kaolinite–gibbsite equilibrium at 25°C,

$$Al_2Si_2O_5(OH)_4 + 5H_2O = 2Al(OH)_3 + 2H_4SiO_4$$

$$\log K = 2\log a_{H_4SiO_4} = -10.4$$

$$\log a_{H_4SiO_4} = 5.2$$

With quartz saturation maintained at $a_{H_4SiO_4} = 10^{-4.0}$, the reaction will be driven to the left, and kaolinite will be more stable than gibbsite. Now let us see how stable pyrophyllite is relative to kaolinite. The reaction equilibrium is

$$Al_2Si_4O_{10}(OH)_2 + 5H_2O = Al_2Si_2O_5(OH)_4 + 2H_4SiO_4$$

$$\log K = 2\log a_{H_4SiO_4} = -6.35$$

$$\log a_{H_4SiO_4} = -3.2$$

Therefore, pyrophyllite is less stable than kaolinite at quartz saturation and, at 25°C and 1 bar, kaolinite is the stable phase we must consider in equilibrium reactions involving muscovite and paragonite. For the kaolinite–muscovite equilibrium,

$$2KAl_3Si_3O_{10}(OH)_2 + 2H^+ + 3H_2O = 3Al_2Si_2O_5(OH)_4 + 2K^+$$

$$2\log \frac{a_{K^+}}{a_{H^+}} = 8.81$$

$$\log \frac{a_{K^+}}{a_{H^+}} = 4.40 \tag{9-50}$$

This equilibrium is represented as a vertical line in Figure 9-20. Similarly, for the

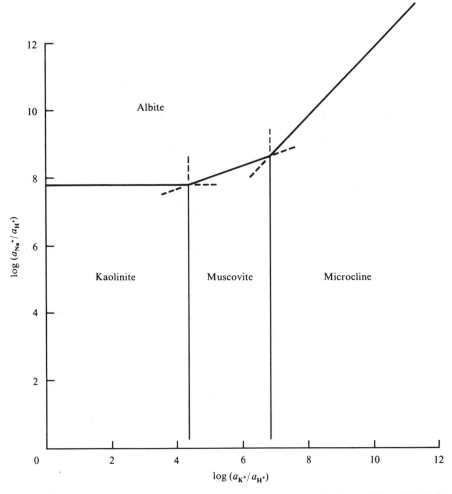

Figure 9-20. Fluid-mineral stability diagram for the system $Na_2O-K_2O-Al_2O_3-SiO_2-HCl-H_2O$ at 25°C.

kaolinite–paragonite equilibrium, we have

$$2\,NaAl_3Si_3O_{10}(OH)_2 + 2\,H^+ + 3\,H_2O = 3\,Al_2Si_2O_5(OH)_4 + 2\,Na^+$$

$$2\log \frac{a_{Na^+}}{a_{H^+}} = 20.0$$

$$\log \frac{a_{Na^+}}{a_{H^+}} = 10.0 \tag{9-51}$$

For muscovite–microcline,

$$3\,KAlSi_3O_8 + 2\,H^+ + 12\,H_2O = KAl_3Si_3O_{10}(OH)_2 + 2\,K^+ + 6\,H_4SiO_4$$

$$2\log\frac{a_{K^+}}{a_{H^+}} = -10.3 - 6\log a_{H_4SiO_4}$$

$$\log\frac{a_{K^+}}{a_{H^+}} = 6.85 \tag{9.-52}$$

For paragonite–albite,

$$3\,NaAlSi_3O_8 + 2\,H^+ + 12\,H_2O = NaAl_3Si_3O_{10}(OH)_2 + 2\,Na^+ + 6\,H_4SiO_4$$

$$2\log\frac{a_{Na^+}}{a_{H^+}} = -10.7 - 6\log a_{H_4SiO_4}$$

$$\log\frac{a_{Na^+}}{a_{H^+}} = 6.66 \tag{9-53}$$

At this point, we discover that paragonite equilibrates with albite at a a_{Na^+}/a_{H^+} ratio that is less than the a_{Na^+}/a_{H^+} ratio where kaolinite equilibrates with paragonite. In other words, we find that paragonite will already transform to albite in the stability field of kaolinite, and therefore paragonite is metastable with respect to kaolinite and albite at 25°C (assuming that the thermodynamic data are reliable). We should next calculate the kaolinite–albite boundary:

$$2\,NaAlSi_3O_8 + 2\,H^+ + 9\,H_2O = Al_2Si_2O_5(OH)_4 + 2\,Na^+ + 4\,H_4SiO_4$$

$$2\log\frac{a_{Na^+}}{a_{H^+}} = -0.45 - 4\log a_{H_4SiO_4}$$

$$\log\frac{a_{Na^+}}{a_{H^+}} = 7.78 \tag{9-54}$$

For the muscovite–albite equilibrium,

$$3\,NaAlSi_3O_8 + K^+ + 3\,H^+ + 12\,H_2O = KAl_3Si_3O_{10}(OH)_2 + 3\,Na^+$$
$$+ H^+ + 6\,H_4SiO_4$$

$$3\log\frac{a_{Na^+}}{a_{H^+}} = \log\frac{a_{K^+}}{a_{H^+}} - 6\log a_{H_4SiO_4} - 5.08$$

$$\log\frac{a_{Na^+}}{a_{H^+}} = \frac{1}{3}\log\frac{a_{K^+}}{a_{H^+}} + 6.31 \tag{9-55}$$

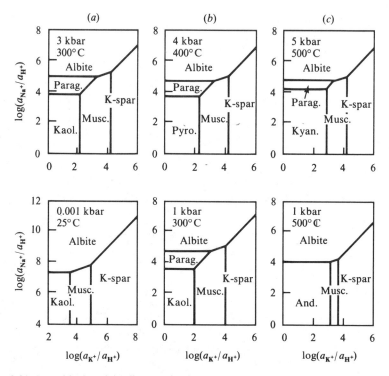

Figure 9-21. Logarithmic activity diagrams for the system $Na_2O-K_2O-Al_2O_3-SiO_2-HCl-H_2O$ at high pressures and temperatures. The positions of the stability-field boundaries were computed for $a_{H_2O} = 1$ at quartz saturation, using thermodynamic equations and data for minerals summarized in the text and those for aqueous species given by Helgeson et al. (1978).

For the albite–microcline equilibrium,

$$NaAlSi_3O_8 + K^+ + H^+ = KAlSi_3O_8 + Na^+ + H^+$$

$$\log \frac{a_{Na^+}}{a_{H^+}} = \log \frac{a_{K^+}}{a_{H^+}} + 1.75 \qquad (9\text{-}56)$$

Note that an extra H^+ was added to each side of equations (9-55) and (9-56) in order to maintain the correct ratio of alkali to proton.

The stabilities of these minerals change markedly with increased temperature. Above 300°C, kaolinite changes to pyrophyllite, and paragonite becomes stable; at 500°C, andalusite and kyanite become stable with respect to pyrophyllite at 1 kbar and 5 kbar, respectively. The progressive changes in the diagram with P and T can be seen in Figure 9-21. The difference between Figures 9-20 and 9-21 in the equilibrium boundaries at 25°C is due to the different data bases. Figure 9-20 is based on Robie et al. (1978), whereas Figure 9-21 is based on Helgeson et al. (1978).

SUMMARY

- Reactions between minerals and an aqueous solution, a gas, or a supercritical fluid phase can be expressed in terms of those molecular or ionic species present in the solution, gas, or fluid that critically affect the equilibrium under the pressure and temperature conditions of interest.

- At constant temperature and pressure, the chemical potentials of these species are linear functions of each other, with slopes determined by the ratios of the stoichiometric coefficients of the solution species (provided that the activities of the mineral phases are always unity). For a reaction controlled by two independent chemical potentials μ_1 and μ_2 each related by stoichiometric coefficients v_1 and v_2,

$$\mu_1 = -\frac{v_2}{v_1} \mu_2 + \text{constant}$$

where the constant is a function of the standard free energy of the reaction. This equation is the basis for the construction of chemical-potential diagrams.

- Alternative axes for chemical potential diagrams may be expressed in terms of the logarithm of the activity, partial pressure, fugacity, or molality of the species in question. One of the most important problems in using these diagrams is to decide which variables are most relevant to solving the problem at hand. A second problem is to decide which kind of diagram will best show the critical relationships between the variables you have chosen and the mineral phases.

- Mineral-solubility diagrams are a special kind of activity diagram in which the activities of all aqueous species present at saturation are plotted against a "master variable" such as pH or P_{CO_2}. The diagrams emphasize the distribution of the various aqueous species as a function of the master variable.

- The saturation index (SI) of an aqueous solution is defined as

$$\text{SI} = \log \frac{\text{IAP}}{K_{sp}}$$

If SI for a given mineral–water system is positive, then the mineral has a tendency to precipitate. If SI is negative, then the water is undersaturated, and the mineral (if present) has a tendency to dissolve. This function is very useful in revealing which mineral or minerals may exert the greatest influence on the chemistry of natural low-temperature waters.

- Low-temperature mineral stability diagrams plot the boundaries between minerals in terms of the logarithm of the activities (or ratios of activities) of the dissolved species that control mineral-solution equilibria. The diagrams may be determined experimentally or may be calculated from thermodynamic data, and they show which minerals should exist in equilibrium with natural waters of known compositions.

- High-temperature activity diagrams for dissolved aqueous species are more difficult to construct quantitatively because the speciation of the high-temperature aqueous phase is often poorly known. Nonetheless, such diagrams are commonly constructed for the gaseous components of a supercritical fluid phase. For the special case when the fluid phase is composed entirely of the gaseous species of interest (e.g., $P_{total} = P_{H_2O} + P_{CO_2}$), only one line on the activity diagram is physically accessible. In such cases, alternative representations such as a $T-X_{CO_2}$ diagram may be more convenient.

PROBLEMS

1. Calculate the equilibrium line between calcite and fluorite at 80°C (cf Figure 9-1) if $\Delta H^\circ_{r(298)}$ for reaction (9-1) is -154.6 kJ and $\log K_{(298)}$ is 22.564. State clearly any assumptions that you need to solve the problem. (*Hint:* You should find $\log K_{(353)} = 18.35$.)

2. Calculate the $\log P_{O_2}$ versus $\log P_{S_2}$ coordinates of the invariant point involving tenorite, cuprite, and chalcocite shown on Figure 9-8. Verify that all slopes shown in the diagram are correct. Check your answer against the figure. Use the following values for $\Delta G^\circ_{f(298)}$ (in kcal·mol^{-1}):

Cu_2O	-34.98
CuO	-30.4
Cu_2S	-20.6
CuS	-11.7
$S_{2(g)}$	19.13

Then recalculate the invariant point in terms of the chemical potentials of O_2 and S_2.

3. Consider the system $MgO-SiO_2-H_2O-CO_2$ at 1 kbar and 700 K. The four mineral phases magnesite ($MgCO_3$), talc $(Mg_3Si_4O_{10}(OH)_2)$, forsterite (Mg_2SiO_4), and quartz, plus a supercritical fluid phase composed only of H_2O and CO_2, make up an isothermally isobarically invariant assemblage with respect to the variables a_{H_2O} and a_{CO_2}, and the divariant areas on the activity diagram will be represented by two-phase

mineral assemblages as opposed to one-phase assemblages. This increase of one phase is a consequence of adding one component to the $MgO-H_2O-CO_2$ system discussed near the beginning of this chapter.

a) Write and balance the four univariant reactions that radiate from the invariant point.

b) Perform a Schreinemakers analysis to determine the sequence of lines around the invariant point.

c) Calculate the slope of each line, choosing $\log f_{H_2O}$ as the ordinate, and draw the diagram.

d) Label each divariant field with the appropriate chemography.

4. Derive equation (9-24):

$$a_{Al^{3+}} = \frac{10^{-pH-7}}{10^{-pH} + 10^{-5}}$$

5. Using the following hydrolysis constants for carbonic acid,

$$K_h = 10^{-1.46} \qquad CO_{2(g)} + H_2O \rightleftharpoons H_2CO_{3(aq)}$$
$$K_1 = 10^{-6.35} \qquad H_2CO_{3(aq)} \rightleftharpoons HCO_{3(aq)}^- + H_{(aq)}^+$$
$$K_2 = 10^{-10.33} \qquad HCO_{3(aq)}^- \rightleftharpoons CO_{3(aq)}^{2-} + H_{(aq)}^+$$

derive the relationships for the acid dissociation of carbonic acid at $pCO_2 = 3.5$, and draw an activity diagram for the three carbon species (H_2CO_3, HCO_3^- and CO_3^{2-}) as a function of pH.

6. Find the pH of distilled water equilibrated with atmospheric CO_2. Acid rain has been defined as rain having a pH less than this value.

7. Draw the solubility diagram for ferric hydroxide as a function of pH from the following data: (cf p. 261 for aluminum hydrolysis)

$$K_1 = 10^{-2.2}$$
$$K_2 = 10^{-3.5}$$
$$K_3 = 10^{-7.3}$$
$$K_4 = 10^{-8.6}$$
$$K_{sp} = 10^{-38}$$

8. Starting with the mass-balance expression for the sum of the dissolved aluminum species, equation (9-33), derive the expression for gibbsite solubility using molalities and activity coefficients for each specie. From this equation, calculate the total dissolved-aluminum concentrations (in molal units) as a function of pH with the Davies equation for activity coefficients at an ionic strength of 0.1 m, and compare to the solubility without activity-coefficient corrections (Figure 9-11).

9. Using data from Robie et al. (1978), construct an activity diagram for the univariant equilibrium between calcite and fluorite in terms of the aqueous species CO_3^{2-} and F^-. Compare and contrast with Figure 9-1.

10. Using data from Robie et al. (1978), construct an activity diagram for log $(a_{Mg^{2-}}/a_{H^+}^2)$ versus log (a_{Na^+}/a_{H^+}), considering the mineral phases kaolinite, albite, brucite, chrysotile, talc, and quartz at 25°C and 1 bar.

10

MINERAL EQUILIBRIA
III. OXIDATION–REDUCTION
REACTIONS

Reduction and oxidation reactions proceeding in geologic environments have long been accepted as factors influencing the formation of rocks and mineral assemblages.

H. P. EUGSTER (1959)

Oxidation–reduction (redox) equilibria are customarily restricted to those chemical reactions that involve a transfer of electrons between species having different valence states—e.g., reactions involving minerals and solutions containing transition metals such as Fe, Mn, and Cr. The natural abundance of iron both in terrestrial rocks and in solutions as diverse as groundwater, sea water, ore-forming hydrothermal fluids, and magmas has always made the study of redox reactions of prime importance in geochemistry.

There is no unique thermodynamic technique for describing redox equilibria, and very different approaches have customarily been taken, depending mostly on the temperature of the system of interest. For instance, consider two alternative ways to express the oxidation of magnetite (Fe_3O_4) to hematite (Fe_2O_3):

$$2\,Fe_3O_4 + \tfrac{1}{2}O_{2(g)} \rightleftharpoons 3\,Fe_2O_3 \tag{10-1}$$

$$2\,Fe_3O_4 + H_2O_{(l)} \rightleftharpoons 3\,Fe_2O_3 + 2\,e^- + 2\,H^+_{(aq)} \tag{10-2}$$

Equation (10-1) is written in terms of equilibrium between both minerals and a gas phase, whereas equation (10-2) assumes both iron oxides to be in equilibrium with liquid water. The former approach has customarily been taken for the description of high-temperature (igneous/metamorphic) redox reactions, whereas the latter approach is commonly adopted to study redox reactions on or near the earth's surface, where liquid water is stable. As we shall see, if H_2O is present in the system, the two approaches are thermodynamically equivalent in terms of expressing the conditions under which magnetite might become oxidized to hematite, and the choice of which formulation to follow is largely a matter of convenience and emphasis.

The first part of this chapter treats redox equilibria in terms of the electrochemistry of aqueous solutions and explains the special kind of activity diagrams known as pε–pH diagrams that are used to display aqueous redox equilibria. The next part approaches the same subject from the perspective of mineral–gas reactions and shows how to use oxygen fugacity (f_{O_2}) as a master variable to describe redox equilibria. The final part ties together many of the different approaches to mineral equilibria discussed in Chapters 8 through 10 and addresses the general problem of selecting thermodynamic variables to solve geochemical problems.

10-1 ELECTROCHEMICAL CONVENTIONS

A thermodynamic description of the large class of reactions involving electron transfers requires the use of certain conventions to maintain a consistency of symbols and equations with those applied in the rest of thermodynamics and with empirical observations. This section and the following two sections cover the basic concepts and conventions for electrochemical reactions and discuss both the construction and significance of pε–pH diagrams. We begin by returning to the example of the Daniell cell introduced in Section 3-8.

In a Daniell cell, one electrode consists of zinc metal in a zinc sulfate solution, and the other electrode is copper metal in a copper sulfate solution. If the two solutions are connected via a porous plug and the wire leads are attached to a galvanometer, we find a current flowing and a polarity indicating that the zinc electrode is negative relative to the copper electrode (Figure 10-1). The negative charge on the zinc electrode suggests to us that the reaction at the zinc-metal surface is producing an excess of electrons in the zinc metal—i.e., an oxidation reaction.

$$Zn_{(s)} \rightleftharpoons Zn^{2+}_{(aq)} + 2e^- \qquad (10\text{-}3)$$

Equation (10-3) represents a *half-cell reaction*; the other half of the galvanic cell reaction takes place at the copper electrode surface, where a positive charge is established by the reduction of copper ions:

$$Cu^{2+}_{(aq)} + 2e^- \rightleftharpoons Cu_{(s)} \qquad (10\text{-}4)$$

By adding the two half-cell reactions, we obtain the overall cell reaction of

$$Zn_{(s)} + Cu^{2+}_{(aq)} \rightleftharpoons Cu_{(s)} + Zn^{2+}_{(aq)} \qquad (10\text{-}5)$$

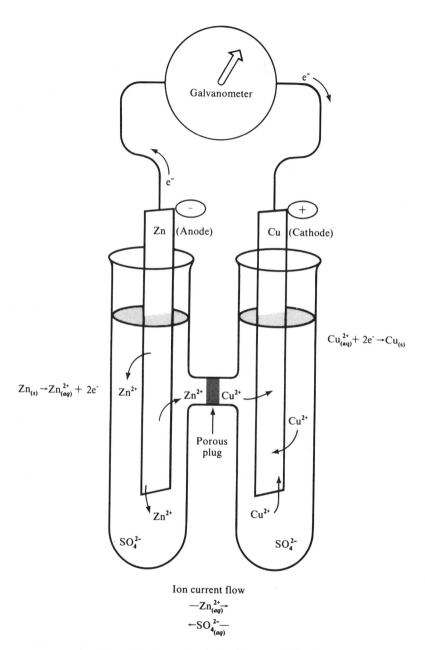

Galvanometer

e⁻

e⁻

Zn (Anode)

Cu (Cathode)

$-$

$+$

$Cu^{2+}_{(aq)} + 2e^- \rightarrow Cu_{(s)}$

$Zn_{(s)} \rightarrow Zn^{2+}_{(aq)} + 2e^-$

Zn^{2+}

Zn^{2+} Cu^{2+}

Cu^{2+}

Porous
plug

Zn^{2+}

Cu^{2+}

SO_4^{2-}

SO_4^{2-}

Ion current flow
$-Zn^{2+}_{(aq)} \rightarrow$
$\leftarrow SO^{2-}_{4(aq)} -$

Figure 10-1. Electrode reactions in the Daniell cell.

An easy mnemonic for remembering redox reactions is this: "**Leo** the Lion **Grr**s!": Loss of Electrons is Oxidation—Gain Refers to Reduction of electrons. The excess electrons produced in the zinc electrode flow spontaneously to the copper electrode through the external circuit when it is closed as shown in Figure 10-1. The negative electrode is called the anode, and the positive electrode is the cathode. Another handy way to keep your electrodes straight is to note that anode and oxidation both begin with vowels whereas cathode and reduction both begin with consonants. The current flow in the external circuit is carried by electrons, whereas the current flow in the cell is carried by ions. Excess zinc ions tend to migrate toward the copper electrode, where a deficiency of positive ions is caused by the reduction of copper. To maintain electroneutrality, sulfate ions migrate in the opposite direction. Thus, positive ions have become known as cations because they migrate to the cathode, and negative ions are called anions because they migrate to the anode.

When a galvanic cell is set up as shown in Figure 10-1 with the zinc electrode on the left and the copper electrode on the right, then the diagrammatic shorthand is

$$Zn \,|\, Zn^{2+} \,|\, Cu^{2+} \,|\, Cu$$

where the vertical lines represent phase boundaries. If the electric potential difference, or cell voltage, ΔV is measured reversibly (with a reverse-current potentiometer or a high-input impedance voltmeter), then that potential will be positive and equal in sign and magnitude to the potential of a metallic conducting lead on the right minus that of an identical lead on the left. This convention is in accordance with the Stockhohm IUPAC convention of 1953 and is similar to the convention of products minus reactants for a standard chemical reaction. The limiting value of the electrical potential difference for zero current through the cell is called the electromotive force (EMF).

Because electrochemical cell measurements always involve two electrodes, the half-cell electrode potential cannot be measured directly. To circumvent this problem, a relative scale of electrode potentials has been designated whereby all potentials are related to the standard hydrogen electrode (SHE). The SHE consists of a platinum electrode in a HCl solution ($a = 1$) with hydrogen gas bubbling across the electrode surface as shown in Figure 10-2. The platinum catalyzes the reaction

$$\tfrac{1}{2}H_{2(g)} = H^+_{(aq)} + e^- \tag{10-6}$$

which is defined as having a standard electrode potential of zero—i.e.,

$$E° = 0$$

The standard state used here refers to pure hydrogen gas at 1 bar and 298.15 K with $a_{HCl} = 1$. Now we can set up the cell

$$Pt_{(s)} \,|\, H_{2(g)} \,|\, H^+_{(aq)} \,|\, Zn^{2+}_{(aq)} \,|\, Zn_{(s)}$$

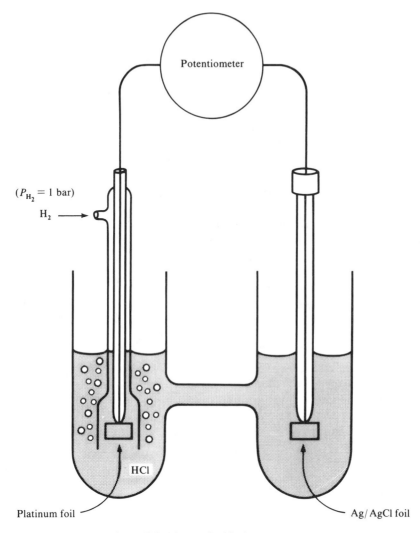

Figure 10-2. The standard hydrogen electrode.

and measure the cell potential for

$$Zn^{2+}_{(aq)} + H_{2(g)} = Zn_{(s)} + 2 H^{+}_{(aq)} \tag{10-7}$$

This procedure allows us to measure

$$Zn^{2+}_{(aq)} + 2 e^{-} = Zn_{(s)}$$

which is really an abbreviation of (10-7). In the standard state, this electrode potential is -0.763 V. By measuring the cell

$$Pt_{(s)} \,|\, H_{2(g)} \,|\, H^+_{(aq)} \,|\, Cu^{2+}_{(aq)} \,|\, Cu_{(s)}$$

we find that $E° = 0.340$ V, so that the Daniell cell potential can be calculated

$$E_{Daniell} = E°_{Cu^{2+}/Cu} - E°_{Zn^{2+}/Zn} = 0.340 - (-0.763) = 1.103 \text{ V}$$

which is the value used in Section 3-8.

Unfortunately, the SHE is difficult to operate (not to mention the dangers of using hydrogen gas), and most electrochemical measurements are made with either a Ag/AgCl or a calomel (Hg/Hg_2Cl_2) reference electrode. Both of these electrodes have been measured carefully and extensively over a range of temperatures and at several molalities of the KCl filling solution with a SHE. One of them is frequently used as the reference electrode in pH measurements, ion-selective electrode measurements, and aqueous redox-potential measurements. One example can be given for measurements of the important electrode potential

$$Fe^{3+}_{(aq)} + e^- = Fe^{2+}_{(aq)} \tag{10-8}$$

that were made by Whittemore and Langmuir (1972) in perchlorate media with a calomel reference electrode. After correcting for hydrolysis and activity coefficients, they were able to calculate a standard electrode potential of $E° = +0.770$ V.

Another important convention is that half-cell reactions are always written as reduction potentials, such as equation (10-8). Because this electrode potential is an abbreviation for

$$Fe^{3+}_{(aq)} + \tfrac{1}{2}H_{2(g)} = Fe^{2+}_{(aq)} + H^+_{(aq)}$$

the equilibrium constant expression is therefore

$$K = \frac{a_{Fe^{2+}} a_{H^+}}{a_{Fe^{3+}} a^{1/2}_{H_{2(g)}}} = \frac{a_{Fe^{2+}_{(aq)}}}{a_{Fe^{3+}_{(aq)}}} \tag{10-9}$$

when $a_{H^+_{(aq)}} = 1$ and $a_{H_{2(g)}} = 1$.

10-2   THE NERNST EQUATION, E_H, AND $p\varepsilon$

Electrochemical energy is simply another form of free energy, as we have discussed in Chapter 3—that is,

$$\boxed{\Delta G = -nFE} \quad \text{and} \quad \boxed{\Delta G° = -nFE°} \quad \text{(cf 3-94 and 3-105)}$$

Note that this sign convention assures that a spontaneous electrochemical reaction is possible only if the electrode potential of the overall cell reaction is positive.

Remembering that

$$\Delta G = \Delta G° + RT \ln \Pi a_i^{\nu_i} \tag{cf 8-8 and 8-9}$$

we can substitute to obtain the *Nernst equation*:

$$E = E^\circ - \frac{RT}{nF} \ln \Pi a_i^{\nu_i}$$ (10-10)

where n is the number of electrons transferred, F is the Faraday constant, and R is the ideal gas constant. At 298.15 K and 1 bar, the equation reduces to

$$E = E^\circ - \frac{0.0592}{n} \log \Pi a_i^{\nu_i}$$ (10-11)

For the iron(II/III) couple, we have

$$E = +0.770 - 0.0592 \log \frac{a_{Fe^{2+}}}{a_{Fe^{3+}}}$$ (10-12)

and $\Delta G^\circ = -nFE^\circ = -(1 \times 96,485 \times (+0.770)) = -74.29 \text{ kJ} \cdot \text{mol}^{-1}$.

In geochemistry, the symbol E_H has been commonly employed instead of E to designate the hydrogen scale (referenced to the SHE) as the basis for electrode potentials. There is no difference in the value between the E_H of geochemistry and the E of electrochemistry. Another variable that is frequently used for the redox state of an aqueous solution instead of E is $p\varepsilon$. This parameter expresses redox as the activity of electrons in solution—i.e.,

$$p\varepsilon = -\log a_{e^-}$$

Returning to the example of the iron(II/III) couple and using equation (10-8) instead of (10-9) for the equilibrium, we have

$$K = \frac{a_{Fe^{2+}}}{a_{Fe^{3+}} a_{e^-}}$$

and

$$\log K = \log \frac{a_{Fe^{2+}}}{a_{Fe^{3+}}} + p\varepsilon$$

or

$$p\varepsilon = \log K - \log \frac{a_{Fe^{2+}}}{a_{Fe^{3+}}}$$

When the activities are in their standard states, then we have

$$p\varepsilon^\circ = \log K$$

or, in general

$$p\varepsilon^\circ = \frac{1}{n} \log K$$ (10-13)

so that

$$p\varepsilon = p\varepsilon^\circ - \log \frac{a_{Fe^{2+}}}{a_{Fe^{3+}}}$$

Comparing equation (10-13) with (10-12), we see that

$$p\varepsilon = \frac{FE}{2.303RT} \quad \text{and} \quad p\varepsilon^\circ = \frac{FE^\circ}{2.303RT} \tag{10-14}$$

The main advantage of $p\varepsilon$ is that it is a direct function of the logarithm of the activities of the redox species in the reaction (Truesdell, 1968), so that the characteristic slopes are proportional to the ratio of the stoichiometric coefficients of electrons and hydrogen ions, which makes $p\varepsilon$–pH diagrams exactly analogous to other activity diagrams. For example, consider the reduction of lead ions,

$$Pb^{2+}_{(aq)} + 2\,e^-_{(aq)} = Pb_{(s)}$$

for which

$$p\varepsilon = p\varepsilon^\circ + \tfrac{1}{2}\log a_{Pb^{2+}_{(aq)}}$$

or the reduction of chromium(VI) to chromium(III),

$$CrO^{2-}_{4(aq)} + 3\,e^-_{(aq)} + 8\,H^+_{(aq)} = Cr^{3+}_{(aq)} + 4\,H_2O_{(l)}$$

for which

$$p\varepsilon = p\varepsilon^\circ - \frac{1}{3}\log \frac{a_{Cr^{3+}}}{a_{CrO^{2-}_4}} - \frac{8}{3}pH$$

This analogy of $p\varepsilon$ ($-\log a_{e^-}$) with pH ($-\log a_{H^+}$) is a mathematical convenience that has been recommended as a better representation of the redox state of an aqueous solution than is E (see, e.g., Stumm and Morgan, 1981). However, the analogy is a mathematical one only and should not imply any chemical analogy. Thorstenson (1984) has amply pointed out that hydrated electrons differ greatly from hydrated protons and, under natural conditions, aqueous electrons do not exist at meaningfully significant concentrations whereas hydrated protons do. Furthermore, emphasis on $p\varepsilon$ (or E_H) as a representation of the redox state of a natural water tends to support the belief that there is a unique redox potential for a natural water, which is generally not true. We shall return to the practical aspects of redox after the development of a typical $p\varepsilon$–pH (or E_H–pH) diagram.

10-3 $p\varepsilon$–pH DIAGRAMS

The $p\varepsilon$–pH diagrams are simply another type of chemical potential diagram such as those derived in the previous chapter with one important difference: they display the redox potential, or pε, as one of the variables. They can show the relative stabilities of several mineral phases on a single two-dimensional plot as a function of pε and pH. The most thorough presentation of these diagrams is given by Garrels and Christ (1965).

The pε–pH diagram is limited to the stability limits of water, and thus the first consideration must be the oxidation and reduction of H_2O. We shall assume standard-state conditions of 298.15 K and 1 bar. For the reduction of oxygen to water, we have

$$\tfrac{1}{2}O_{2(g)} + 2\,e^- + 2\,H_{(aq)}^+ = H_2O_{(l)}$$

$$K = \frac{a_{H_2O}}{P_{O_2}^{1/2}\, a_{e^-}^2\, a_{H^+}^2}$$

$$\Delta G_r^\circ = \Delta G_{f,H_2O_{(l)}}^\circ - 2\,\Delta G_{f,H_{(aq)}^+}^\circ - 2\,\Delta G_{f,e^-}^\circ - \tfrac{1}{2}\Delta G_{f,O_{2(g)}}^\circ$$

$$= \Delta G_{f,H_2O_{(l)}}^\circ = -237.141 \text{ kJ}\cdot\text{mol}^{-1}$$

$$\log K = \frac{-\Delta G_r^\circ}{2.303RT} = -\left(\frac{-237{,}141}{2.303 \times 8.3141 \times 298.15}\right) = 41.56$$

$$\log K = \log a_{H_2O} - \tfrac{1}{2}\log P_{O_2} + 2\,p\varepsilon + 2\,pH = 41.56$$

Because $P_{O_2} = 1$ bar and $a_{H_2O} = 1$ at standard-state conditions,

$$p\varepsilon = 20.78 - pH \tag{10-15}$$

Equation (10-15) plots as a straight line with a slope of -1 on the pε–pH diagram (Figure 10-3). Lower partial pressures of oxygen will move the equilibria line toward more reducing conditions.

The lower stability limit of water is the reduction of protons to hydrogen gas:

$$H_{(aq)}^+ + e^- \rightleftarrows \tfrac{1}{2}H_{2(g)}$$

$$\log K = \tfrac{1}{2}\log P_{H_2} + pH + p\varepsilon$$

Because $\Delta G_r^\circ = 0$ and $\log K = 0$, we have

$$p\varepsilon = -pH \tag{10-16}$$

because $P_{H_{2(g)}} = 1$ bar. Equation (10-16) is shown in Figure 10-3 along with lines showing the affect of lower partial pressures of hydrogen. Note that lowering the partial pressure of either O_2 or H_2 decreases the stability field of liquid water. One of the most important examples of pε–pH relationships is the iron system. The first step is to identify the most important aqueous species of iron: Fe^{2+}, $FeOH^+$, Fe^{3+}, $FeOH^{2+}$, $Fe(OH)_2^+$, $Fe(OH)_3^\circ$, and $Fe(OH)_4^-$. We shall begin with the oxidized species and derive all the appropriate boundaries that separate regions in which the various hydrolysis products predominate. The first

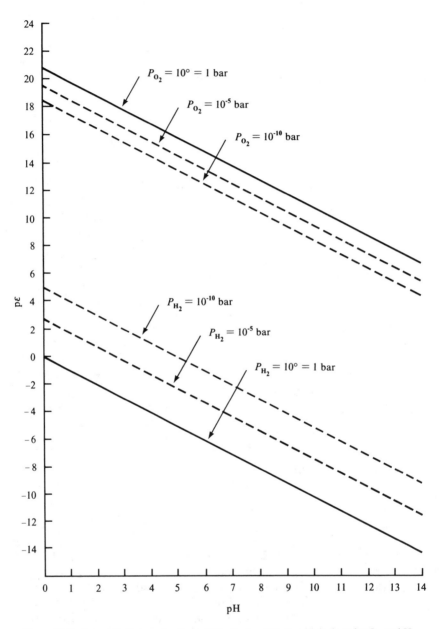

Figure 10-3. A pε–pH diagram for the stability limits of H_2O, with isobars for O_2 and H_2.

hydrolysis reaction is the conversion of Fe^{3+} to $FeOH^{2+}$:

$$Fe^{3+} + H_2O = FeOH^{2+} + H^+ \qquad K = 10^{-2.2}$$

for which

$$\log K_1 = \log a_{FeOH^{2+}} - pH - \log a_{Fe^{3+}} = -2.2 \qquad (10\text{-}17)$$

When the activities of Fe^{3+} and $FeOH^{2+}$ are equal, then

$$pH = 2.2 = pK_1$$

which is plotted as a vertical line at a pH of 2.2 on the pε–pH diagram (Figure 10-4). Likewise, for every hydrolysis product of Fe^{3+}, there will be a vertical line whose position is determined by the condition

$$pH = pK_n$$

This relationship results in

$$pK_1 = 2.2 = pH \text{ for } Fe^{3+}\text{–}FeOH^{2+}$$

$$pK_2 = 3.5 = pH \text{ for } FeOH^{2+}\text{–}Fe(OH)_2^+$$

$$pK_3 = 7.3 = pH \text{ for } Fe(OH)_2^+\text{–}Fe(OH)_3^\circ$$

$$pK_4 = 8.6 = pH \text{ for } Fe(OH)_3^\circ\text{–}Fe(OH)_4^-$$

where the hydrolysis constants are from Baes and Mesmer (1976). Next, we must consider the reduction of $Fe^{3+}_{(aq)}$ to $Fe^{2+}_{(aq)}$ and the hydrolysis of $Fe^{2+}_{(aq)}$. For the iron(II/III) redox couple, we have

$$p\varepsilon^\circ = \frac{FE^\circ}{2.303RT} = \frac{96,485 \times 0.770}{2.303 \times 8.3141 \times 298.15} = 13.0$$

Substituting this value into equation (10-13), we obtain

$$p\varepsilon = 13.0 - \log \frac{a_{Fe^{2+}}}{a_{Fe^{3+}}} \qquad (10\text{-}18)$$

and $p\varepsilon = 13.0$ when $a_{Fe^{2+}} = a_{Fe^{3+}}$. Equation (10-18) plots as a straight horizontal line that intersects the $Fe^{3+}\text{–}FeOH^{2+}$ line at an invariant point at $pH = 2.2$. Then for the equilibrium between $FeOH^{2+}$ and Fe^{2+}, we have

$$FeOH^{2+}_{(aq)} + e^- + H^+_{(aq)} \rightleftharpoons Fe^{2+}_{(aq)} + H_2O_{(l)}$$

which is derived from the algebraic sum of the two reactions

$$Fe^{3+}_{(aq)} + e^- = Fe^{2+}_{(aq)} \qquad\qquad p\varepsilon = 13.0$$

$$-[Fe^{3+}_{(aq)} + H_2O_{(l)} = FeOH^{2+}_{(aq)} + H^+_{(aq)}] \quad -[\log K = -2.2 + pH]$$

$$\overline{FeOH^{2+}_{(aq)} + H^+_{(aq)} + e^- = Fe^{2+}_{(aq)} + H_2O \qquad\qquad p\varepsilon = 15.2 - pH}$$

In similar fashion, we may derive the slope conditions and intercepts for the

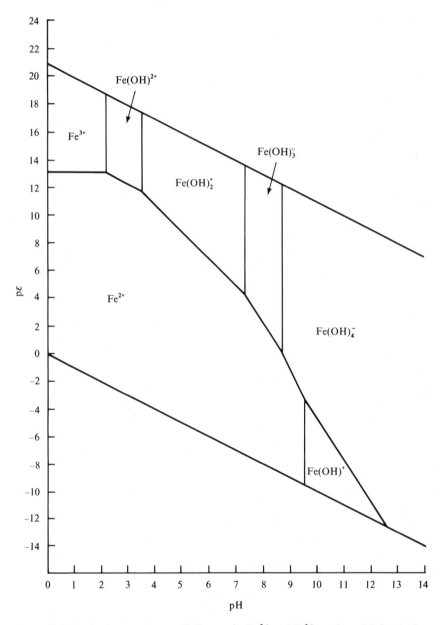

Figure 10-4. Predominance-area $p\varepsilon$–pH diagram for $Fe^{2+}_{(aq)}$ and $Fe^{3+}_{(aq)}$ species at $25°C$ and 1 bar.

remaining equiactivity boundaries. For the $Fe^{2+}-Fe(OH)_2^+$ equilibrium,

$$Fe^{3+}_{(aq)} \quad + e^- \quad = Fe^{2+}_{(aq)} \qquad\qquad p\varepsilon = 13.0$$

$$-[Fe^{3+}_{(aq)} \quad + 2\,H_2O_{(l)} \quad = Fe(OH)^+_{2(aq)} + 2\,H^+_{(aq)}] \quad -[\log K = -5.7 + 2\,pH]$$

$$\overline{Fe(OH)^+_{2(aq)} + 2\,H^+_{(aq)} + e^- = Fe^{2+}_{(aq)} + 2\,H_2O_{(l)}} \qquad p\varepsilon = 18.7 - 2\,pH$$

For the $Fe^{2+}-Fe(OH)_3^\circ$ equilibrium,

$$Fe^{3+}_{(aq)} \quad + e^- \quad = Fe^{2+}_{(aq)} \qquad\qquad p\varepsilon = 13.0$$

$$-[Fe^{3+}_{(aq)} \quad + 3\,H_2O_{(l)} \quad = Fe(OH)^\circ_{3(aq)} + 3\,H^+_{(aq)}] \quad -[\log K = -13.0 + 3\,pH]$$

$$\overline{Fe(OH)^\circ_{3(aq)} + 3\,H^+ + e^- = Fe^{2+}_{(aq)} + 3\,H_2O_{(l)}} \qquad p\varepsilon = 26 - 3\,pH$$

For the $Fe^{2+}-Fe(OH)_4^-$ equilibrium,

$$Fe^{3+}_{(aq)} \quad + e^- \quad = Fe^{2+}_{(aq)} \qquad\qquad p\varepsilon = 13.0$$

$$-[Fe^{3+}_{(aq)} \quad + 4\,H_2O_{(l)} \quad = Fe(OH)^-_{4(aq)} + 4\,H^+_{(aq)}] \quad -[\log K = -21.6 + 4\,pH]$$

$$\overline{Fe(OH)^-_{4(aq)} + 4\,H^+_{(aq)} + e^- = Fe^{2+}_{(aq)} + 4\,H_2O_{(l)}} \qquad p\varepsilon = 34.6 - 4\,pH$$

For the $FeOH^+-Fe(OH)_4^-$ equilibrium,

$$Fe^{3+}_{(aq)} + e^- \quad = Fe^{2+}_{(aq)} \qquad\qquad p\varepsilon = 13.0$$

$$-[Fe^{3+}_{(aq)} + 4\,H_2O_{(l)} \quad = Fe(OH)^-_{4(aq)} + 4\,H^+_{(aq)}] \quad -[\log K = -21.6 + 4\,pH]$$

$$Fe^{2+}_{(aq)} + H_2O_{(l)} \quad = FeOH^+_{(aq)} + H^+_{(aq)} \qquad \log K = -9.5 + pH$$

$$\overline{Fe(OH)^-_{4(aq)} + 3\,H^+_{(aq)} + e^- = FeOH^+_{(aq)} + 3\,H_2O_{(l)}} \qquad p\varepsilon = 25.1 - 3\,pH$$

Every invariant point where three univariant lines meet provides a convenient check for arithmetic errors. If the third line does not meet at the intersection of the other two, then the calculations are in error, and they should be rechecked.

It is important to remember that this pε–pH diagram (Figure 10-4) is really a "predominance-area" diagram because the outlined divariant fields are those areas containing where the designated species make up more than about 50% of the total concentration. Each line represents the pε–pH conditions where the two species are present in equal proportions, and every intersection of three lines represents the pε–pH point where the three species are present in equal proportions. Every specie is, in fact, present at some activity value everywhere on the diagram, but only inside its designated field is it the predominant specie. Using the same equations just derived, you can also construct contour lines that have the same slopes but depict different activity levels for a given specie.

Figure 10-4 is useful precisely because at a quick glance you can see which type of species will be predominant over a given range of pε and pH. For example, reduced groundwaters and anaerobic-lake-sediment and marine-sediment pore waters are dominated by $Fe^{2+}_{(aq)}$, whereas acid mine waters are

TABLE 10-1

SOME COMMOM IRON MINERALS

Mineral name	Formula	$\log K_{sp}$ at 298 K	Reference
Jarosite	$KFe_3(SO_4)_2(OH)_6$	-93	Chapman et al. (1983)
Ferrihydrite	$Fe(OH)_3$	-39	Norvell and Lindsay (1982)
Siderite	$FeCO_3$	-10.5	Langmuir (1969)

dominated by Fe^{3+} and $FeOH^{2+}$. Alkaline lake waters are dominated by the $Fe(OH)_4^-$ ion.

Equilibria involving solid phases also can be shown on a $p\varepsilon$–pH diagram by very similar derivations. We shall consider the iron system further by introducing the minerals jarosite, ferrihydrite, siderite, and pyrite. These minerals represent a wide range of compositions and origins, and they are among the most common iron minerals that form under low-temperature conditions. Because jarosite, siderite, and pyrite contain the additional compositional variables of potassium, sulfate, carbonate, and sulfide, these additional variables must be arbitrarily fixed at some value so that their aqueous equilibria can be represented on a $p\varepsilon$–pH diagram. Typical concentrations of the aqueous ions in natural waters where these minerals seem to be stable can be used as guidelines to fix the appropriate ionic activities. Taking advantage of these guidelines, we shall fix the ion activities at the following values: $a_{K^+} = 10^{-4}$, $a_{SO_4^{2-}} = 10^{-2}$, $P_{CO_2} = 10^{-2}$, and $a_{S^{2-}} = 10^{-2}$.

The two most-oxidized minerals are jarosite and ferrihydrite, which is a poorly crystalline form of ferric hydroxide. These two minerals take part in the following dissolution equilibria:

$$KFe_3(SO_4)_2(OH)_{6(s)} \rightleftarrows K^+_{(aq)} + 3\,Fe^{3+}_{(aq)} + 2\,SO^{2-}_{4(aq)} + 6\,OH^-_{(aq)}$$

$$\log K = -93$$

$$Fe(OH)_{3(s)} \rightleftarrows Fe^{3+}_{(aq)} + 3\,OH^-_{(aq)}$$

$$\log K = -39$$

When the two phases are in equilibrium, we have

$$KFe_3(SO_4)_2(OH)_{6(s)} + 3\,OH^-_{(aq)} \rightleftarrows 3\,Fe(OH)_{3(s)} + K^+_{(aq)} + 2\,SO^{2-}_{4(aq)}$$

We can convert this expression to one that uses hydrogen ions by adding the dissociation constant expression for water:

$$H_2O_{(l)} \rightleftarrows H^+_{(aq)} + OH^-_{(aq)}$$

$$\log K = -14$$

The reaction expression now becomes

$$KFe_3(SO_4)_2(OH)_{6(s)} + 3H_2O_{(l)} \rightleftarrows 3Fe(OH)_{3(s)} + K^+_{(aq)} + 2SO^{2-}_{4(aq)} + 3H^+_{(aq)}$$

$$\log K = -18$$

The equilibrium-constant expression is

$$\log a_{K^+_{(aq)}} + 2\log a_{SO^{2-}_{4(aq)}} - 3\,pH$$

$$= \log K_{sp(jarosite)} - 3\log K_{sp(ferrihydrite)} + 3\log K_w = -18$$

or

$$pH = 6 + \tfrac{1}{3}\log a_{K^+} + \tfrac{2}{3}\log a_{SO_4^{2-}}$$

Substituting the activity values previously chosen for the ionic species, we have

$$pH = 6 - \tfrac{4}{3} - \tfrac{4}{3} = 3.33$$

Note that this equilibrium is not a redox reaction, so that it must be represented by a vertical line at a pH of 3.33 on a pε–pH diagram (see Figure 10-5). However, it is important to emphasize that the exact position of this equilibrium boundary is not well known for several reasons.

1. The values of a_{K^+} and $a_{SO_4^{2-}}$ must be held constant to draw the diagram, but in real environments this may not happen.
2. The K_{sp} of the ferric hydroxide may vary considerably with grain size. Langmuir and Whittemore (1971) found values of $\log(a_{Fe^{3+}} a_{OH^-}^3)$ to vary from -37 to -43.5 for laboratory solutions and natural groundwaters. Substituting these values into the ferrihydrite–jarosite equilibrium results in an equilibrium pH ranging from -1.16 to 5.33. Well-aged crystalline ferric hydroxide (goethite) removes jarosite from the stability field for natural systems at 25°C.
3. The $\log K_{sp}$ for jarosite also is not well established. Values range from -93.21 to -98.56, which changes the equilibrium pH from 3.33 to 5.19.

From Figure 10-4 we see that, under acidic conditions, ferric iron is reduced at pε values below 13. Therefore, jarosite and ferrihydrite would also be reduced and solubilized. The reduction of jarosite may be written

$$KFe_3(SO_4)_2(OH)_{6(s)} + 3e^- + 6H^+_{(aq)} \rightleftarrows K^+_{(aq)} + 3Fe^{2+}_{(aq)} + 2SO^{2-}_{4(aq)} + 6H_2O_{(l)}$$

The log of our equilibrium-constant expression for unit activities of liquid water and jarosite is

$$\log K = 3\,pε + 6\,pH + \log a_{K^+} + 3\log a_{Fe^{2+}} + 2\log a_{SO_4^{2-}}$$

By combining this expression with the solubility-product constant for jarosite, we obtain

$$\log K = \log K_{sp(jarosite)} + 3\,pε^\circ_{(Fe^{2+}/Fe^{3+})} - 6\log K_w$$

which is equivalent to

$$\log K = -93 + 39 + 84 = 30$$

and

$$p\varepsilon = 10 - \tfrac{1}{3}\log a_{K^+} - \log a_{Fe^{2+}} - \tfrac{2}{3}\log a_{SO_4^{2-}} - 2\,pH$$

Substituting the activity values, we obtain

$$p\varepsilon = 12.66 - 2\,pH - \log a_{Fe^{2+}}$$

Now we must choose an activity value for $Fe^{2+}_{(aq)}$; this is a difficult task because the activity varies considerably in natural waters as a function of pH. However, it is possible to show how the equilibrium boundary shifts with different activities of $Fe^{2+}_{(aq)}$. For this example, we shall use $a_{Fe^{2+}} = 10^{-4}$, so that

$$p\varepsilon = 16.66 - 2\,pH$$

This equilibrium line goes above $p\varepsilon = 13$ for pH values below 1.83 and, because dissolved iron is predominantly in the trivalent state under those conditions, the jarosite–$Fe^{2+}_{(aq)}$ equilibrium must terminate at that point. What happens to the jarosite stability field? It also terminates because the value of $a_{Fe^{2+}}$ that we fixed must also be equal to $a_{Fe^{3+}}$ when iron is in that form. Fixing $a_{Fe^{3+}}$ defines the solubility limits for jarosite under extremely acid conditions:

$$KFe_3(SO_4)_2(OH)_{6(s)} + 6\,H^+_{(aq)} \rightleftarrows K^+_{(aq)} + 3\,Fe^{3+}_{(aq)} + 2\,SO_4^{2-}_{(aq)} + 6\,H_2O$$

$$\log a_{K^+} + 3\log a_{Fe^{3+}} + 2\log a_{SO_4^{2-}} + 6\,pH = \log K_{sp(jarosite)} - 6\log K_w$$

$$= -93 + 84 = -9$$

Because

$$6\,pH = -9 - \log a_{K^+} - 3\log a_{Fe^{3+}} - 2\log a_{SO_4^{2-}} = 11$$

we have

$$pH = 1.83$$

which is the minimum pH for jarosite stability, given the assumed levels of ion activity.

The equilibrium line for the reduction of ferrihydrite is easier to derive than that for jarosite. The reaction is

$$Fe(OH)_{3(s)} + e^- + 3\,H^+_{(aq)} \rightleftarrows Fe^{2+}_{(aq)} + 3\,H_2O_{(l)}$$

and

$$\log a_{Fe^{2+}_{(aq)}} + 3\,pH + p\varepsilon = \log K_{sp(ferrihydrite)} - 3\log K_w + p\varepsilon^\circ_{(Fe^{2+}/Fe^{3+})} = 16$$

$$p\varepsilon = 16 - \log a_{Fe^{2+}} - 3\,pH$$

$$= 20 - 3\,pH$$

Next, we shall look at the stability field for siderite:

$$FeCO_{3(s)} \rightleftarrows Fe^{2+}_{(aq)} + CO^{2-}_{3(aq)}$$

$$\log a_{Fe^{2+}} + \log a_{CO^{2-}_3} = -10.5$$

Because $a_{Fe^{2+}} = 10^{-4}$, we have

$$\log a_{CO^{2-}_3} = -6.5$$

If P_{CO_2} is fixed and $a_{CO^{2-}_3}$ is fixed, then the pH is fixed through the carbonate equilibria:

$$CO_{2(g)} + H_2O_{(l)} \rightleftarrows H_2CO_{3(aq)} \qquad K_H = 10^{-1.47}$$

$$H_2CO_{3(aq)} \rightleftarrows H^+_{(aq)} + HCO^-_{3(aq)} \qquad K_1 = 10^{-6.35}$$

$$HCO_{3(aq)} \rightleftarrows H^+_{(aq)} + CO^{2-}_{3(aq)} \qquad K_2 = 10^{-10.33}$$

and we have

$$\log a_{CO^{2-}_3} = \log K_H + \log K_1 + \log K_2 + \log P_{CO_2} + 2\,pH = -6.5$$

$$2\,pH = -6.5 + 2 + 10.33 + 6.35 + 1.47 = 13.65$$

$$pH = 6.83$$

Thus, at a pH of 6.83 and above, siderite becomes a stable phase. It dissolves at lower pH values because of the decreasing activity of $CO^{2-}_{3(aq)}$.

The ferrihydrite–siderite equilibrium,

$$Fe(OH)_{3(s)} + e^- + CO^{2-}_{3(aq)} \rightleftarrows FeCO_{3(s)} + 3\,H_2O_{(l)}$$

leads to the expression

$$p\varepsilon + 3\,pH - \log a_{CO^{2-}_3} = \log K_{sp(ferrihydrite)} - \log K_{sp(siderite)}$$

$$- 3\log K_w + p\varepsilon^\circ_{(Fe^{2+}/Fe^{3+})}$$

$$= -39 + 10.5 + 42 + 13 = 26.5$$

or

$$p\varepsilon = 20 - 3\,pH$$

which means that this boundary will be congruent with the $Fe(OH)_{3(s)}$–$Fe^{2+}_{(aq)}$ boundary.

Finally, the stability field for pyrite will be delineated using an alternative approach with free-energy data. Pyrite stability is more complex because it can form in predominance areas of $SO^{2-}_{4(aq)}$, $H_2S_{(aq)}$, and $HS^-_{(aq)}$. The sulfate–pyrite equilibrium is given by the reaction

$$8\,H_2O_{(l)} + FeS_{2(s)} \rightleftarrows Fe^{2+}_{(aq)} + 2\,SO^{2-}_{4(aq)} + 14\,e^- + 16\,H^+_{(aq)}$$

The free energy of this reaction can be calculated from data in Robie et al. (1978):

$$\Delta G^\circ_r = 489.227 \ kJ \cdot mol^{-1}$$

and

$$\log K = -\frac{489{,}227}{(2.303)(8.3144)(298.15)} = -85.7$$

so that

$$\log a_{Fe^{2+}} + 2 \log a_{SO_4^{2-}} - 14 \, p\varepsilon - 16 \, pH = -85.7$$

Substituting activities and rearranging, we obtain

$$p\varepsilon = 5.55 - 1.14 \, pH$$

For the reaction

$$FeS_{2(s)} + 4 H^+_{(aq)} + 2 e^- \rightleftarrows Fe^{2+}_{(aq)} + 2 H_2S_{(aq)}$$

we have

$$\Delta G_r^\circ = 25.7 \, kJ \cdot mol^{-1}$$

$$\log K = -4.5 = \log a_{Fe^{2+}} + 2 \log a_{H_2S} + 2 \, p\varepsilon + 4 \, pH$$

$$p\varepsilon = 1.75 - 2 \, pH$$

For the reaction

$$FeS_{2(s)} + 2 H^+_{(aq)} + 2 e^- \rightleftarrows Fe^{2+}_{(aq)} + 2 HS^-_{(aq)}$$

we have

$$\Delta G_r^\circ = 105.5$$

$$\log K = -18.5 = \log a_{Fe^{2+}} + 2 \log a_{HS^-} + 2 \, p\varepsilon + 2 \, pH$$

$$p\varepsilon = -5.25 - pH$$

which is below the H_2O–H_2 stability limits for water and therefore is not plotted. Finally, for the pyrite–ferrihydrite equilibrium,

$$FeS_{2(s)} + 11 H_2O_{(l)} \rightleftarrows Fe(OH)_{3(s)} + 2 SO_{4(aq)}^{2-} + 15 e^- + 19 H^+_{(aq)}$$

we have

$$\Delta G_r^\circ = 583$$

$$\log K = -102 = 2 \log a_{SO_4^{2-}} - 15 \, p\varepsilon - 19 \, pH$$

$$p\varepsilon = 6.53 - 1.27 \, pH$$

This completes the $p\varepsilon$–pH diagram for a few common minerals in the Fe–C–S–H_2O system. Figure 10-5 graphically displays the relative-stability fields and equilibrium solubilities of jarosite, ferrihydrite, siderite, and pyrite as a function of redox potential and pH. A dashed line indicating the boundary conditions for $a_{Fe^{2+}_{(aq)}} = 10^{-2}$ is also shown. As the iron activities increase, the

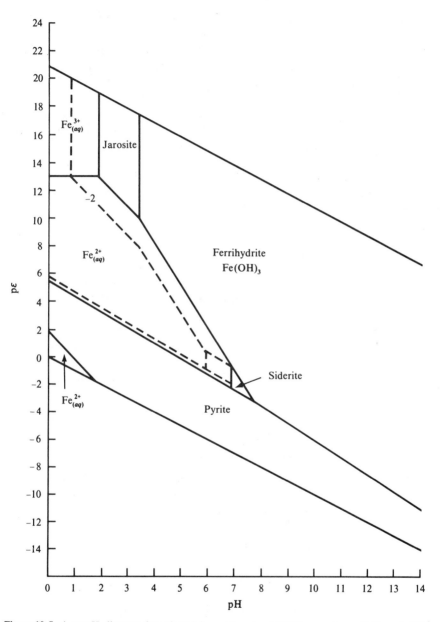

Figure 10-5. A pε-pH diagram for selected iron minerals at 25°C, assuming activities of 10^{-4} for dissolved iron, 10^{-4} for dissolved potassium, and 10^{-2} for dissolved sulfur, and assuming $P_{CO_2} = 10^{-2}$.

stability area of each solid phase increases. Several qualitative features are readily apparent. Most oxidizing groundwaters should be very low in dissolved iron, being limited by the solubilities of ferrihydrite and jarosite except under strongly acidic conditions. Reduced groundwaters should have increased concentrations of iron until sulfate reduction occurs, and then the concentrations should decrease as pyrite is formed. Siderite has a very limited stability field and should occur only in environments containing high P_{CO_2}, low dissolved sulfur, and moderate reducing conditions containing abundant organic matter. Although there are several complications in the application of such a diagram to natural waters, iron does behave very much in accordance with these predictions. It is important to remember that $p\varepsilon$–pH diagrams can give an excellent perspective on the relative stabilities of several minerals in a single diagram, but they cannot be used for quantitative chemical modeling because

1. they are constructed under conditions of constant activities of ions whereas, over a reasonable range of $p\varepsilon$ and pH, ionic composition can vary greatly;
2. direct comparison of solubilities with natural water compositions cannot be made because nonideality corrections have not been made;
3. it may often be difficult to decide which phases should be included in the diagram (e.g., should you choose ferrihydrite, goethite, or hematite for the dominant form of iron oxide?);
4. the redox potential of a natural water is not a clearly defined quantity like pH and, in fact, a range of redox values for different couples (e.g., Fe(II/III), As(III/V), O_2/H_2O, NH_4/NO_3, and SO_4^{2-}/S^{2-}) may coexist in any given water sample.

10-4 SOME PRECAUTIONARY NOTES ON REDOX EQUILIBRIA IN NATURAL WATERS

Redox conditions of natural waters are essential to our understanding of geochemical processes, but the theory and practice of redox is fraught with difficulties. Thorstenson (1984) has made a thorough review of the concept of $p\varepsilon$ and its application, in which he makes the following important points:

1. the analogy between $p\varepsilon$ and pH as master variables is generally carried much further than is justified;
2. a single, thermodynamically meaningful redox potential cannot be assigned to a natural water;
3. the most useful approach is the study of multiple redox couples from individual determinations of redox species.

Traditionally, platinum electrodes have been used to measure $p\varepsilon$ (or E_H) in natural systems and they have, in a few instances, been related to a specific redox couple such as the $S_{(aq)}^{2-}/S_{(s)}$ couple in reduced marine sediments (Berner, 1963) and the $Fe_{(aq)}^{2+}/Fe_{(aq)}^{3+}$ couple in acid mine waters (Nordstrom, Jenne, and Ball, 1979). More commonly, however, redox species are present in concentrations too low to give an adequate electrode response. Furthermore, there are strong kinetic barriers to the establishment of equilibrium between different redox couples. In a survey of over 700 groundwaters having E_H and pH measurements, many with one or more redox species analyzed, Lindberg (1983) found that, where different redox couples could be directly compared through the Nernst equation, they did not agree. Consequently, it is clear that redox reactions in natural waters are dominated by disequilibrium.

10-5 REDOX EQUILIBRIA AT HIGH TEMPERATURES

Equilibria in the system Fe−Si−O−H

We now turn our attention away from the redox potentials of aqueous solutions and consider how oxidation−reduction reactions in the Fe−Si−O−H system can be studied using oxygen fugacity (f_{O_2}) as a master variable. As we shall soon see, the content of free molecular oxygen in equilibrium with either supercritical aqueous fluids or all but the most-oxidizing magmas is negligible, which results in very low values of f_{O_2}. Thus, species other than molecular oxygen must be called upon to transfer the O^{2-} between the oxidized and reduced assemblages. For instance, the oxidation of magnetite to hematite mentioned in the opening of this chapter could be written in several alternative ways, all involving gaseous species:

$$4\,Fe_3O_4 + O_{2(g)} \rightleftharpoons 6\,Fe_2O_3 \qquad (10\text{-}19)$$

or

$$2\,Fe_3O_4 + H_2O_{(g)} \rightleftharpoons 3\,Fe_2O_3 + H_{2(g)} \qquad (10\text{-}20)$$

or

$$2\,Fe_3O_4 + CO_{2(g)} \rightleftharpoons 3\,Fe_2O_3 + CO_{(g)} \qquad (10\text{-}21)$$

or

$$8\,Fe_3O_4 + CO_{2(g)} + 2\,H_2O_{(g)} \rightleftharpoons 12\,Fe_2O_3 + CH_{4(g)} \qquad (10\text{-}22)$$

Reactions (10-19) through (10-22) are thermodynamically equivalent in terms of magnetite oxidation, but they differ in the gas species that oxidizes magnetite. Reaction (10-19), written only in terms of O_2, is the simplest but, if it is important to consider the proportions of gas species actually present in a system (i.e., X_{H_2} or X_{CO_2}), then it may prove more convenient to choose one of the other three representations.

The equilibrium constant for reaction (10-19) is exactly equal to the inverse of the oxygen fugacity of an equilibrium assemblage of magnetite and hematite

(MH):

$$K_{MH} = f_{O_2}^{-1} \tag{10-23}$$

assuming, of course, that both iron oxides are stoichiometric and have unit activity. At 1000 K and 1 bar,

$$-\log K = \log f_{O_2} = \left(\frac{6 \, \Delta G^{\circ}_{f(Fe_2O_3,1000)} - 4 \, \Delta G^{\circ}_{f(Fe_3O_4,1000)}}{2.303RT} \right)$$

or

$$\log f_{O_2} = \frac{-207.98 \text{ kJ}}{2.303 \times 0.008314 \text{ kJ} \cdot \text{K}^{-1} \times 1000 \text{ K}} = -10.86$$

When the total pressure on the system is greater than 1 bar, it is easy to correct the oxygen fugacity by choosing standard states of P and T for the solids and 1 bar and T for O_2 gas. The pressure correction to the equilibrium constant is then proportional to ΔV_s for the reaction—i.e., $6\bar{V}_{Fe_2O_3} - 4\bar{V}_{Fe_3O_4} = +0.3548 \text{ J} \cdot \text{bar}^{-1}$ in this case. The complete pressure correction term for MH equilibrium is $\Delta V_s(P - 1)/2.303RT$, or $0.0185(P - 1)/T$ (cf equation (8-44)); thus, at 5 kbar and 1000 K, we have $\log f_{O_2} = -10.77$. For most oxidation–reduction reactions involving oxide or silicate phases, the effect of total pressure on f_{O_2} is small, but it should never be neglected.

The MH equilibrium is only one reaction in the geologically important system Fe–Si–O. The other significant reactions, arranged in order of decreasing oxygen fugacity, are

$$3\,Fe_2SiO_4 + O_2 \rightleftharpoons 2\,Fe_3O_4 + 3\,SiO_2 \qquad \text{(FMQ)}$$
$$\text{fayalite} \qquad\qquad \text{magnetite} \quad \text{quartz}$$

$$\frac{6}{4x - 3}\,Fe_xO + O_2 \rightleftharpoons \frac{2x}{4x - 3}\,Fe_3O_4 \qquad \text{(WM)}$$
$$\text{wuestite} \qquad\qquad \text{magnetite}$$

$$\tfrac{3}{2}Fe + O_2 \rightleftharpoons \tfrac{1}{2}Fe_3O_4 \qquad \text{(IM)}$$
$$\text{iron} \qquad \text{magnetite}$$

$$2x\,Fe + O_2 \rightleftharpoons 2\,Fe_xO \qquad \text{(IW)}$$
$$\text{iron} \qquad\quad \text{wuestite}$$

$$SiO_2 + 2\,Fe + O_2 \rightleftharpoons Fe_2SiO_4 \qquad \text{(QIF)}$$
$$\text{quartz} \quad \text{iron} \qquad\quad \text{fayalite}$$

Wuestite is a nonstoichiometric phase with an Fe/O ratio that depends on f_{O_2} and T, which accounts for the noninteger coefficients in the WM and IW reactions. The x shown in these reactions is a measure of cation vacancies resulting from the presence of both Fe^{2+} and Fe^{3+} in wuestite and ranges

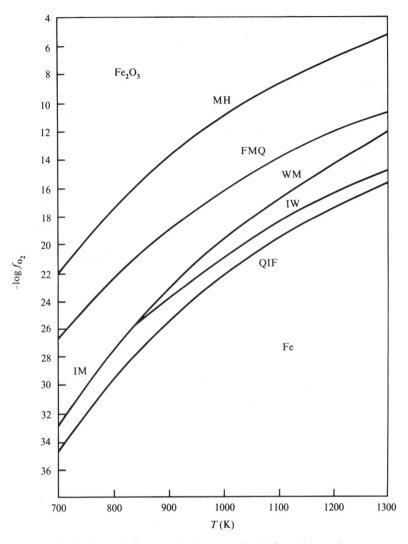

Figure 10-6. Oxygen buffer curves in the system Fe–Si–O at 1 bar total pressure.

between 0.85 and 0.95. Wuestite becomes unstable relative to magnetite and iron below about 835 K.

The temperature dependence of these reactions is shown in Figure 10-6. Note that the stability of magnetite is restricted to the area between the MH and FMQ lines in the presence of quartz but can extend down to the lower oxygen fugacities of the WM line in the absence of quartz. The wuestite field is confined to the narrow wedge between WM and IW. Because both metallic iron and wuestite are extremely rare phases in terrestrial rocks, most oxygen fugacities encountered in igneous or metamorphic environments lie above wuestite stability.

The precise locations of many of the lines shown in Figure 10-6 have been the subject of debate in the literature. Although we have already shown how these lines can be calculated from thermodynamic data, more commonly the f_{O_2} values of these reactions are measured indirectly in the laboratory. One of the more modern techniques uses a high-temperature thermogravimetric balance to detect very tiny weight changes resulting from gain or loss of oxygen in solid phases. A solid electrolyte cell is used to monitor the changes in oxygen fugacity. The electrolyte cell is fashioned from solid zirconia (ZrO_2) with small amounts of CaO or Y_2O_3 present in solid solution; the lower valence cations substituting for Zr^{4+} create anion vacancies in the crystal structure. Under the influence of an electrical field, O^{2-} anions exchange with the vacant anion sites and move towards the more positive potential. Thus, the electrical potential of such a cell is sensitive to the activity of oxygen anions, and it can be calibrated in terms of f_{O_2} (Sato, 1971). The f_{O_2} in the furnace can be varied by controlling the partial pressures of two gas species related by an oxidation–reduction reaction, such as $CO–CO_2$ or $CO_2–H_2$ (Nafziger et al., 1971), and can be monitored by the zirconia electrode. Myers and Eugster (1983) used this method to locate the IW, WM, MH, and QIF curves at temperatures above 1100 K.

The oxidation–reduction reactions shown in Figure 10-6 are commonly referred to as examples of "oxygen buffers." The term "buffer" is appropriate because the oxygen fugacities of these assemblages are single-valued at fixed temperature and total pressure, a direct consequence of the fact that the equilibrium constants for these reactions are written as functions of oxygen fugacity only. Thus, as long as both the reduced and oxidized assemblages are in equilibrium, the oxygen fugacity is rigorously defined. A system of magnetite + hematite + gas can tolerate small amounts of oxidation or reduction (by converting some of the magnetite to hematite, or vice versa) *without changing the oxygen fugacity* of the system. The buffering capacity is achieved by changing the Fe/O ratio of the mineral assemblages because, so long as both magnetite and hematite are present, the oxygen fugacity is fixed by the equilibrium constant (Figure 10-7). This principle was used by H. P. Eugster to control the oxygen fugacity of mineral assemblages in experimental systems, revolutionizing the study of oxidation–reduction mineral reactions at high temperatures and pressures (Eugster, 1957; Eugster and Wones, 1962; Huebner, 1971).

It is important to see how these oxygen-buffer systems are affected by the addition of gaseous H_2O to the mineral assemblage. Clearly, the addition of H_2O cannot change the oxygen fugacities of any of the buffers, because its presence does not alter the thermodynamic properties of any of the solid buffer phases. In fact, the same point can be made for the addition of any number of components to either the mineral assemblage or to the gas phase: so long as the thermodynamic properties of magnetite and hematite are not altered by these additions (e.g., by solid solution), then f_{O_2} of the system must be unchanged. However, at

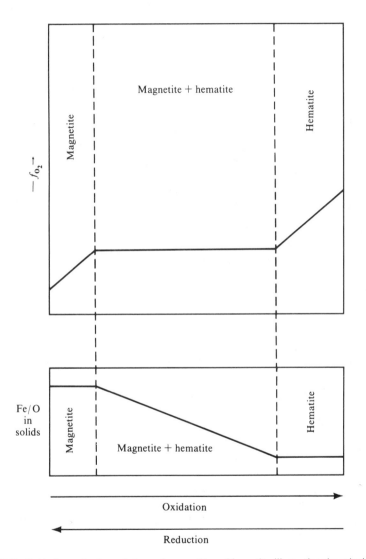

Figure 10-7. Oxidation-reduction relations of magnetite and hematite, illustrating the principles of the oxygen buffer and buffer capacity. With both magnetite and hematite present, a small increase in the amount of oxygen in the system is compensated by conversion of magnetite to hematite (decreasing the Fe/O ratio in the oxide assemblage) while the oxygen fugacity remains fixed. The buffering capacity of the system is exhausted when all the magnetite is gone and the local f_{O_2} is once again free to increase.

high temperature, thermal dissociation of H_2O is very rapid, so we have

$$H_2 + \tfrac{1}{2}O_2 \rightleftharpoons H_2O \tag{10-24}$$

and

$$K_w = \frac{f_{H_2O}}{f_{H_2} f_{O_2}^{1/2}} \tag{10-25}$$

which is the equilibrium constant for the formation of steam from the elements. Thus, at equilibrium, a high-temperature gas in the system O–H will be composed of H_2O, O_2, and H_2 molecules, and the total pressure of the system will be

$$P_{total} = \frac{f_{H_2O}}{\Gamma_{H_2O}} + \frac{f_{H_2}}{\Gamma_{H_2}} + \frac{f_{O_2}}{\Gamma_{O_2}} \tag{10-26}$$

assuming ideal mixing in the gaseous solution (cf equation (9-18)). The last term in (10-26) is the partial pressure of oxygen gas, and it can be conveniently dropped from the summation because it is so small in common oxygen-buffer reactions that it does not measurably add to the total pressure (cf Figure 10-6). Substituting (10-25) into (10-26), we obtain

$$P_{total} = \frac{K_w f_{H_2} f_{O_2}^{1/2}}{\Gamma_{H_2O}} + \frac{f_{H_2}}{\Gamma_{H_2}} \tag{10-27}$$

If an oxygen-buffer assemblage is present in the system, then the oxygen fugacity in (10-27) is fixed by the equilibrium constant of the buffer reaction. The only unknown quantity remaining in (10-27) is the hydrogen fugacity, which can be found by substitution and rearrangement:

$$f_{H_2} = P_{total} \left(\frac{K_w K_{bu}^{1/2}}{\Gamma_{H_2O}} + \frac{1}{\Gamma_{H_2}} \right)^{-1} \tag{10-28}$$

The partial pressures of hydrogen for three of the oxygen buffers in the system Fe–Si–O–H at a total pressure of 1 kbar are shown in Figure 10-8. It is clear from the figure that the fluid phase is a H_2–H_2O mixture, with the proportion of H_2 increasing as f_{O_2} decreases. For all oxygen fugacities above the stability of fayalite, the equilibrium hydrogen partial pressures are small relative to the total pressure. These calculations show that, under such conditions, the assumption of ideal mixing is justifiable because the gas is overwhelmly composed of H_2O molecules.

Redox equilibria in the C–O and C–O–H systems

In addition to H_2O and H_2, carbon-bearing species such as CO_2 and CH_4 are the most geochemically abundant components in a high-temperature gas phase. These species also take part in redox equilibria, and they introduce some significant complications. We look first at the C–O system.

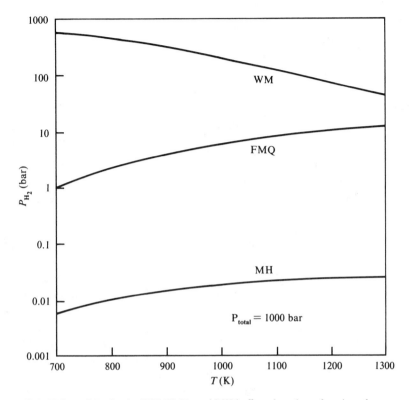

Figure 10-8. Values of P_{H_2} for the WM, FMQ, and MH buffers plotted as a function of temperature for an O–H gas at a total pressure of 1 kbar. The balance of the gas phase is almost entirely composed of H_2O, because P_{O_2} is so small (cf Figure 10-6). Note that H_2 is a significant component ($> 1\%$) only for conditions that are more reducing than the FMQ buffer.

The f_{O_2} of the C–O system is buffered if graphite (gr) is present. The relevant equations are

$$C_{(gr)} + O_2 \rightleftharpoons CO_2$$

$$K_{CO_2} = \frac{f_{CO_2}}{f_{O_2}} \tag{10-29}$$

$$C_{(gr)} + \tfrac{1}{2}O_2 \rightleftharpoons CO$$

$$K_{CO} = \frac{f_{CO}}{f_{O_2}^{1/2}} \tag{10-30}$$

$$P_{total} = \frac{f_{CO_2}}{\Gamma_{CO_2}} + \frac{f_{CO}}{\Gamma_{CO}} \tag{10-31}$$

By substitution of (10-29) and (10-30) into (10-31), we may obtain a quadratic equation in f_{O_2} (see Problem 7):

$$\left(\frac{K_{CO_2}}{\Gamma_{CO_2}}\right)^2 f_{O_2}^2 - \left(2P_T\frac{K_{CO_2}}{\Gamma_{CO_2}} + \left(\frac{K_{CO}}{\Gamma_{CO}}\right)^2\right)f_{O_2} + P_{total}^2 = 0 \qquad (10\text{-}32)$$

The coefficients in (10-32) can be calculated at any total pressure (P_{total}) and temperature by substituting the appropriate equilibrium constants and fugacity coefficients, and f_{O_2} can be calculated. The oxygen fugacity defined by graphite equilibrium in this system is extremely dependent on total pressure. Figure 10-9 shows the differences between the 1-bar and 2-kbar cases, which amounts to more than five orders of magnitude in f_{O_2} at 1200 K. The dashed lines in the figure are the QIF and FMQ buffers, which are shown for comparison. The upper section of Figure 10-9 shows the mole fraction of CO_2 as a function of temperature. Only at temperatures in excess of 1100 K does CO appear in significant concentrations. Below that temperature, the assumption of ideal mixing in the gas phase is justifiable.

The solid lines in Figure 10-9 define the maximum stability of graphite at 1 bar and at 2 kbars. At higher values of f_{O_2} relative to an isotherm, graphite is oxidized. However, at constant temperature, lower f_{O_2} values are possible while still maintaining equilibrium with graphite if H is added as a component to the gas phase. The addition of H has the effect of increasing the thermodynamic variance of the system by one, so f_{O_2} is no longer constrained to the solid curves in Figure 10-9. This is a significant addition, because the increase in variance assures that the oxygen fugacity is no longer buffered. Thus, the presence of graphite in a metamorphic rock does not imply that the oxygen fugacity of the metamorphic fluid was buffered, unless there is some evidence that the fluid phase was restricted to the binary system C–O. It does, however, define a *maximum* f_{O_2}, although that maximum is strongly dependent on the total pressure of the fluid phase.

The system C–O–H contains an overwhelming number of different organic molecular species that are gases at high temperatures and pressures. French and Eugster (1965) showed that the only significant specie that need concern us in addition to CO_2, CO, H_2O, and H_2 is methane (CH_4). The use of the word "significant" has a special meaning relating to the calculation of gas compositions: a significant species must be capable of exerting a partial pressure that is large enough at some geologically realistic combination of f_{O_2} and T materially to affect the total-pressure equation. Thus, for the C–O–H system we have

$$P_{total} = \frac{f_{H_2O}}{\Gamma_{H_2O}} + \frac{f_{CO_2}}{\Gamma_{CO_2}} + \frac{f_{CO}}{\Gamma_{CO}} + \frac{f_{CH_4}}{\Gamma_{CH_4}} + \frac{f_{H_2}}{\Gamma_{H_2}} \qquad (10\text{-}33)$$

The compositions of a C–O–H gas in equilibrium with graphite can be approximately calculated by substituting (10-25), (10-29), (10-30), and the

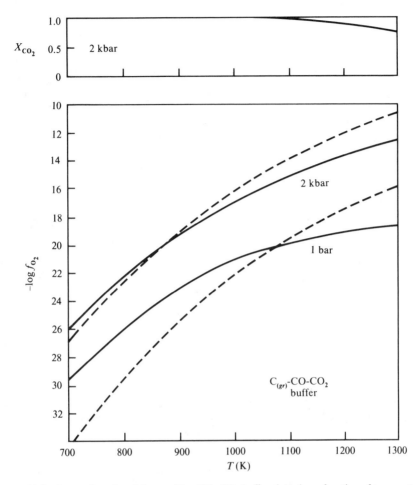

Figure 10-9. Oxygen fugacity of the graphite–CO–CO$_2$ buffer plotted as a function of temperature, showing the extreme dependence on total pressure (solid lines). The dashed lines enclose the stability field of fayalite in the Fe–Si–O system (bottom = QIF; top = FMQ) and are shown for reference. The upper graph shows the mole fraction of CO$_2$ in the gas phase, revealing that the proportion of CO is negligible below 1000 K.

equilibrium constant for methane,

$$C_{(gr)} + 2\,H_{2(g)} \rightleftharpoons CH_{4(g)}$$

$$K_{CH_4} = \frac{f_{CH_4}}{f_{H_2}^2} \tag{10-34}$$

into (10-33), expressing all fugacities in terms of either f_{O_2} or f_{H_2}. In order to solve the resulting equation, either f_{H_2} or f_{O_2} must be specified. For example, Figure 10-10 shows the gas composition as a function of temperature for oxygen

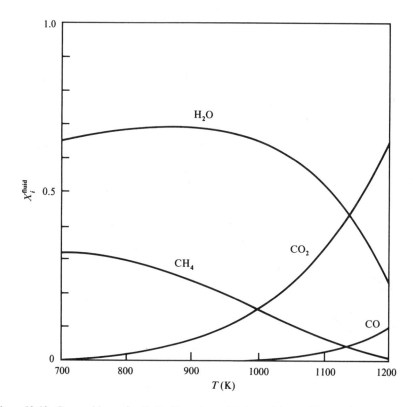

Figure 10-10. Compositions of a C–O–H gas in equilibrium with graphite, calculated for a total pressure of 2 kbar and an oxygen fugacity exactly 2 log units more reducing than the FMQ buffer of Myers and Eugster (1983).

fugacities arbitrarily taken as two log units less than the FMQ buffer. Note that methane is an important species at low temperature and tends to be replaced by CO_2 at higher temperatures. This is largely a reflection of the fact that the lines that define f_{O_2} of the graphite–CO–CO_2 equilibrium have a slope lower than those of the oxygen-buffer curves. Thus, following a path parallel to an oxygen-buffer curve produces a fluid that becomes more oxidizing (relative to graphite) with increasing temperature. As oxygen fugacities fall to still lower values, the gas phase becomes overwhelmingly composed of $CH_4 + H_2 + H_2O$, with vanishly small amounts of CO_2 and CO. At oxygen fugacities more oxidizing than graphite–gas, the fluid phase is essentially a binary H_2O–CO_2 mixture.

You should recall that the form of equation (10-33) implies the ideal mixing of all species. Because of the significant variations in the amount of all species present, this assumption can be viewed as no more than an approximation. Thus, the accuracy of Figure 10-10 can be materially improved by incorporating activity coefficients for all gas species. The calculation is considerably more cumbersome and requires a computer, but it is absolutely necessary in order to ensure more accurate results, especially for temperature less than 600°C.

Determination of oxygen fugacities in igneous rocks

The redox state of a rock that formed at high temperatures can sometimes be determined from a study of the mineral assemblages present and an analysis of the compositions of mineral solid solutions. Suitable minerals for analysis are those that take part in oxidation–reduction reactions. The oxygen fugacity calculated from such an assemblage is assumed to be a "fossil" intensive variable that records the oxygen fugacity that prevailed at high temperatures when the rock formed. For this reason, it is important to exclude rocks that obviously have been affected by later events such as hydrothermal or supergene alteration. Another important point is that an oxygen-fugacity value by itself is meaningless without temperature information. For example, the HM buffer spans more than 10 orders of magnitude in $\log f_{O_2}$ between 800 K and 1200 K (Figure 10-6). However, oxygen-buffer reactions tend to be roughly parallel in $(\log f_{O_2})$–T space, and implied *gradients* in oxygen fugacity or chemical potential may be meaningful in the absence of temperature data.

Perhaps the most successful oxygen geobarometer, and certainly the most widely used assemblage in igneous rocks, is based on the compositions of two *coexisting* Fe–Ti oxides (Buddington and Lindsley, 1964). The composition of each phase is restricted ideally to the binary joins Fe_3O_4–Fe_2TiO_4 (magnetite (mt)–ulvospinel (usp), or $mt_{(ss)}$) and Fe_2O_3–$FeTiO_3$ (hematite (hem)–ilmenite (ilm), or $ilm_{(ss)}$), and is controlled by both the exchange reaction

$$\begin{array}{cccc} Fe_3O_4 & + \ FeTiO_3 & \rightleftharpoons Fe_2TiO_4 & + \ Fe_2O_3 \\ \text{in } mt_{(ss)} & \text{in } ilm_{(ss)} & \text{in } mt_{(ss)} & \text{in } ilm_{(ss)} \end{array} \tag{10-35}$$

and the oxidation reaction

$$\begin{array}{ccc} 4\,Fe_3O_4 & + \ O_2 \rightleftharpoons & 6\,Fe_2O_3 \\ \text{in } mt_{(ss)} & & \text{in } ilm_{(ss)} \end{array} \tag{10-36}$$

The compositions of the solid-solution phases depend on the free energies of each reaction, temperature, and f_{O_2}:

$$-\frac{\Delta G^{\circ}_{(10\text{-}35)}}{RT} = \ln \frac{X_{\text{usp}}^{mt(ss)}(1 - X_{\text{ilm}}^{ilm(ss)})}{(1 - X_{\text{usp}}^{mt(ss)})X_{\text{ilm}}^{ilm(ss)}} + \ln \frac{\lambda_{\text{usp}}^{mt(ss)}\lambda_{\text{hem}}^{ilm(ss)}}{\lambda_{\text{mt}}^{mt(ss)}\lambda_{\text{ilm}}^{ilm(ss)}} \tag{10-37}$$

and

$$-\frac{\Delta G^{\circ}_{(10\text{-}36)}}{RT} = \ln \frac{(X_{\text{hem}}^{ilm(ss)})^6}{(X_{\text{mt}}^{mt(ss)})^4} + \ln \frac{(\lambda_{\text{hem}}^{ilm(ss)})^6}{(\lambda_{\text{mt}}^{mt(ss)})^4} - \ln f_{O_2} \tag{10-38}$$

Spencer and Lindsley (1981) developed a solution model for the coexisting $mt_{(ss)}$ and $ilm_{(ss)}$ phases and reviewed the results of experiments performed by a number of investigators on the Fe–Ti–O system under known f_{O_2}–T conditions.

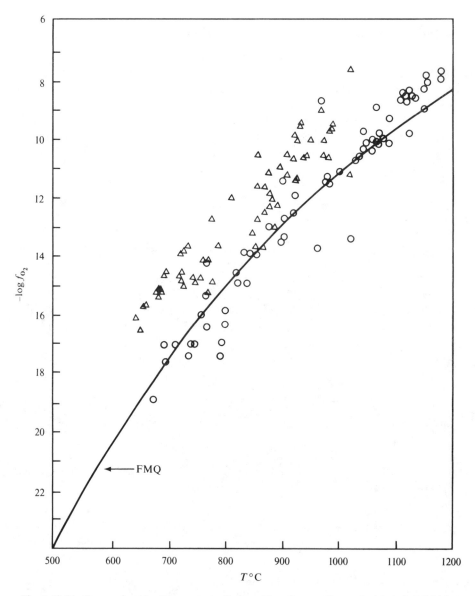

Figure 10-11. Oxygen fugacity–temperature relations of mafic extrusive rocks (circles) and felsic extrusive rocks (triangles), determined by the magnetite-ilmenite geobarometer. FMQ buffer curve shown for reference. (Adapted from Haggerty, 1976)

It is important to emphasize that the calculations would not be possible without these experimental data. As a result of such calculations, the standard free energies shown in (10-37) and (10-38) can be determined, making it possible to retrieve both an f_{O_2} and a temperature from the measured compositions of the coexisting oxide phases. Spencer and Lindsley present a graph from which $\log f_{O_2}$ and T can be read directly from the intersection of $mt_{(ss)}$ and $ilm_{(ss)}$ isopleths. Some natural compositions extend beyond the limits of the solution model and cannot be used to estimate T or f_{O_2}. Haggerty (1976) has summarized the data obtained from the magnetite–ilmenite geobarometer for various igneous rock compositions. His compilation of mafic and felsic extrusive suites is reproduced in Figure 10-11. The dashed line is the fayalite–magnetite–quartz buffer, shown for reference.

It is also possible to measure oxygen fugacity at high temperatures indirectly by using the zirconia electrochemical cell. Sato and Wright (1966) used such a cell to measure f_{O_2} of a cooling Hawaiian lava lake, and they reported oxygen fugacities near, but slightly greater than, the FMQ buffer. Another use of the ZrO_2 electrochemical cell has been the monitoring of the "intrinsic" oxygen fugacities of natural igneous rocks heated to high temperatures under controlled laboratory conditions (Sato, 1972). The results of many such investigations are summarized by Haggerty (1976, fig. Hg-171). Most of the oxygen fugacities recorded in this way are in the same f_{O_2}–T ranges as those shown in Figure 10-11.

Determination and significance of oxygen fugacities in metamorphic rocks

Several different methods have been used to deduce the oxygen fugacities recorded by metamorphic assemblages. Some methods are qualitative, such as study of changes in the Fe^{2+}/Fe^{3+} or Mg^{2+}/Fe^{2+} ratio of silicates. The latter ratio is significant because oxidation involves the conversion of Fe^{2+} to Fe^{3+}; thus, a positive correlation between oxidation state and Mg^{2+}/Fe^{2+} ratio has been widely noted in metamorphic rocks (e.g., Chinner, 1960).

The oxide phases present in metamorphic rocks provide the greatest amount of quantitative redox information, although the magnetite–ilmenite geobarometer is relatively useless at low to intermediate grades of metamorphism because the composition of magnetite in these rocks is pure Fe_3O_4 (Rumble, 1976a). Nonetheless, graphical interpretations of phase equilibria in the Fe–Ti–O system may be very revealing. Figure 10-12 plots isothermal-isobaric oxygen fugacity versus composition for intermediate to high metamorphic grades, showing how the trivariant (P,T,f_{O_2}) oxide assemblage changes as a function of f_{O_2}. In a study of kyanite-grade metaquartzites from western New Hampshire, Rumble (1973) reported the presence of hematite + magnetite and rutile + ilmeno-hematite assemblages in the same thin section, which corresponds to points a and b in Figure 10-12. At 600°C, this difference corresponds to $\Delta f_{O_2} \cong 10^{-4}$ bar, or four orders of magnitude in f_{O_2}.

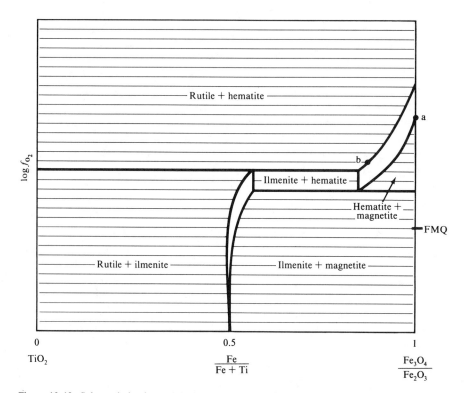

Figure 10-12. Schematic isothermal, isobaric diagram showing the distribution of tie lines in the Fe–Ti–O system at temperatures corresponding to intermediate grades of metamorphism. The relative position of the FMQ buffer corresponds to 600°C. Points a and b are mentioned in the text. (Adapted from Rumble, 1976)

Another technique for determining gradients in fugacity or chemical potentials is algebraic and is based on the solution of a set of simultaneous equations relating P, T, μ_i, and X_i for i components of interest (Rumble, 1976b, Spear et al., 1982). The method assumes that Gibbs–Duhem equations apply for all the intensive variables in mineral solution phases at equilibrium, and it is useful for either igneous or metamorphic rocks. When applied specifically to μ_{O_2} in metamorphic rocks, significant gradients are very commonly inferred. For a recent summary of specific examples, see Rice and Ferry (1982, table 1, pp. 298–299).

The inference of both large-scale and small-scale variations in μ_{O_2} suggests that the redox state of many metamorphic rocks is inherited from premetamorphic conditions—e.g., from variations in the Fe^{2+}/Fe^{3+} ratio of the parent rocks. Much of the evidence for this view was summarized by Thompson (1972).

In contrast, Eugster (1972) suggested that most metamorphic rocks undergo progressive reduction during metamorphism as a result either of the oxidation of organic matter buried with sediments or of the diffusion of mantle-derived H_2 through the crust. The two cases are not necessarily exclusive. Certain oxygen buffers such as magnetite–hematite have a large buffering capacity. During reduction, a relatively small amount of hydrogen is used for conversion of hematite to magnetite. Thus, if f_{O_2} of the reducing fluid is not too far removed from the HM buffer, some hematite will persist for a long time, which fixes the *local* μ_{O_2} equal to that of the HM buffer (cf Figure 10-7). Such buffers could theoretically maintain local gradients in μ_{O_2} even if the diffusing metamorphic fluid is relatively reducing.

In this section, we have discussed the role of oxygen in metamorphic processes, but this discussion leads to a more general question of just how the chemical potential of any component in a geochemical system becomes fixed. As suggested in the previous paragraph, there are two distinct possibilities: either the chemical potential is fixed by the compositions and relative masses of minerals present in the system, or the chemical potential is imposed by a component that flows into the system from a large external reservoir with which it is in equilibrium. The gradients in f_{O_2} that were inferred from the compositions of the Fe–Ti oxides shown in Figure 10-12 must be an example of the first case. A contrasting example could be found in a turbulent, well-mixed mountain stream, where f_{O_2} is in equilibrium with the atmosphere and is not determined by river–sediment reactions. If any chemical potential is fixed by equilibrium with an external reservoir, then one degree of freedom is removed from the system, because the system is incapable of exerting any control over that chemical potential. Of course the difference between an internally-fixed and an externally-fixed component may be largely a matter of scale and the boundaries that are chosen for the system.

10-6   OVERVIEW: CHOOSING VARIABLES, REACTIONS, AND SYSTEM BOUNDARIES

The application of thermodynamics to the interpretation of geologic processes requires considerable skill in identifying the system, choosing the dominant or controlling reactions and their most appropriate variables, and finding the clearest graphical and analytical techniques to represent the system. This skill, like any other, is gained through practice, but a summary here of some important points from the first ten chapters should make it easier to acquire.

The choice of a geochemical system is often guided by the investigator's particular interest. This choice is not as important as the proper identification of the system after having chosen it. That is, it should be clear whether the system is applicable to microenvironments with a coordinate scale of micrometers, millimeters, or centimeters or to macroenvironments with a scale of meters to kilometers. Likewise, the time scale is important. Is the system relevant to a time scale of minutes to years, or is it considered in terms of geologic time such as thousands to billions of years? The chosen spatial and temporal scale may well determine whether equilibrium is likely to prevail or not. In many cases, the larger the scale, the less likely it is that the system will be at or near equilibrium. For very large-scale systems, such as global processes, equilibrium has relatively little application, and fluxes and residence times are more important factors. Likewise, for very small-scale systems, equilibrium may not apply. If the appropriate scale in time and space is selected and the correct variables are chosen, then the degree of equilibrium in the system usually can be determined.

If it can be established that the system is of small to moderate size, then the system should be examined to see whether it is open or closed and whether or not it is in partial or local equilibrium within the chosen time interval. Very commonly it is worthwhile to measure or calculate several system parameters to see whether or not the system is in partial equilibrium. For example, natural groundwaters are likely to be in disequilibrium with respect to redox reactions, but they may be in equilibrium with respect to gypsum or calcite saturation. The best method is to carefully analyze the groundwaters, examine the minerals present, and find out what the field relations indicate. Even when you know that the system is not at equilibrium, it can be very helpful to determine what the equilibrium conditions are, so that the distance from equilibrium—i.e., the driving force for reaction rates—can be estimated.

Choosing the dominant or controlling reactions in a given geochemical system may be relatively straightforward for readily observable processes such as weathering, riverine, lacustrine, and estuarine reactions. Similarly the important variables may be easy to identify for those processes where a multitude of parameters can all be directly measured, including both the mineral phases and the aqueous phases. Although it may not be possible to observe directly the minerals reacting in groundwater systems, the composition of the water can be used along with appropriate chemical models to anticipate what minerals should be dissolving and precipitating (see Appendix E).

In contrast, a metamorphic rock, which may have been in equilibrium with a fluid phase at one time, no longer contains the original fluid phase (with the possible exception of fluid inclusions). Only the existing mineral assemblages in the metamorphic rock can be used to infer relative changes in the composition or redox state of the fluid that existed at some remote time in the evolution of the rock. This situation is quite the opposite of that in groundwater investigations.

In those cases where the solution phase is available for study, it is almost always advantageous to consider the thermodynamic properties of dissolved constituents as potentially important variables, leading to the profusion of ion activity diagrams and $p\varepsilon$–pH diagrams used in low-temperature geochemistry. However, even if the solution phase has long since vanished, it may still be equally relevant to interpret mineral equilibria in terms of variations in the activities of dissolved constituents that presumably were in equilibrium with some observed mineral assemblage. For example, $(\log f_{O_2})$–pH diagrams have long been used to interpret the redox state of supercritical aqueous fluids implicated in the formation of hydrothermal ore deposits, and activity diagrams of all types have found increasing use in the interpretation of mineral reactions in metamorphic rocks (cf Eugster, 1981). One obvious difference between the use of such diagrams to describe surface conditions (ca 1 bar, 25°C) and their use to describe high-temperature geochemical environments is that the temperature in the latter case not only is unknown but commonly is variable. This uncertainty in temperature has a strong effect not only on the positions of univariant lines on activity diagrams but also on the identity of the mineral phases that can equilibrate with the aqueous fluid (cf Figure 9-22). Thus, many graphical analyses of high-temperature mineral equilibria have chosen temperature as an independent variable (e.g., $\log f_{O_2}$ versus T or $X_{CO_2}^{fluid}$ versus T), in combination with some function of the activities of the molecular constituents of the fluid phase (CO_2, H_2O, HCl, etc.). This choice is partly a reflection of the fact that, at elevated temperature, molecular species tend to be dominant relative to ionic species.

With these generalizations in mind, once you know what system you are interested in, your next job is to decide how to formulate the most relevant mineral or fluid reactions. This task requires an analysis of the number of components in the system, which amounts to deciding which chemical species and which intensive parameters are the most critical ones in effecting changes in the mineral assemblage. Next, you must decide whether to express the system variables that you have chosen in terms of chemical potentials, activities, partial pressures. fugacities, mole fractions, molalities, or ratios of these parameters. All these decisions depend to some extent on what data are available, but they also depend very much on what kinds of questions you are trying to answer in the first place. Because this problem may be especially complex at elevated temperatures and pressures, we particularly recommend the reviews of Ferry and Burt (1982) and Eugster (1981), both of which discuss in detail the choice and manipulation of variables in metamorphic fluids.

SUMMARY

- Oxidation–reduction (redox) reactions involve transfer of electrons between species having different valence states.

- In an electrochemical cell, oxidation occurs at the anode. The excess electrons produced by oxidation flow through an external circuit to the cathode, where reduction occurs. Within the electrochemical cell, positive charges are carried by cations that migrate toward the cathode, and negative charges are carried by anions that migrate toward the anode.

- The potential of a half-cell reaction is measured relative to the standard hydrogen electrode ($H_{2(g)} | H_{(aq)}^+$, or SHE), which is defined to have a standard potential of 0 V at 298.15 K. In practice, secondary reference electrodes such as $Ag | AgCl$ and $Hg | Hg_2Cl_2$ (calomel) are used in place of the SHE to measure half-cell potentials. By convention, half-cell reactions are written as reduction potentials.

- The Gibbs free energy of an electrochemical reaction is related to the potential E of an electrochemical cell by

$$\Delta G = -nFE$$

where n is the number of electrons transferred, and F is the Faraday constant. The Nernst equation describes the relationship between the electrochemical potential and the activities of ions participating in redox equilibria:

$$E = E^\circ - \frac{RT}{nF} \ln \Pi a_i^{v_i}$$

where E° is the standard-state potential of the reaction.

- E_H (electrochemical potential measured relative to SHE), $p\varepsilon$ (the negative logarithm of the activity of the electron), and f_{O_2} are three variables commonly used to describe redox equilibria in geochemistry. When combined with pH, either $p\varepsilon$ or $\log f_{O_2}$ is a more-useful variable than E_H because both $p\varepsilon$–pH and $(\log f_{O_2})$–pH diagrams obey the conventions and rules of construction common to all activity diagrams.

- The measured redox potentials of most natural waters cannot be related to any single meaningful half-cell potential. For this reason, the use of $p\varepsilon$ as a master variable is not always justified.

- At higher temperatures and pressures, mineral–fluid and mineral–melt redox equilibria are commonly expressed in terms of f_{O_2}, even though species more abundant than O_2 molecules are responsible for transfer of oxygen in the system.

- Redox reactions that are isobarically univariant with respect to T and f_{O_2} are called oxygen buffers if crystalline phases take part in redox equilibrium. The oxygen fugacities of buffer assemblages can be measured at high temperatures by a solid-state electrochemical cell, or they can be calculated from standard thermodynamic data.

- Many mineral buffer assemblages are theoretically capable of controlling the f_{O_2} of a natural system at elevated temperatures. In practice, this can happen only (1) if the relative mass of the buffer assemblage is large compared to the amount of fluid phase in contact with it, and (2) if the buffering capacity of the mineral assemblage is high. It is much more likely that most naturally occurring buffer assemblages serve only as indicators of relative oxidation state.

- The bulk composition of a supercritical fluid in the system C–O–H depends on f_{O_2}. Under relatively oxidizing conditions, CO_2 and H_2O predominate. At lower f_{O_2} values, graphite may precipitate, and CH_4 and H_2 (and, to a lesser extent, CO) may become present in significant amounts.

PROBLEMS

1. Using the standard electrode potentials given in Section 10-1 for the Cu^{2+}/Cu and Fe^{3+}/Fe^{2+} couples, calculate the standard-state free energy change for the oxidation of 1 mol of solid copper by aqueous ferric ion. (*Ans:* -231.56 kJ·mol^{-1})

2. Sato (1960) postulated that, in the absence of other redox couples, the oxygen/peroxide couple

$$O_2 + 2H^+ + 2e^- \rightleftharpoons H_2O_2$$

may govern the redox potential of natural waters. Given a standard potential of 0.682 V, a pH of 6.0, and a partial pressure of oxygen of 10^{-1}, find the equilibrium concentration of hydrogen peroxide when the measured potential is 0.505 V. (*Ans:* 10^{-7} *m*) If the other variables remain constant, what is the measured potential at pH = 5.0? (*Ans:* 0.564 V)

3. Acid mine waters commonly contain both $Fe^{2+}_{(aq)}$ and $Fe^{3+}_{(aq)}$ and have redox potentials that are fairly high. For a total dissolved iron concentration of 40 mg·L^{-1} and $E_H = 0.700$ V, calculate the concentrations of $Fe^{2+}_{(aq)}$ and $Fe^{3+}_{(aq)}$ at 25°C, assuming ideal solution behavior (no ion pairing, and activity coefficients are unity). (*Ans:* $Fe^{2+}_{(aq)} = 67.2 \times 10^{-5}$ *m*, and $Fe^{3+}_{(aq)} = 4.4 \times 10^{-5}$ *m*) If this solution is in equilibrium with ferrihydrite, find the pH of the solution. (*Ans:* 2.78)

4. Reconstruct the pε–pH diagram for iron using goethite (log $K_{sp} = -43$) instead of ferrihydrite, and demonstrate the major changes in the equilibrium boundaries. What are the dominant changes when compared to Figure 10-5?

5. Choose an element that has more than one redox state, and construct a pε–pH diagram at 25°C and 1 bar for that element from the available data (e.g., Robie et al., 1978). If molar volume and enthalpy data are available, compare and contrast the changes in the equilibrium boundaries at a higher temperature and pressure.

6. At 1000 K and 5 kbar, the oxygen fugacity of the nickel–bunsenite buffer

$$2\,Ni + O_2 = \quad 2\,NiO$$
$$\text{bunsenite}$$

is $10^{-15.360}$ bar. Use this information to calculate the standard free energy of formation of bunsenite at 1 bar and 1000 K. The molar volumes are $\bar{V}_{Ni} = 6.588$ cm^3·mol^{-1}, $\bar{V}_{NiO} = 10.970$ cm^3·mol^{-1}. (*Ans:* -149.22 kJ·mol^{-1})

7. Derive equation (10-32).

8. Using equation (10-32), calculate the oxygen fugacity and the mole fraction of CO in a C–O gas in equilibrium with graphite at a total pressure of 250 bars for isotherms 600, 800, 1000, and 1200 K. Use the following data.

T(K)	Γ_{CO_2}	Γ_{CO}	log K_{CO_2}	log K_{CO}
600	0.948	1.08	34.401	14.319
800	1.020	1.07	25.826	11.914
1000	1.039	1.06	20.676	10.459
1200	1.044	1.05	17.239	9.479

$$\bar{V}_{(graphite)} = 5.298 \text{ cm}^3 \cdot \text{mol}^{-1}$$

The equilibrium constants are given for 1 bar and T. Don't forget to correct them to 250 bars. Assume ideal mixing of CO and CO_2. (*Sample answer*: At 600 K, $\log f_{O_2} = -32.14$ and $X_{CO} = 8.56 \times 10^{-5}$)

9. The oxidation of siderite to magnetite can be written as

$$3\,FeCO_3 + \tfrac{1}{2}O_2 = Fe_3O_4 + 3\,CO_2$$
$$\text{siderite} \qquad\qquad \text{magnetite}$$

Calculate the oxygen fugacity at which magnetite and siderite are in equilibrium with a gas phase in the system C–O, for a total pressure of 2 kbar and 700 K. Take $\Delta G_{r(1,700)}$ for the reaction as written as -298.20 kJ. Log K for $CO + \tfrac{1}{2}O_2 = CO_2$ is 16.556 under the stated conditions, and the fugacity coefficients for CO_2 and CO are 1.777 and 2.70, respectively. The molar volumes of siderite and magnetite are 29.378 $cm^3 \cdot mol^{-1}$ and 44.524 $cm^3 \cdot mol^{-1}$. (*Ans*: $\log f_{O_2} = -24.50$)   Would your answer be different if the gas phase were in the system C–O–H? Explain.

10. Why do the oxygen fugacities of many mafic igneous rocks shown in Figure 10-11 cluster roughly around the FMQ buffer curve?

11

THERMODYNAMIC DATA: MEASUREMENTS AND ESTIMATIONS

*Thermodynamics is an experimental science, and not a branch of meta-
physics. It consists of a collection of equations, and also some inequalities,
which inter-relate certain kinds of measurable physical quantities. In any
thermodynamic equation every quantity is thermodynamically measurable.
What can such an equation "tell one" about one's system or process? Or, in
other words, what can we learn from such an equation about the microscopic
explanation of a macroscopic change? Nothing whatever. What then is the
use of thermodynamic equations? They are useful precisely because some
quantities are easier to measure than others.*

M. L. McGlashan (1979)

If you are like most students who are coming to grips with thermodynamics for
the first time, you may take the whole subject of thermodynamic data pretty
much for granted. After all, there is so much to learn about how to approach a
problem and how to choose the "right" equation to suit the problem at hand that
there is an unfortunate tendency to run to any table of data and plug in some
magical number for ΔH or $\log K$ and hope that the answer which comes flashing
up on your calculator will be meaningful. Up to this point in the book, we may
have reinforced that impression by supplying whatever data were required for
problem solving without explaining where they came from. It is now time to
address the subject of data selection directly.

Let's suppose for a minute that we had not presented any data for the
problems in the preceding chapters but instead had provided a list of standard
sources of thermodynamic data, with the idea that you should fend for yourself. If
you spent the time to consult all these sources, you would probably find either

that the substance you were looking for is not listed in any of the tables, or that the substance appears in more than one source, with the values for enthalpy or heat capacity or whatever you seek being different in each table. In the latter case, the solution to your problem obviously depends on which values you select as input data. Thus, it is important to understand where these numbers come from, how it is that substantially different numbers can be generated, and how to choose between conflicting data sets. This chapter discusses the most important methods for generating thermodynamic data, and Chapter 12 describes strategies for the compilation and evaluation of thermodynamic data obtained from such experiments.

11-1 CALORIMETRIC MEASUREMENTS

Calorimetry is the direct measurement of either (1) the heat capacity or heat content of individual phases or (2) the heat associated with some chemical reaction. From these data, it is possible to calculate third-law entropies, standard enthalpies and free energies, and the temperature effect on these functions. Most of the thermodynamic data for minerals are based, at least to some degree, on calorimetric measurements.

Heat-capacity measurements

A heat-capacity experiment measures the amount of energy necessary to raise the temperature of a substance by a certain number of degrees. Heat capacities rank among the most essential of all thermodynamic data, because they provide two very different kinds of information. First, they are a measure of the temperature dependence of the enthalpy because

$$\left(\frac{\partial H}{\partial T}\right)_P = C_P \tag{3-109}$$

More importantly, because of the third law of thermodynamics, heat-capacity measurements provide a means of calculating the principal contribution to the third-law entropy of phases through the relation

$$S^{\circ}_{298} = \int_0^{298.15} C_P \frac{dT}{T} + \frac{\Delta H_p}{T_p} + S_0 \tag{cf 3-125}$$

where ΔH_p and T_p are the enthalpies of inversion and the temperatures of any phase changes that occur between 0 and 298.15 K, and S_0 is the residual or zero-point entropy at 0 K.

To perform the necessary measurements, investigators have used several fundamentally different methods. Measurements from near absolute zero to a maximum temperature of about 400 K allow precise evaluation of the integral in

(3-125). Heat-capacity measurements at higher temperatures are used to extend the measurements of finite entropy above 400 K and to define the temperature effect on both the entropy and enthalpy. Both low-temperature and high-temperature heat-capacity measurements depend on the ability to measure very small heat changes with very great accuracy, and the specific design of calorimeters is very complex. Although a detailed description of the instrumentation involved is unwarranted here, you might profit by studying the design of a low-temperature calorimeter that has been in use, with continual modification, at the U.S. Geological Survey laboratories since the late 1960s (Robie and Hemingway, 1972; Robie et al., 1976). The calorimeter itself is adiabatically shielded and uses liquid helium to cool the substance under investigation to near 4.2 K. Each measurement consists of monitoring the voltage and calculating the current that passes through a heater in the calorimeter for a fixed time. The temperature rise resulting from the heating interval is measured with a platinum resistance thermometer. The size of the heating interval is determined largely by the magnitude of the heat-capacity change anticipated; temperature increases range from a fraction of a degree to several degrees. All observations are stored on magnetic tape for subsequent computer processing; third-law entropies accurate to $\pm 0.1\%$ to $\pm 0.2\%$ can then be extracted from C_P versus T curves similar to Figure 3-3.

Some low-temperature calorimeters do not operate below 50 K. Extrapolations from the lowest temperature achieved experimentally to 0 K could result in significant errors in third-law entropy because (as mentioned in previous chapters) certain phases at very low temperatures show transition or ordering effects that produce anomalous peaks in the heat-capacity curve and add to the third-law entropy of the phase (cf Figure 4-12, a lambda transition resulting from the ordering of magnetic moments of Fe^{2+} in fayalite at 64.88 K). If the rate of transformation is very slow compared to the rate of cooling (for example, Al/Si ordering in feldspars and felspathoids), then no effect on the heat capacity can be measured, and a residual, or zero-point, entropy contribution must be added to the calorimetric term.

For heat-capacity measurements above room temperature, a differential scanning calorimeter (DSC) has recently proved useful. This instrument compares the different amounts of power required to keep two sample holders at the same temperature. Successive scans are made over each temperature interval for the unknown, a sapphire (or similar) heat-capacity standard, and an empty pan. Because the heat capacity of the standard phase is known very precisely, the measured power differences can be used to calculate the heat capacity of the unknown phase. Krupka et al. (1979) describe the method in more detail and report DSC measurements for a number of minerals and glasses. The uncertainties involved in DSC measurements ($\pm\frac{1}{2}\%$ to $\pm 1\%$) are considerably larger than uncertainties associated with low-temperature calorimeters. The principal advantage of the DSC is that only 20 mg of sample is required, as compared with more than 1 g for cryogenic calorimeters. This is a very important factor that adds to the appeal of the DSC, because it has opened the possibility of measuring the heat capacities of synthetic minerals that can be grown only in very small quantities.

Drop calorimetry and heat-content measurements

The drop calorimeter measures the difference in enthalpy for a known mass of substance at two temperatures—i.e., $H_T - H_{T_{ref}}$ where T_{ref} is the operating temperature, or reference temperature, of the calorimeter. Drop calorimetry has been the major technique for determining high-temperature enthalpies or heat contents of minerals and glasses, although the advent of the differential scanning calorimeter has somewhat reduced its influence in recent years. The principle of drop calorimetry is simple: the sample is heated to the desired temperature T and dropped into a calorimeter operating near room temperature T_{ref}, where the heat evolved is measured as the sample cools to the operating temperature. One of the most common receiving devices is the Bunsen *ice calorimeter*, which operates on the principle that the mass of ice melted isothermally at $0°C$ as a result of the sample drop is a direct measure of the amount of heat liberated during the cooling of the sample from T to T_{ref}. In practice, it is the volume change resulting from melting the ice that is measured. Details of the design and operation of drop-calorimeter experiments are described by Douglas and King (1968).

Enthalpy-of-solution measurements

Direct measurements of heats of solution have long provided the great majority of standard enthalpy data available to geochemists. In commonly tabulated form, most standard enthalpy data are presented in terms of the formation of a compound either from its component elements or from oxides. Consider albite as an example; $\Delta H^\circ_{f(1,298)}$ refers to the enthalpy change that would occur for

$$Na_{(c)} + Al_{(c)} + 3\,Si_{(c)} + 4\,O_{2(g)} = NaAlSi_3O_{8(c)} \qquad (11\text{-}1)$$

at 1 bar and 298.15 K, where subscripts (c) and (g) refer to the crystalline and gaseous states, respectively. As you may well imagine, bringing metallic sodium, aluminum, silicon, and gaseous oxygen together would not result in the formation of albite at any temperature—these elements would form instead a metastable mixture of metallic oxides, because the oxides are the phases that would nucleate most rapidly. For these reasons, the determination of standard enthalpies of formation of many rock-forming minerals requires the construction of elaborate reaction schemes that may include more than 20 individual reactions. These schemes are the potential sources of a great deal of trouble, and it is important for you to appreciate how and why they are constructed.

Many different kinds of calorimeters have been used over the years to measure reaction heats. Probably most of the enthalpy data for minerals have come from various designs of the *acid-solution calorimeter*, which produced an impressive body of data for the Bureau of Mines subsequent to World War II and for the U.S. Geological Survey in later years. The principle involves measuring the heat of solution associated with dissolving crystalline phases in aqueous solutions such as HCl or HF. These calorimeters operate at temperatures ranging from 300 K to 400 K, depending on design. A measurement is divided into (1) a calibration event that measures the amount of electrical energy required to raise the temperature of the calorimeter by a small temperature interval and (2) a solution event that measures the temperature change caused by dissolving a known mass of a phase in the calorimeter solvent. Elaborate corrections are required for small but significant heat exchange between the calorimeter and a large constant-temperature bath. The precision achievable in these measurements is about $\pm 0.1\%$ for heats of solution greater than 100 kJ.

There are a number of experimental difficulties associated with acid-solution calorimetry. One of the most prominent is that highly refractory phases—such as corundum (Al_2O_3) or certain aluminosilicates—do not dissolve or dissolve very slowly, even in the strongest acid. Under some conditions, a gel or precipitate might form; either of these could invalidate the measurements. Perhaps the most critical concern is that the reaction scheme must be precisely stoichiometric, which requires (1) that all reactions in the scheme be precisely defined with respect to stoichiometric coefficients and (2) that all dilutions be performed with the utmost care. To explain why this stoichiometry is so significant, we shall consider the case of one set of measurements for determining the standard enthalpy of formation of low albite (Hemingway and Robie, 1977a). Three of the reactions studied involve dissolution of low albite (l-ab), α-quartz (α-qz), and crystalline $NaAlO_2$ in 20.1% (by weight) HF solution at 333.15 K. The reactions and their measured enthalpies of solution* are

$$3\,SiO_{2(\alpha\text{-}qz)} + [1849\,HF + 8163\,H_2O]_{(aq)}$$

$$= [3\,H_2SiF_6 + 1831\,HF + 8169\,H_2O]_{(aq)} - (413{,}211 \pm 627)\,J \quad (11\text{-}2)$$

$$NaAlO_{2(c)} + [3\,H_2SiF_6 + 1831\,HF + 8169\,H_2O]_{(aq)}$$

$$= [3\,H_2SiF_6 + NaF + AlF_3 + 1827\,HF + 8171\,H_2O]_{(aq)}$$

$$- (282{,}884 \pm 280)\,J \quad (11\text{-}3)$$

$$NaAlSi_3O_{8(l\text{-}ab)} + [1849\,HF + 8163\,H_2O]_{(aq)}$$

$$= [3\,H_2SiF_6 + NaF + AlF_3 + 1827\,HF + 8171\,H_2O]_{(aq)}$$

$$- (628{,}980 \pm 1000)\,J \quad (11\text{-}4)$$

where the species enclosed in brackets are dissolved in the calorimeter soution. If we subtract (11-4) from the sum of (11-2) and (11-3), all aqueous species will cancel

---

* The important subject of propagation of uncertainties in calorimetric experiments is deferred until the following subsection.

exactly, so that we can obtain the desired enthalpy of formation for low albite from the heats of solution. This requirement that aqueous species cancel underlines the necessity for taking scrupulous care to ensure that stoichiometry is maintained. Assuming that this requirement is met, we are left with

$$3 \, SiO_{2(\alpha\text{-}qz)} + NaAlO_{2(c)} = NaAlSi_3O_{8(l\text{-}ab)} - (67,115 \pm 1045)J \quad (11\text{-}5)$$

which is the enthalpy of the indicated reaction at 333.15 K.

A calorimetric method that uses molten salts instead of aqueous acids to measure enthalpies of solution has recently produced a considerable body of data for rock-forming minerals. The apparatus currently in widest use is a twin microcalorimeter of the Tian–Calvet type (Navrotsky, 1977). One of the appealing advantages of this calorimeter is that only 20 to 100 mg of sample is required (versus 1 to 2 g for an acid calorimeter). The sample is dropped into a molten salt ($2 \, PbO \cdot B_2O_3$ is the solvent of choice but is inappropriate for some materials). Once the solution reaction is initiated, a thermopile registers the change in temperature difference between the reaction chamber and a twin chamber that acts as a control. The total heat effect is proportional to the area under a heat-flow-versus-time curve, and the calorimeter must be calibrated with known phases. The operating temperature ranges from 600°C to 1200°C, and heat effects can be measured with a standard deviation of about $\pm 1\%$. This precision is worse than that encountered with acid calorimeters, but this difference is partly offset because the heats of solution in molten salts are about an order of magnitude smaller than comparable heats of solution in acid calorimeters. This method is appropriate for highly refractory phases that are reluctant to dissolve in strong acids. Molten-salt calorimetry is subject to many of the same uncertainties discussed for acid-calorimeter experiments and has some additional disadvantages. Careful corrections must be applied to account for the water content of the molten salts, and it is difficult to assure complete stirring of the sample.

We return to the determination of the standard enthalpy of formation of albite as an example of the complexity of some calorimetric schemes. This would be an easier task were it possible to measure the heat of reaction for

$$\tfrac{1}{2}Na_2O_{(c)} + \tfrac{1}{2}Al_2O_{3(c)} + 3 \, SiO_{2(\alpha\text{-}qz)} = NaAlSi_3O_{8(l\text{-}ab)} \quad (11\text{-}6)$$

directly, based on heat-of-solution measurements for each of the crystalline phases listed in (11-6). This approach is not feasible, however, because Na_2O cannot be obtained as a stable stoichiometric compound. For this reason, Hemingway and Robie (1977a) chose to conduct a series of solution experiments to determine the standard enthalpy of formation for $NaAlO_2$. By combining this value with the enthalpy of reaction for (11-5), they would be able to calculate the desired value for albite. This task required measuring the heat of solution of metallic Al, crystalline NaCl, crystalline $NaAlO_2$, $HCl \cdot 12.731 \, H_2O$ (or $4N$ HCl), and 14.73 mol of liquid water in the 20.1% HF solution. The reactions are

$$Al_{(c)} + [1849\,HF + 8163\,H_2O]_{(aq)}$$
$$= [AlF_3 + 1846\,HF + 8163\,H_2O]_{(aq)} + \tfrac{3}{2}H_{2(g)} \quad (11\text{-}7)$$

$$14.73\,H_2O_{(l)} + [AlF_3 + 1846\,HF + 8163\,H_2O]_{(aq)}$$
$$= [AlF_3 + 1846\,HF + 8177.73\,H_2O]_{(aq)} \quad (11\text{-}8)$$

$$NaCl_{(c)} + [AlF_3 + 1846\,HF + 8177.73\,H_2O]_{(aq)}$$
$$= [NaF + HCl + AlF_3 + 1845\,HF + 8177.73\,H_2O]_{(aq)} \quad (11\text{-}9)$$

$$NaAlO_{2(c)} + [1849\,HF + 8163\,H_2O]_{(aq)}$$
$$= [NaF + AlF_3 + 1845\,HF + 8165\,H_2O]_{(aq)} \quad (11\text{-}10)$$

$$HCl \cdot 12.731\,H_2O_{(l)} + [NaF + AlF_3 + 1845\,HF + 8165\,H_2O]_{(aq)}$$
$$= [NaF + AlF_3 + HCl + 1845\,HF + 8177.73\,H_2O]_{(aq)} \quad (11\text{-}11)$$

Subtracting the sum of (11-10) and (11-11) from the sum of (11-7), (11-8), and (11-9), we obtain

$$Al_{(c)} + NaCl_{(c)} + 14.73\,H_2O_{(l)} = NaAlO_{2(c)} + HCl \cdot 12.731\,H_2O_{(l)} + \tfrac{3}{2}H_{2(g)} \quad (11\text{-}12)$$

From the heats of solution for reactions (11-7) through (11-11), Hemingway and Robie (1977a) obtained $\Delta H_r = -(315{,}507 \pm 1255)\,J$ at 298.15 K for reaction (11-12). This value may then be combined with previously determined standard enthalpies of formation for NaCl, 4 N HCl, and H_2O (Wagman et al., 1968, 1976) to obtain the standard enthalpy of formation for $NaAlO_2$:

$$Na_{(c)} + \tfrac{1}{2}Cl_{2(g)} = NaCl_{(c)} \quad (11\text{-}13)$$

$$14.73\,H_{2(g)} + 7.365\,O_{2(g)} = 14.73\,H_2O_{(l)} \quad (11\text{-}14)$$

$$HCl \cdot 12.731\,H_2O_{(l)} = \tfrac{1}{2}Cl_{2(g)} + 13.231\,H_{2(g)} + 6.365\,O_{2(g)} \quad (11\text{-}15)$$

$$\underline{Al_{(c)} + NaCl_{(c)} + 14.73\,H_2O_{(l)} = NaAlO_{2(c)} + HCl \cdot 12.731\,H_2O_{(l)} + \tfrac{3}{2}H_{2(g)}} \quad (11\text{-}16)$$

$$Na_{(c)} + Al_{(c)} + O_{2(g)} = NaAlO_{2(c)} \quad (11\text{-}17)$$

Addition of reactions (11-13) through (11-16) gives $\Delta H_r^\circ = -(1{,}135{,}990 \pm 1255)\,J \cdot mol^{-1}$ for the formation of $NaAlO_2$ from the elements at 1 bar and 298.15 K. We are not yet home free, however, because we still need to know the standard enthalpy of formation of α-quartz. This value can be determined by direct reaction of α-quartz and metallic silicon with gaseous F_2 in a calorimeter of different design (Wise et al., 1963):

$$SiO_{2(\alpha\text{-}qz)} + 2\,F_{2(g)} = SiF_{4(g)} + O_{2(g)} \quad (11\text{-}18)$$

$$Si_{(c)} + 2\,F_{2(g)} = SiF_{4(g)} \quad (11\text{-}19)$$

Subtracting (11-18) from (11-19), we obtain

$$Si_{(c)} + O_{2(g)} = SiO_{2(\alpha\text{-}qz)} \tag{11-20}$$

which carries a recommended value of $\Delta H_r = -(910{,}700 \pm 1000)\,\text{J} \cdot \text{mol}^{-1}$ (CODATA Task Group, 1976). Finally, by adding the reactions

$$3\,SiO_{2(\alpha\text{-}qz)} + NaAlO_{2(c)} = NaAlSi_3O_{8(l\text{-}ab)} \tag{11-5}$$

$$Na_{(c)} + Al_{(c)} + O_{2(g)} = NaAlO_{2(c)} \tag{11-17}$$

$$\underline{3\,(Si_{(c)} + O_{2(g)}) = 3\,SiO_{2(\alpha\text{-}qz)}} \tag{11-20}$$

$$Na_{(c)} + Al_{(c)} + 3\,Si_{(c)} + 4\,O_{2(g)} = NaAlSi_3O_{8(l\text{-}ab)} \tag{11-1}$$

we obtain the desired standard heat of formation of low albite from the elements: $-(3935.12 \pm 3.42)\,\text{kJ} \cdot \text{mol}^{-1}$.

Such an intricate scheme is by no means unique, but it is required here because Na_2O happens to be an inappropriate phase for calorimetric study. The same holds true for K_2O, and for all minerals containing K_2O.

Error accumulation and correlation

Every calorimetric datum carries an experimental uncertainty. Traditionally, this uncertainty has been taken to be twice the standard deviation of the mean value calculated from a set of individual measurements. Because calculations of ΔH_r°, ΔS_r°, and ΔG_r° are commonly obtained by combining standard thermodynamic properties, each with a different uncertainty, it is important to know how to combine the individual uncertainties to obtain the propagated uncertainty associated with the calculated property. This important subject has been treated in some detail by Anderson (1977b). However, for most calorimetric measurements, the data set is too small for statistics to be truly valid.

The customary approach follows the conventional rules of error propagation (Bevington, 1969). For $x = f(u, v, \ldots)$, we have for the case of addition or subtraction $x = au \pm bv \pm \cdots$, where $a, b, \ldots$ are constants. If $u, v, \ldots$ are standard thermodynamic properties and x is the change in those properties resulting from a chemical reaction, then $a, b, \ldots$ are the stoichiometric coefficients of reaction. The uncertainty σ_x in x is simply

$$\sigma_x = (a^2\sigma_u^2 + b^2\sigma_v^2 + \cdots)^{1/2} \tag{11-21}$$

provided that the uncertainties in $u, v, \ldots$ are uncorrelated (i.e., independent of one another). The corresponding expression for multiplication or division is

$$\frac{\sigma_x^2}{x^2} = \frac{\sigma_u^2}{u^2} + \frac{\sigma_v^2}{v^2} \tag{11-22}$$

for $x = \pm auv$ or $x = \pm au/v$—assuming, once again, that the uncertainties in u and v are uncorrelated.

Because of the complexity of many calorimetric schemes, it commonly turns out that the uncertainties in the standard thermodynamic functions that are combined in a reaction are not independent. The reason for this noninde-pendence is that two or more phases involved in a reaction may share part of the same calorimetric scheme, and many of the uncertainties associated with reactions in this scheme may cancel. For instance, consider the enthalpy of the low-albite–analbite polymorphic transition. This value could be calculated from the enthalpies of formation of low albite and analbite (Robie et al., 1978):

$$\text{analbite} \qquad \Delta H^\circ_{f(298)} = -(3924.24 \pm 3.64)\,\text{kJ}$$

$$\underline{-[\text{low albite} \qquad \Delta H^\circ_{f(298)} = -(3935.12 \pm 3.42)\,\text{kJ}]}$$

$$\Delta H^\circ_{r(298)} = \quad +10.88 \pm 4.99\,\text{kJ}$$

The uncertainty in ΔH°_r for the transition is here determined according to equation (11-21) as

$$[(3.64)^2 + (3.42)^2]^{1/2} = 4.99$$

This value is much too large, however, because the calorimetric schemes used to calculate the standard enthalpies of formation for both polymorphs of $NaAlSi_3O_8$ are the same, and most of their accumulated uncertainties cancel. The *true* uncertainty of ± 1.26 kJ is calculated by combining, in accordance with (11-21), the uncertainties in the heat-of-solution measurements for analbite and low albite that were obtained in aqueous HF at 323 K (Waldbaum and Robie, 1971).

As a second example, consider the dehydration of analcime:

$$NaAlSi_2O_6 \cdot H_2O \rightleftharpoons NaAlSi_2O_6 + H_2O_{(l)}$$
$$\text{analcime} \qquad\quad \text{dehydrated} \qquad\qquad\qquad\qquad (11\text{-}23)$$
$$\text{analcime}$$

Johnson et al. (1982) measured the heat of solution in aqueous HF of a well-characterized natural analcime and that of the same analcime after dehydration. Their abbreviated results (shown here without specifically accounting for the stoichiometry of the acid solution) are

$$\text{analcime + acid} \qquad \Delta H_{\text{soln}(298)} = -(500.11 \pm 0.88)\,\text{kJ} \quad (11\text{-}24)$$

$$\text{dehydrated analcime + acid} \qquad \Delta H_{\text{soln}(298)} = -(540.50 \pm 1.41)\,\text{kJ} \quad (11\text{-}25)$$

$$H_2O_{(l)} + \text{acid} \qquad \Delta H_{\text{soln}(298)} = -(0.42 \pm 0.04)\,\text{kJ} \quad (11\text{-}26)$$

Subtracting (11-24) from the sum of (11-25) and (11-26) and combining the uncertainties according to (11-21), we determine the enthalpy of reaction (11-23) to be $-(40.81 \pm 1.66)\,\text{kJ} \cdot \text{mol}^{-1}$.

Johnson et al. (1982) also reported standard enthalpies of formation for both analcime and dehydrated analcime. The calculation of these values required the construction of an elaborate calorimetric scheme for the same reason discussed in the preceding subsection. Using their standard enthalpy values and the associated uncertainties,

analcime	$\Delta H_{f(298)}^{\circ} = -(3296.9 \pm 3.3)\,kJ$
dehydrated analcime	$\Delta H_{f(298)}^{\circ} = -(2970.2 \pm 3.5)\,kJ$
$H_2O_{(l)}$	$\Delta H_{f(298)}^{\circ} = -(285.83 \pm 0.04)\,kJ$

and combining the uncertainties according to (11-21), we obtain a value of $-(40.87 \pm 4.81)\,kJ$ for the enthalpy of reaction (11-23). We note that the uncertainty calculated here is much larger than that calculated earlier from the heats of solution ($\pm 1.66\,kJ$). Once again, the reason is that analcime and dehydrated analcime share many common dissolution reactions. In our second computation, we added together uncertainties for many values that in fact are canceled during the computation.

The point to be made here is that the true thermodynamic uncertainty of any reaction calculated from calorimetric measurements can be evaluated only by careful examination of the original data. This is the only sure way to account for internally correlated measurements.

11-2 PHASE EQUILIBRIUM STUDIES

Over the last twenty years, a large number of very careful experimental studies of mineralogical phase equilibria have been published. These data are commonly used to retrieve standard thermodynamic data for minerals. Some of the methods used to extract these data from the experiments are discussed in Appendix D.

All retrieval methods depend on a fundamental principle of experimental petrology: to demonstrate equilibrium, a reaction must be *reversed*. To understand this concept, recall that every point on a univariant line represents the condition $\Delta G_r = 0$, because the univariant line is the projection of the intersection of two free-energy surfaces onto the $P-T$ plane. However, the univariant curve can never be located precisely by experimental methods because, as the free energy of the reaction approaches zero, reaction rates decrease dramatically. Consider once again the all-purpose schematic reaction $A \rightleftharpoons B$, where A is any reactant assemblage, and B is any product assemblage. Suppose that an isobaric sequence of experiments, each of which started as assemblage A, gives the following results for temperatures T_1 through T_4, where $T_1 < T_2 < T_3 < T_4$:

T_1	no change
T_2	no change
T_3	a trace of assemblage B forms
T_4	100% B

Another sequence of experiments using assemblage B as starting material gives these results:

T_1 B + A in equal amounts

T_2 a trace of assemblage A forms

T_3 no change

T_4 no change

Only those experiments that show some reaction (i.e., A → B, or B → A) are significant, because they unequivocally demonstrate the relative positions of free-energy surfaces for assemblages A and B (Figure 11-1). The extent of reaction

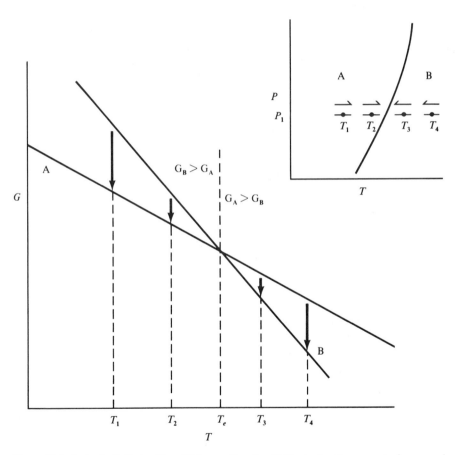

Figure 11-1. Isobaric G–T plot ($P = P_1$) for reaction A = B, illustrating the concept of a reversal bracket. Because B → A at temperature T_2 and A → B at temperature T_3, the equilibrium temperature T_e must lie between T_2 and T_3. Thus T_2 and T_3 taken together form a reversal bracket. The inset is a P–T diagram showing isobar P_1, temperatures T_1 through T_4, and the univariant line A = B. The arrows over the four experimental points indicate the direction from which equilibrium was approached.

depends on kinetic factors and is immaterial so long as the investigator is convinced that some reaction has occurred. The formation of a trace amount of assemblage B is just as informative as the complete conversion of A to B. From the experimental data just given, it is apparent that at T_2 we have B → A $(G_B > G_A)$, whereas at T_3 we have A → B $(G_A > G_B)$. At some intermediate temperature, G_A must equal G_B, so the equilibrium temperature T_e of the univariant line must be located somewhere between T_2 and T_3. The pair of experiments at T_2 and T_3 define a *reversal bracket* for the isobar. Experiments at T_1 and T_4 give further support to the bracket, but they add no new information.

Many different techniques have been used to measure directly mineral reactions of petrologic significance. Details of the variety of experimental techniques common in petrology are important, but the subject is lengthy and complex. An excellent book edited by Ulmer (1971) reviews most of the important methods in some detail. For retrieval purposes, the most accurate data can be obtained from hydrothermal experiments, in which a fluid pressure is applied to a small reactant sample sealed in a cylindrical, metallic reaction vessel known in experimental jargon as a bomb. The reactant assemblage itself is commonly sealed with a small amount of water in a thin-walled tube of noble metal such as gold or platinum; the tube isolates the system under study from the bomb environment. In effect, the result is a sealed metallic test tube with the reactant assemblage located at the bottom of the tube (Figure 11-2). The fluid pressure in such apparatus can be measured directly by means of a Bourdon tube gauge up to 7.7 kbar with a precision of $\pm 1.5\%$. Heating is commonly achieved with an external furnace. Although the temperature of the sample is seldom measured directly (a thermocouple is inserted into the bomb wall adjacent to the sample chamber), many experimenters report an accuracy of $\pm 5°C$ for experiments spanning several weeks. The external furnace limits the practical maximum temperature to about 850°C, depending on bomb composition and pressure. Higher temperatures can be achieved by placing the heating assembly inside the bomb and using a chemically inert gas such as nitrogen or argon to apply the pressure. Such internally heated bombs are complex in design and notoriously cantankerous in operation.

To obtain higher pressures (in the range of 10 to 50 kbar), solid-media apparatus may be used. In the most popular design of "piston–cylinder" apparatus, pressure is transmitted to the sample by solids or softened glasses and cannot be measured directly. Pressure is applied by forcing a hardened piston against the solid media, which are contained in a cylindrical bomb (Figure 11-3). Calibration of piston–cylinder apparatus is usually performed by measuring certain standardized reactions such as the pressure-sensitive conversion of bismuth-I to bismuth-II or the change in the melting of gold with pressure. Friction between piston, cylinder, and the solid media surrounding the sample is significant and adds to pressure uncertainty. Temperature is measured with a thermocouple in contact with the sample container, but thermal gradients may be severe, adding to uncertainty. For these reasons, both pressures and temperatures

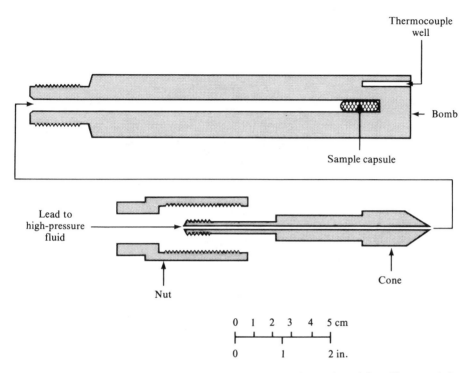

Figure 11-2. Cross section of cold-seal bomb assembly. The stainless steel core is forced into a conical seal at the bomb opening and held in place by a hexagonal nut. The softer cone is slightly deformed against the harder bomb metal, making a tight seal. (After Huebner, 1971)

of experiments using solid-media apparatus have uncertainties much larger than those of hydrothermal experiments.

Regardless of apparatus, most retrieval experiments are quench experiments. This means that the reactant assemblage is held at the desired P and T for a period of time ranging from a fraction of an hour to several months, and it is then rapidly brought to ambient lab condition ("quenched") where it can be examined. The assumption of a quench experiment is that the rate of cooling greatly exceeds the rate of reaction, so that the assemblage examined at room temperature is taken to represent the assemblage that existed at high T and P. The reaction rates of most silicates are slow enough that the quenching process itself does not generally lead to new reaction products, but one must be alert to the possibility. Usually the quenching principle works too well—especially at relatively low pressures and temperatures. In countless hydrothermal experiments, reactant assemblages have been allowed to stew at the desired conditions for many weeks, only to emerge unscathed with no detectable reaction. These are frustrating experiments because they are completely equivocal: the reactant assemblage could be stable, in which case it will never react, or it could be

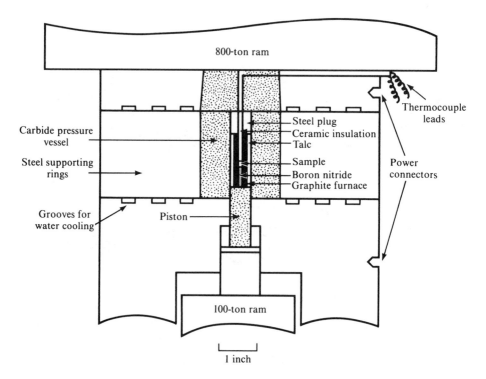

Figure 11-3. Piston-cylinder apparatus, according to the design of Boyd and England (1960). In practice, details of the sample and furnace assembly may vary widely (see, for example, Johannes et al., 1971).

unstable but recalcitrant to react owing to unfavorable kinetics such as lack of product nuclei.

For these reasons, the detection of small amounts of reaction is an important art. One popular technique uses mixtures of product and reactant assemblages as starting material. This method ensures that seed nuclei are available for the reaction to proceed in either direction, but only fairly large changes in the product/reactant ratio (equivalent to 10% to 30% reaction) can be detected routinely. Relative peak heights of critical product and reactant phases on X-ray diffractograms are a common but potentially misleading tool for detection direction. In contrast, some experimenters have found success using the weight changes of a seed crystal surrounded by a fine powder of both product and reactant assemblages as an indicator of very tiny degrees of reaction. The idea is that a weight loss corresponds to dissolution of some of the seed crystal, suggesting that the assemblage that contains the seed phase is unstable. A weight gain has the opposite significance. Great care is required in handling the seed crystal, where weight changes as small as 5 μg are considered significant.

A fundamental problem associated with retrieval experiments is that of completely characterizing all the phases present in the experiment. It should be apparent by now that, in order to retrieve accurate thermodynamic data for a reaction, both the chemical compositions and the ordering parameters of all phases must be known. The former is particularly important for phases that can significantly change their composition during an experiment. Many experimenters use synthetic minerals that are too fine-grained for routine microprobe examination, and indirect methods for estimating compositions (such as percentage yield from a known synthetic composition) have been routinely used. As a result, mineral compositions reported in some experiments are only best guesses.

Consider an even more important problem: even if the run products could be accurately characterized at room temperature, there is no guarantee that these properties pertained to the phases under the *experimental* conditions. This uncertainty is potentially a significant drawback in using phase-equilibrium experiments to calculate standard thermodynamic data. It is worth pointing out that, although the problem of characterization is no less critical in calorimetry experiments, it is a much easier matter to carefully measure the properties of individual phases before dissolving them in a solvent than it is to measure the same properties in a mixture of phases sealed in a gold capsule under a pressure of 2 kbar and at a temperature of 600°C!

11-3 VOLUME MEASUREMENTS

Determinations of the molar volumes of pure phases and of species in solution are needed in order to calculate the effect of pressure on the free energy of reactions (cf equation (4-4)) and are essential for precise calculations of mineral-gas equilibria. As examples of the range of experimental methods encountered, we shall briefly discuss volume determinations of crystalline phases, gases, and aqueous species.

Solids

The molar volume of a crystalline solid can be calculated by dividing the gram formula weight of the substance by the measured density. In practice, however, it is more precise to calculate the molar volume directly from the volume of the unit cell (as calculated from X-ray data). The relationship is simply

$$\text{molar volume (cm}^3 \cdot \text{mol}^{-1}) = \frac{V \times 0.6023}{Z}$$

where V is the unit cell volume in cubic angstroms (Å^3), and Z is the number of formula units present in the X-ray cell. The constant 0.6023 in the expression is Avogadro's number divided by $(10^{24} \text{Å}^3)/(1 \text{ cm}^3) = 1$. The bulk of molar-volume data for crystals has been obtained from X-ray measurements at or near

25°C (e.g., Robie et al., 1978). Under these conditions, the accuracy of the measurement for most rock-forming minerals is in the range of $\pm 0.05\%$ to $\pm 0.25\%$ of the reported volume, and the accuracy depends on the success of the refinement of X-ray data that is required to calculate the lattice constants of the unit cell. If measurements are obtained on a solid solution rather than a stoichiometric compound, the uncertainty in the composition of the solid phase may well exceed the uncertainty in its measured volume.

Gases and supercritical fluids

As shown in Chapter 6, the fugacities of gases and supercritical fluids are determined by means of volume measurements. These experiments rank among the most difficult and challenging of the physicochemical measurements required in thermodynamics, especially when carried out at very high temperatures and pressures. Also, great differences exist in both apparatus and technique; Presnall (1971) reviews most of the important methods. Regardless of technique, the experimental principle is very simple: the molar volume of a pure gas can be calculated as a function of pressure and temperature by measuring the number of moles of gas that occupy a container of known volume at the pressure and temperature of interest. The procedure used by Presnall (1969) to determine the fugacity coefficients of hydrogen at 1800 bar is instructive of the principles involved. The apparatus consists of a small-volume, high-pressure portion (a pressure vessel contained in a furnace) and a large-volume, low-pressure portion (a 10-L bottle connected to a mercury manometer). The pressure vessel is heated to the desired temperature, evacuated, and filled with hydrogen until a predetermined pressure is reached. After thermal equilibration, the gas is expelled into the evacuated low-pressure part of the system, where the mass of hydrogen present can be calculated from the manometer reading and the final gas temperature. Thus, the experimental measurements are the pressure and temperature of the hydrogen in the pressure vessel and the number of moles present at the end of the experiment. It is essential to know precisely the volume of the pressure vessel under the experimental conditions, which must include corrections for thermal expansion and elastic stretch of the alloy and for the small volume of associated capillary tubing. Also, the procedure is tedious because each measurement requires a separate filling of the pressure vessel.

A very different procedure is particularly well suited to gases that liquefy at room temperature and low to moderate pressures (e.g., steam and carbon dioxide). In this case, a screw press or bellows system can be used to inject very accurately known increments of fluid into the pressure vessel. Such a procedure allows continuous measurements of changes in fluid pressure along an isotherm. The apparatus required tends to be rather complex (e.g., Burnham et al., 1969).

Similar procedures can be used to study gas mixtures, but the composition of the gas phase must be determined, as well as the total number of moles of gas present. Greenwood (1969) presents an experimental method for H_2O–CO_2

mixtures. Such experiments are correspondingly more difficult than those that use pure gases, and relatively few measurements at high temperatures and pressures are available for gas mixtures of geologic interest.

Aqueous electrolytes

The partial molal volume of an electrolyte species in aqueous solution can be obtained from the *apparent molal volume* ϕ_v, which is defined by

$$\phi_v = \frac{V - n_1 \bar{V}_1^\circ}{n_2} \qquad (T,P \text{ constant}) \tag{11-27}$$

where V is the volume of a solution containing n_1 moles of H_2O and n_2 moles of electrolyte, and $\bar{V}_1^\circ$ is the molar volume of pure H_2O at constant T and P. The partial molal volume $\bar{V}_2$ of electrolyte is then

$$\bar{V}_2 = \left(\frac{\partial V}{\partial n_2}\right)_{T,P} = \phi_v + n_2 \left(\frac{\partial \phi_v}{\partial n_2}\right)_{T,P} \tag{11-28}$$

Because the second term in (11-28) vanishes as $n_2 \to 0$, the apparent molal volume is equal to the partial molal volume at infinite dilution. A graphical method commonly is used to determine $\bar{V}_2$. The method relies on the observation that the apparent molar volume is a linear function of molar concentration, according to

$$\phi_v = \phi_v^\circ + S_v^* \sqrt{c} \tag{11-29}$$

where ϕ_v° is the intercept (equivalent to the apparent molar volume at infinite dilution), c is the molarity, and S_v^* is a slope expression that depends on the electrolyte type and charge (Millero, 1972). Thus, if experimental values of apparent molal volume are plotted against the square root of solution molarity, the intercept at infinite dilution will be equal to the partial molal volume of the solute at the T and P of interest, referred to an infinite-dilution standard state.

Apparent molal volumes can be calculated directly from measurements of the densities of aqueous electrolyte solutions obtained over a range of temperatures and molalities. Many very accurate methods are available for the determination of solution densities at nearly ambient temperatures. For example, Millero (1967) describes an ingenious magnetic float densimeter that operates somewhat like the hygrometer commonly used to check antifreeze in your car but has a precision of 2 ppm. Measurements at elevated temperatures are necessarily much less precise. Ellis (1966) developed a method to measure the densities of alkali halides in aqueous solution up to 200°C. The method uses twin stainless-steel vessels, each containing the electrolyte solution, connected by capillary tubing. One vessel is kept at 25°C, and the second one is heated in an oil bath to the desired temperature. As the second vessel is heated, some of the solution is transferred through the capillary system to the colder vessel. The amount of solution transferred is proportional to the *difference* between the density of the solution at the oil bath temperature and the density of the solution at 25°C. The latter density can be measured very accurately by more routine procedures. The system is maintained at a constant vapor pressure that is just high enough to prevent boiling.

11-4 SOLUBILITY MEASUREMENTS

The solubility of a mineral is the total amount of the mineral that has dissolved at equilibrium. Solubilities can be useful for obtaining thermodynamic properties including solubility product constants, stability or hydrolysis constants, and activity coefficients. Several examples of solubility calculations are explained in other chapters: calcite (Chapters 8 and 12), gypsum (Chapter 12), and gibbsite (Chapter 9). In these examples the solubilities are corrected for nonideal behavior by calculating activity coefficients of the type shown in Chapter 7. We shall now illustrate another procedure, using the solubility of barite as an example.

If s moles of $BaSO_4$ have dissolved per liter of solution, then, for the reaction

$$BaSO_{4(s)} \rightleftharpoons Ba^{2+}_{(aq)} + SO^{2-}_{4(aq)}$$

we have

$$s = m_{Ba^{2+}} = m_{SO_4^{2-}} \tag{11-30}$$

The solubility-product constant K_{sp} is defined as

$$\begin{aligned}
K_{sp} &= (a_{Ba^{2+}})(a_{SO_4^{2-}}) \\
&= (m_{Ba^{2+}} \gamma_{Ba^{2+}})(m_{SO_4^{2-}} \gamma_{SO_4^{2-}}) \\
&= m_{Ba^{2+}} m_{SO_4^{2-}} \gamma^2_{\pm BaSO_4} \\
&= s^2 \gamma^2_{\pm BaSO_4}
\end{aligned} \tag{11-31}$$

Taking logarithms, we obtain

$$\log K_{sp} = 2 \log s + 2 \log \gamma_{\pm BaSO_4} \tag{11-32}$$

or

$$\log s = \tfrac{1}{2} \log K_{sp} - \log \gamma_{\pm BaSO_4}$$

Using the Debye–Hückel limiting law, we get

$$\log s = \tfrac{1}{2} \log K_{sp} + A_{|z_+ z_-|} \sqrt{I} \tag{11-33}$$

Thus, from equation (11-33), we can see that, if barite solubility has been measured over a range of ionic strength, then a $\log m$ versus $\sqrt{I}$ plot should give a straight line at low concentrations, which would allow extrapolation to the reference state of infinite dilution. The intercept for $I = 0$ will be proportional to the thermodynamic solubility-product constant ($\tfrac{1}{2} \log K_{sp}$ in this example). Neumann (1933) has measured barite solubility over a range of concentrations of several added electrolytes. The data for barite solubility in KCl solutions at low

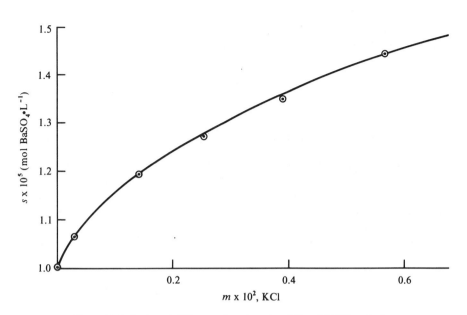

Figure 11-4. Barite solubility as a function of molality of KCl in solution.

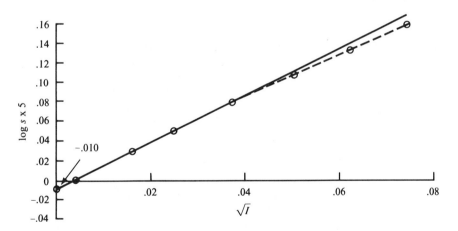

Figure 11-5. Barite solubility data from Figure 11-4 plotted in terms of $\log s$ versus $\sqrt{I}$.

concentrations is shown in Figure 11-4. The data are replotted as $\log s$ versus $\sqrt{I}$ in Figure 11-5. At low ionic strength, the points form a perfect straight line that extrapolates to -5.010, and thus

$$\tfrac{1}{2}\log K_{sp} = -5.010$$

$$\log K_{sp} = -10.020$$

$$K_{sp} = 10^{-10.02}$$

This procedure is a bit easier than calculating activity coefficients and ion-pair constants for every point, but it works only at extremely low concentrations. If ion pairs exist in the system, then an optimal best fit can be made for both the K_{sp} and any ion-pair constants iterating over the whole range of conditions. This procedure requires the use of a reliable model for activity coefficients.

11-5 CONDUCTIVITY MEASUREMENTS

The conductivity of electrolytic solutions played a very important role in the development of electrolyte theory. This type of data helped to formulate many of our current ideas about hydrolysis and the dissociation of dissolved aqueous species. The basic method is quite simple. Two platinum electrodes of 1 cm² surface area each, held 1 cm apart, are placed into the solution to be measured. The electrodes are connected to an appropriate resistance bridge, traditionally a Wheatstone bridge, and the specific resistance R is measured. The resistance is converted to the specific conductance K by reciprocation:

$$K = \frac{1}{R} \, (\text{ohm}^{-1} \cdot \text{cm}^{-1}) \tag{11-34}$$

The equivalent conductance Λ is then obtained by multiplying K by the volume V of solution containing 1 gram-equivalent of the solute:

$$\Lambda = KV \tag{11-35}$$

If an electrolyte shows very little dissociation into ions, then the equivalent conductance will be quite low compared to an electrolyte that fully dissociates because the greater the degree of dissociation, the greater the concentration of charge-conducting ions. The degree of dissociation is expressed by

$$\alpha = \frac{\Lambda}{\Lambda_e} \tag{11-36}$$

where Λ_e represents the equivalent conductance of the fully dissociated electrolyte.

Conductivity measurements can be used to determine stability constants for ion pairs and complexes and to determine solubility-product constants. Shedlovsky and MacInnes (1935) determined the first dissociation constant of carbonic acid by this method in the following manner. For the reaction

$$H_2CO_{3(aq)} \rightleftharpoons H^+_{(aq)} + HCO^-_{3(aq)}$$

the equilibrium constant is

$$K_1 = \frac{a_{H^+} a_{HCO_3^-}}{a_{H_2CO_3}}$$

Substituting the degree of dissociation, α, we find that the concentration of H^+ is

αm, that of HCO_3^- is αm, and that of H_2CO_3 is $(1 - \alpha)m$, so that

$$K_1 = \frac{\alpha^2 m^2 \gamma_{H^+} \gamma_{HCO_3^-}}{(1 - \alpha)m\gamma_{H_2CO_3}} = \frac{\alpha^2 m \gamma_{\pm H_2CO_3}^2}{(1 - \alpha)} \tag{11-37}$$

assuming that $\gamma_{H_2CO_3} \approx 1$. Substituting equation (11-36), we have

$$K_1 = \frac{\Lambda^2 m \gamma_\pm^2}{(1 - \Lambda/\Lambda_e)\Lambda_e^2} = \frac{\Lambda^2 m \gamma_\pm^2}{\Lambda_e(\Lambda_e - \Lambda)} \tag{11-38}$$

We find Λ_e by combining the measured conductances of the ions. Thus, Λ_e for carbonic acid can be obtained from conductivity data on H^+Cl^-, K^+Cl^-, and $K^+HCO_3^-$ measured at the same concentration:

$$\Lambda_{e(H_2CO_3)} = \Lambda_{e(HCl)} - \Lambda_{e(KCl)} + \Lambda_{e(KHCO_3)} \tag{11-39}$$

Equation (11-39) is also plotted as a function of $\sqrt{m}$ to obtain the limiting equivalent conductance Λ_0:

$$\Lambda_e = \Lambda_0 - A\sqrt{m} \tag{11-40}$$

This equation can then be used to calculate Λ_e at low dilution and can be substituted into (11-38) to obtain an apparent dissociation constant. By correcting with an appropriate form of Debye–Hückel equation, one can derive the thermodynamic dissociation constant.

Conductivity measurements are not as reliable as other methods available at low temperatures. The equivalent conductance can give the degree of association only if the ionic mobilities are assumed constant or can be calculated. This assumption is approximately valid for relatively few systems, and the calculations are not as quantitative as one needs to arrive at highly precise equilibrium constants. These measurements work best as solutions approach infinite dilution, but large experimental errors can then arise from sensitivity losses, which are inescapable at such low concentrations.

In contrast, conductivity measurements are one of the best ways to obtain data on the degree of association in aqueous salt solutions at high pressures and temperatures, where difficulties associated with instrumentation and procedures preclude other methods. For example, conductivity measurements have shown that dilute NaCl solutions are very weakly ionized above $400°C$, due primarily to the sharp decline in the dielectric constant of H_2O at high temperatures (Quist and Marshall, 1968). Franck (1973) gives a more general review.

11-6 ELECTROMOTIVE FORCE MEASUREMENTS

Data derived from carefully constructed electrochemical cells can be the most reliable, in terms of both accuracy and precision, of any type of thermodynamic measurements. These measurements are limited, however, to certain types of reactions—i.e., those that transfer electrons at a fast enough rate (and with a sufficient potential difference) that they can be measured under essentially reversible conditions. Most electrochemical measurements are made on aqueous solutions for this reason, but an increasing number have been successfully done using solid-state electrolytes at high temperatures (e.g., Sato, 1971).

The electromotive force of a chemical reaction is measured as electrical energy by a suitable potentiometer, but it must be related to the chemical transfer of electrons in a reaction. We saw in Chapter 3 that reversible electrochemical work is equivalent to the Gibbs free energy change for a reaction:

$$-W_{rev} = \Delta G_r = -nFE$$

By measuring electrode potentials over a range of temperature, we can determine entropies, enthalpies, and heat capacities. Applying the temperature derivative to the equation above, we find

$$\left(\frac{\partial(\Delta G_r)}{\partial T}\right)_P = -nF\left(\frac{\partial E}{\partial T}\right)_P = -\Delta S_r \qquad (11\text{-}41)$$

or

$$\Delta S_r = nF\left(\frac{\partial E}{\partial T}\right)_P$$

Combining (11-41) with (3-105), we obtain

$$\Delta H = \Delta G + T\,\Delta S = -nFE - nFT\left(\frac{\partial E}{\partial T}\right)_P$$

or

$$\Delta H = -nF\left[E - T\left(\frac{\partial E}{\partial T}\right)_P\right] \qquad (11\text{-}42)$$

Cell potentials are commonly of high enough precision to determine the heat capacity:

$$\Delta C_P = \left(\frac{\partial \Delta H}{\partial T}\right)_P = nFT\left(\frac{\partial^2 E}{\partial T^2}\right)_P \qquad (11\text{-}43)$$

The reference cell for all electrode potentials is the standard hydrogen electrode (SHE, cf Section 10-1), but it is safer and more convenient to use other half-cell reactions as practical reference cells for routine laboratory measurements. If a silver chloride electrode is connected to a SHE, we can measure the potential of the silver chloride electrode directly because $E^\circ_{(SHE)} = 0$. We then have constructed the cell

$$Pt|H_2, HCl_{(aq)}|AgCl, Ag$$

The lefthand electrode reaction is

$$\tfrac{1}{2}H_{2(g)} \rightleftarrows H^+_{(aq)} + e^- \qquad (11\text{-}44)$$

and the righthand electrode reaction is

$$AgCl_{(s)} + e^- \rightleftarrows Ag_{(s)} + Cl^-_{(aq)} \tag{11-45}$$

which gives a total cell reaction of

$$AgCl_{(s)} + \tfrac{1}{2}H_{2(g)} \rightleftarrows Ag_{(s)} + H^+_{(aq)} + Cl^-_{(aq)} \tag{11-46}$$

When measured under standard-state conditions, the electrode potential is 0.222 V. Rewriting equation (11-45) as an oxidation reaction, we have

$$Ag_{(s)} + Cl^-_{(aq)} \rightleftarrows AgCl_{(s)} + e^- \qquad E° = -0.222 \text{ V}$$
$$\Delta G° = 21.422 \text{ kJ} \cdot \text{mol}^{-1} \tag{11-47}$$

Now we can use the far more convenient Ag–AgCl reference electrode for the measurement of other potentials and half-cell reactions. For example, the Ag–AgCl electrode is commonly used in ion-selective and combination pH electrodes. For further information, see Bates (1973).

The dissociation constant (K_w) and enthalpy of dissociation of water provide a good example of important thermodynamic properties that can be measured using EMF methods. Covington et al. (1977) used several mixed alkali halide electrolytes of the type MX, where M = Li, Na, K, Cs and X = Cl, Br to study water dissociation. The general electrochemical cell was

$$Pt_1|H_2, MOH(m_1), MX(m_2)|AgX, Ag$$

which is similar to the SHE/Ag–AgCl cell except for the electrolytes. The cell reaction is

$$\tfrac{1}{2}H_{2(g)} \rightleftarrows H^+_{(aq)} + e^-$$
$$\underline{AgX_{(s)} + e^- \rightleftarrows Ag_{(s)} + X^-_{(aq)}}$$
$$\tfrac{1}{2}H_{2(g)} + AgX_{(s)} \rightleftarrows Ag_{(s)} + H^+_{(aq)} + X^-_{(aq)}$$

The Nernst equation is then

$$E = E° - \frac{2.303RT}{F} \log a_{H^+_{(aq)}} a_{X^-_{(aq)}} \tag{11-48}$$

The dissociation constant for water is

$$K_w = \frac{a_{H^+_{(aq)}} a_{OH^-_{(aq)}}}{a_{H_2O_{(l)}}} = \frac{m_{H^+_{(aq)}} \gamma_{H^+_{(aq)}} m_{OH^-_{(aq)}} \gamma_{OH^-_{(aq)}}}{a_{H_2O_{(l)}}}$$

Substituting this expression into (11-48), we find

$$E = E° - \frac{2.303RT}{F} \log \frac{K_w a_{H_2O} m_{X^-_{(aq)}} \gamma_{X^-_{(aq)}}}{m_{OH^-_{(aq)}} \gamma_{OH^-_{(aq)}}} \tag{11-49}$$

Separating terms to

$$E - E^\circ = \frac{-2.303RT}{F} \left[\log K_w + \log a_{H_2O} + \log \frac{m_{X^-}}{m_{OH^-}} + \log \frac{\gamma_{X^-}}{\gamma_{OH^-}} \right] \quad (11\text{-}50)$$

and rearranging, we obtain

$$\frac{F(E - E^\circ)}{2.303RT} + \log \frac{m_{X^-}}{m_{OH^-}} = pK_w - \log \frac{\gamma_{X^-}}{\gamma_{OH^-}} - \log a_{H_2O} \quad (11\text{-}51)$$

in which the left-hand side of the equation contains measurable or known quantities and the right-hand side contains unknowns. At low concentrations, $a_{H_2O} = 1$. Thus, if the left-hand side of (11-51) is calculated from measurements made over a range of ionic strengths, and if it can be extrapolated to infinite dilution, then the intercept will be equal to pK_w if the activity coefficient ratio present on the right-hand side can be evaluated.

As a first approximation, the activity coefficients were estimated by the Brønsted–Guggenheim method, and a series of straight lines resulted that converged to the pK_w at $m = 0$. This plot is shown in Figure 11-6. A better result was obtained by using the Pitzer equations for single electrolytes, and the resulting plot (Figure 11-7) gives a single straight line at $k\,pK_w = 828.45$ mV, or

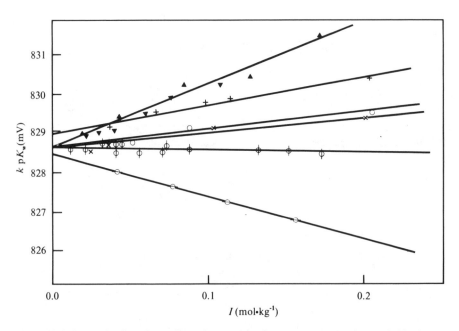

Figure 11-6. Determination of pK_w from electromotive force measurements (see text). (Covington et al., 1977)

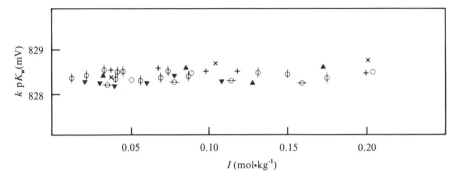

Figure 11-7. Determination of pK_w from electromotive force measurements (see text). (Covington et al., 1977)

$pK_w = 14.004 \pm 0.005$ (where $k = RT\ln(10/F)$). The straight-line results mean that the improved SI method of Pitzer (combined with mixed electrolyte theory) completely accounts for nonideality within the experimental uncertainty and ionic strengths of these measurements ($I < 0.20$). One mixed-electrolyte system was chosen to redetermine the infinite dilution intercept at several temperatures, going through the same procedure. These measurements then provided a value for the enthalpy and heat capacity of dissociation by differentiating $pK_w(T)$ via equations (11-42) and (11-43).

Electrochemical measurements can be used to determine not only dissociation constants as just described, but also activity coefficients and, of course, free energies.

11-7 MOLECULAR DATA

A wide variety of spectroscopic and spectrophotometric methods are available that can be used to measure the atomic, ionic, molecular (vibrational, rotational, translational, and electronic), and magnetic structure of solids, liquids, and gases. Many of these techniques can be used to measure equilibrium constants, although their most valuable function is to elucidate the structure and energetic properties of substances at the atomic scale. Because the techniques and procedures are rather complex and go beyond the intended scope of this book, we simply indicate some of the techniques that have been used and note a few examples here.

Table 11-1 lists the various spectroscopic and spectrophotometric techniques that are useful in deriving thermodynamic quantities. As an example, we note that the thermodynamic functions C_P° and S° for $F_{2(g)}$ reported in the JANAF Tables are completely derived from Raman spectral measurements and vapor-pressure dissociation measurements. In a series of papers on the thermodynamics of minerals based on vibrational spectra, Kieffer (1979*a,b,c*, 1980, 1982)

TABLE 11-1
SPECTROSCOPIC/SPECTROPHOTOMETRIC
TECHNIQUES

Raman spectroscopy
Infrared (IR) spectroscopy
Ultraviolet–visible spectrophotometry
Nuclear magnetic resonance (NMR)
Electron spin resonance (ESR)
X-ray spectroscopy
Microwave spectroscopy

has shown how C_V and C_P values at temperatures above ~ 100 K can be estimated for about 30 major rock-forming minerals. Perusal of the literature of aqueous stability constants shows that a great many have been measured by Raman, IR, or UV–visible spectral methods.

11-8   ESTIMATION TECHNIQUES

A precise and accurate thermodynamic measurement is always preferred over an estimation, but very often a particular property has not been measured or cannot be measured. In these situations, it may be possible to estimate the value, especially if it is readily correlated with reliable chemical analogs. Estimations are typically semiempirical and semitheoretical, having a basis in the molecular properties of the substance. The sounder the theory and the more factors that are known to contribute to the thermodynamic property, the better will be the estimation.

Heat capacity and entropy

Perhaps the earliest thermodynamic estimation was the law of Dulong and Petit (1819), who found that the molal heat capacity of the solid elements is nearly constant at 26 J·mol^{-1}·K^{-1} at ambient temperatures. This observation was followed in 1831 by Neumann and later by Kopp (1864), who proposed that the molal heat capacity of a compound can be estimated from the sum of the molal heat capacities of the constituent elements—i.e.,

$$C_P = \sum v_i C_{P,i} \tag{11-52}$$

where v_i is taken to mean the stoichiometric proportion of element i in the compound.

For covellite at 25°C, we find

$$C_{P,\text{CuS}} = C_{P,\text{Cu}} + C_{P,\text{S}} = 31.80 + 33.15 = 64.95 \text{ J·mol}^{-1}\cdot\text{K}^{-1}$$

which compares well with the experimental value of $66.65 \, \text{J} \cdot \text{mol}^{-1} \cdot \text{K}^{-1}$. Equation (11-52), as we have noted before, assumes that $\Delta C_{P,r} = 0$, which is a good approximation for solid phases. If a gas phase is involved, then Kopp's rule no longer holds. For example, the estimated C_P for fluorite is $56.62 \, \text{J} \cdot \text{mol}^{-1} \cdot \text{K}^{-1}$, compared to the measured value of $68.87 \, \text{J} \cdot \text{mol}^{-1} \cdot \text{K}^{-1}$.

If Kopp's rule of additivity works for heat capacities of some solids, then it should also work for entropies. Latimer (1921, 1951, 1952) argued from the principles of quantum theory and the known relationship between mass and entropy that the entropy of an element in a compound can be approximated by

$$S_i^\circ = \tfrac{3}{2} \ln m_i + S_0' \qquad (11\text{-}53)$$

where m is the elemental atomic mass, and S_0' is a constant for all elements. He then adapted Kopp's heat-capacity law to the estimation of entropies as

$$S^\circ = \sum_i v_i S_i^\circ \qquad (11\text{-}54)$$

which gave reasonable estimates in many cases if he separated anion contributions according to charge as shown in Table 11-2. Now, if we consider fluorite again, the difference becomes: calculated $S^\circ = 75.7 \, \text{J} \cdot \text{mol}^{-1} \cdot \text{K}^{-1}$ compared to experimental $S^\circ = 68.87 \, \text{J} \cdot \text{mol}^{-1} \cdot \text{K}^{-1}$ because Latimer has derived a S_{F^-} contribution from solid compounds rather than from the gaseous element.

Kellogg, Kubaschewski, and Unal (Kubaschewski and Alcock, 1979) have used Latimer's technique for heat capacities (without any valence dependency) with quite good results. From the tables in Kubaschewski and Alcock (1979), we find that the calculated C_P for fluorite is $70.29 \, \text{J} \cdot \text{mol}^{-1} \cdot \text{K}^{-1}$, which compares quite favorably with the experimental value. Estimates of the Maier–Kelley heat-capacity coefficients can also be made from the Kubaschewski tables (Kubaschewski and Alcock, 1979, p. 183).

Because the distribution of bonding energy in a mineral is very difficult to analyze on an elemental basis, and because most rock-forming mineral formulas can be broken down into stoichiometric proportions of oxide components, it makes much more sense to formulate constituent heat-capacity and entropy groups in terms of oxides. This approach was employed by Helgeson (1969) to obtain necessary data for the solubilities and other thermodynamic properties of rock-forming minerals.

Let us consider an example of the use of oxides. Fyfe, Turner, and Verhoogen (1958) suggested the relationship

$$S_i^\circ = \sum v_i S_{i(\text{oxides})} + \hat{K} \, \Delta V_r \qquad (11\text{-}55)$$

where $\sum v_i S_{i(\text{oxides})}$ is the sum of the entropies of the constituent oxides, ΔV_r is the volume change for the reaction $\sum$ oxides $\rightarrow$ compound, and $\hat{K}$ is an empirical fitting constant that has an average value around 2.5 for common silicates if the entropy is in $J \cdot K^{-1} \cdot mol^{-1}$ and the volume change in cm^3. Accordingly, the entropy of akermanite $(Ca_2MgSi_2O_7)$ can be estimated from

$$S^\circ_{akermanite} = 2S^\circ_{CaO} + S^\circ_{MgO} + 2S^\circ_{SiO_2} + 2.5\,\Delta V_r$$

which is equal to $192.95\ J \cdot K^{-1} \cdot mol^{-1}$. Better estimates can be obtained if the summation reaction includes one phase of the same structural class as the unknown mineral. For akermanite,

$$\begin{array}{cccccc}
Ca_2MgSi_2O_7 & + & \alpha\text{-}Al_2O_3 & = & Ca_2Al_2SiO_7 & + & MgO \\
\text{akermanite} & & \text{corundum} & & \text{gehlenite} & & \text{periclase}
\end{array} + \begin{array}{c} SiO_2 \\ \alpha\text{-quartz} \end{array} \quad (11\text{-}56)$$

for which

$$S^\circ_{akermanite} + \Delta S^\circ_r = S^\circ_{gehlenite} + S^\circ_{periclase} + S^\circ_{\alpha\text{-quartz}} - S^\circ_{corundum} \quad (11\text{-}57)$$

The entropy of reaction (ΔS°_r) approaches zero if the righthand side of (11-57) is corrected for the volume change. The algorithm recommended by Helgeson et al. (1978) is

$$S^\circ_i = \frac{S^\circ_{s,i}(V^\circ_{s,i} + V^\circ_i)}{2V^\circ_{s,i}} \quad (11\text{-}58)$$

TABLE 11-2

LATIMER'S CATION/ANION ENTROPY CONTRIBUTIONS TO
SOLID COMPOUNDS AT 298 K (in $J \cdot mol^{-1} \cdot K^{-1}$)

Cations							
Ag	53.6	Dy	60.3	Mn	43.1	Se	48.5
Al	33.5	Br	60.7	Mo	51.5	Si	33.9
As	47.91	Eu	59.0	N	24.3	Sm	59.0
Au	64.0	F	28.9	Na	31.4	Sn	54.8
B	20.5	Fe	43.5	Nd	58.2	Sr	50.2
Ba	57.3	Ga	46.9	Ni	43.9	Ta	62.3
Be	18.0	Gd	59.8	Os	63.2	Tb	59.8
Bi	65.3	Ge	47.3	Pb	64.9	Te	56.1
Br	49.0	Hf	61.9	Pd	53.1	Th	66.5
C	21.8	Hg	64.4	Pr	57.7	Ti	41.0
Ca	39.0						
Cb	51.0	Ho	60.7	Pt	63.6	Tl	64.4
Cd	54.0	I	56.1	Ra	66.1	Tm	61.1
Ce	57.7	In	54.4	Rb	49.8	U	66.9
Cl	36.8	Ir	63.6	Re	62.8	V	42.3
Co	44.4	K	38.5	Rh	52.3	W	62.8
Cr	42.7	La	57.7	Rn	52.3	Y	50.2
Cs	56.9	Li	14.6	S	35.6	Yb	61.5
Cu	45.2	Lu	61.9	Sb	55.2	Zn	45.6
		Mg	31.8	Sc	40.6	Zr	50.6

TABLE 11-2 (Continued)

Anions	Charge on positive ion			
	+1	+2	+3	+4
F^-	23.0	20.0	16.7	20.9
Cl^-	41.8	33.9	28.9	33.9
Br^-	54.4	45.6	38	42
I^-	61.1	56.9	52.3	54.4
CN^-	30.1	25	——	——
OH^-	20.9	18.8	12.6	——
ClO^-	58.6	42	33	——
ClO_2^-	80.3	71.1	58.6	——
ClO_3^-	104.2	83.7	——	——
ClO_4^-	109	92	——	——
BrO_3^-	111	95.8	79.5	——
IO_3^-	107	92	——	——
$H_4IO_6^-$	141.8	126	——	——
NO_2^-	74.5	62.8	——	——
NO_3^-	90.8	74.1	62.8	58.6
VO_3^-	83.7	75.3	——	——
MnO_4^-	133	117	——	——
O^{2-}	10.0	2.1	2.1	4.2
S^{2-}	34.3	21	5.4	10.5
Se^{2-}	67	47.7	33	——
Te^{2-}	69.0	50.6	38	——
CO_3^{2-}	63.6	47.7	33	——
SO_3^{2-}	79.5	62.3	46	——
$C_2O_4^{2-}$	92	74.1	58.6	——
SO_4^{2-}	92	72	57.3	50.2
CrO_4^{2-}	109	87.9	——	——
SiO_4^{2-}	79.5	57.7	38	33.0
SiO_3^{2-}	70.3	43.9	29.3	——
PO_4^{3-}	100	71.1	50.2	——
HCO_3^-	72.8	54.4	42	——
$H_2PO_4^-$	95.4	75.3	——	——
$H_2AsO_4^-$	105	87.9	——	——

Source: Adapted from Latimer (1952), p. 361–363.

where S_i° and V_i° refer to the entropy and volume of the phase to be estimated, $S_{s,i}^\circ$ refers to the sum of the righthand side of (11-57), and $V_{s,i}^\circ$ refers to an analogous summation written for volumes. When equation (11-57) is applied to akermanite, the new estimate for S° is 209.5 $J \cdot mol^{-1} \cdot K^{-1}$, which is virtually identical to the measured calorimetric value (209.3 $\pm$ 2.1 $J \cdot mol^{-1} \cdot K^{-1}$, Weller and Kelley, 1963). This procedure gives remarkably good estimates for nonferrous compounds.

A further improvement was obtained by Robinson and Haas (1983), who considered estimates of heat capacity, relative enthalpy, and calorimetric entropy for silicate minerals based on heat capacities and entropies for fictive oxide components of specific structural groups. For example, Mg may occur in fourfold, sixfold, or eightfold coordination, so that MgO-4, MgO-6, and MgO-8 would each be a separate fictive component. The fictive components were obtained by simultaneous least-squares regression on weighted data. The fitting procedure reflects a precision better than 2% for heat capacity and relative enthalpy and better than 5% for entropy, a significant improvement over other techniques. Heat-capacity, calorimetric-entropy, and relative-enthalpy estimates are made by summing the fictive components in the appropriate stoichiometric proportions. Estimates for acmite and illite (two minerals not used in the fitting procedure) deviate less than 3% from the measured calorimetric values.

Heat capacities and entropies of gases are readily estimated. Monatomic and diatomic gases have nearly constant heat capacities at 298 K that can be calculated from kinetic theory. For monatomic gases, $C_P \simeq 21 \text{ J} \cdot \text{mol}^{-1} \cdot \text{K}^{-1}$; for diatomic gases, $C_P \simeq 29\text{--}38 \text{ J} \cdot \text{mol}^{-1} \cdot \text{K}^{-1}$, increasing with increasing molecular mass. The same general relations hold for the entropies of monatomic and diatomic gases (see Kubaschewski and Alcock, 1979, p. 192). For polyatomic gases, missing data on vibrational frequencies, bond lengths and angles, and electronic states can often be estimated in order to calculate C_P or S. A simpler and highly effective method is that of Latimer (1921), who used for diatomic gases the equation

$$S_i^\circ = \frac{3}{2} R \ln M_i + \frac{R}{2} \ln M_a + \frac{R}{2} \ln M_b + S_0 \qquad (11\text{-}59)$$

where S_0 is an empirical constant, and a and b represent the elements constituting the diatomic gas i of molecular weight M.

Heat capacities and entropies of aqueous ions and aqueous reactions have also been estimated by several techniques. A number of equations for estimating the entropy of aqueous ions have been proposed (see Lewis and Randall, 1961, pp. 522–525). Most of these equations are based on the molecular mass, the charge, and the radius of the ion. For example, Powell and Latimer (1951) used the equation

$$S^\circ = \frac{3R}{2} \ln M + 37 - 270 \frac{z}{r^2} \qquad (11\text{-}60)$$

where z is the ionic charge, r is an effective radius in angstroms which is defined to be 1.0 Å greater than the crystal radius for negative ions or 2.0 Å greater than the crystal radius for positive ions, and S° is expressed in $\text{cal} \cdot \text{mol}^{-1} \cdot \text{K}^{-1}$.

A powerful method for estimating the entropies and heat capacities of aqueous ions at high temperatures has been proposed by Criss and Cobble

(1964a,b) using a correspondence principle. They found a linear correlation between the "absolute" entropy of an ion at 25°C and the absolute entropy at any given higher temperature. The absolute entropy, S_{abs}°, is related to the conventional entropy, S_{conv}° by the equation

$$S_{abs}^\circ = S_{conv}^\circ - 5Z \quad \text{(in cal} \cdot \text{mol}^{-1} \cdot \text{K}^{-1}) \tag{11-61}$$

where Z is the ionic charge that results from assigning the absolute entropy for $H_{(aq)}^+$ a value of -5.0, an essential requirement for the linear correlation between $S_{T,abs}^\circ$ and $S_{298,abs}^\circ$:

$$S_{T,abs}^\circ = a_T + b_T S_{298,abs}^\circ \tag{11-62}$$

The correspondence principle (equation (11-62)) works very well up to 200°C and may be extrapolated to higher temperatures by assuming that the liner correlation of the a and b parameters with temperature continues to be linear above 200°C. Tables of these parameters are given by Criss and Cobble (1964a). Average ionic heat capacities can also be calculated from the following equation (derived from the correspondence principle):

$$C_P]_{298}^T = \alpha_T + \beta_T S_{298,abs}^\circ \tag{11-63}$$

In fact, a linear correlation exists between absolute ionic entropies and absolute ionic heat capacities at 25°C, although more scatter is inherent in the data than is the case for the other correlations.

Enthalpy and free energy

The basic premises that were pioneered by Kopp and Latimer have been considerably revised and extended in subsequent literature, particularly in the field of geochemistry. As a result, estimates of heat capacities and entropies of phases and ionic species have become fairly commonplace. It is unfortunately a much more difficult matter to obtain estimates for the standard enthalpies and free energies of formation for compounds, especially for chemically and structurally complex phases such as silicates. Even when such estimates are made, they tend to be much less reliable than either heat-capacity or entropy estimates for the same phase. Nonetheless, it is possible to derive enthalpies of very simple compounds by setting up an appropriate cyclic reaction scheme.

One of the best known thermodynamic cycles that can be used to derive enthalpy data is the Born–Haber cycle. Consider the enthalpy of formation of halite from the elements at standard-state conditions. A separate pathway can be derived from the initial state of $Na_{(s)}$ and $Cl_{2(g)}$ to solid, crystalline halite, as shown in Figure 11-8. Elemental sodium can be vaporized, and the enthalpy of

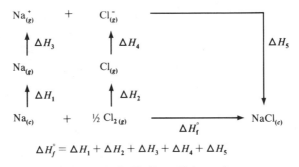

$$\Delta H_f^\circ = \Delta H_1 + \Delta H_2 + \Delta H_3 + \Delta H_4 + \Delta H_5$$

Figure 11-8. The Born–Haber cycle.

vaporization (or sublimation) is ΔH_1. The dissociation enthalpy of

$$\tfrac{1}{2}Cl_{2(g)} \rightarrow Cl_{(g)}$$

is ΔH_2. Then the elements are ionized, and the enthalpy changes in that process are ΔH_3 and ΔH_4. Finally, the gaseous ions are combined to form the crystal lattice of halite, and ΔH_5 is known as the lattice energy. Data for the energies of vaporization, dissociation, ionization, and electron affinity for $Cl_{(g)} \rightarrow Cl_{(g)}^-$ can be found in the literature. Lattice energies can be calculated from theory and from knowledge of the crystal structure as long as the lattice is predominantly ionic (see Evans, 1966, or Swalin, 1972). Thus the Born–Haber cycle can be useful for calculating enthalpies of formation of simple ionic compounds but not for calculating those of complex covalent structures such as silicate minerals. Enthalpies of formation for the alkali halides that are estimated from the Born–Haber cycle differ from experimentally determined values by 15 to 40 $kJ \cdot mol^{-1}$. The differences undoubtedly are due to nonionic contributions to the lattice energy of the crystals. Of course, lattice energies can be calculated by this cycle if the enthalpy of formation is known. Because of the poor agreement between observed and estimated enthalpies, this path is the more reliable one.

A well-behaved relationship exists between the volume of contraction and the enthalpy of formation for structurally simple compounds (Kubaschewski and Alcock, 1979). The volume of contraction is given by

$$\Delta V = \frac{100(k\,\text{MV} - \sum \text{AV})}{\sum \text{AV}}$$

where MV is the molal volume, AV is the atomic volume, and k is an empirical factor related to the lattice structure. However, the uncertainties in this correlation can be rather large for some compounds, and nonionic or complex compounds do not correlate well.

Wilcox and Bromley (1963) found correlations for enthalpies and free energies of inorganic compounds, and they treated complex compounds as equal to the sum of their simplest component parts (similar to the idea of estimating the entropy of a silicate mineral in terms of the sum of its component oxides). When Parks and Nordstrom (1979) applied this rule to a select group of compounds, they found calculated values to be consistently lower than the experimental ones. They dealt with this problem by applying a correction factor derived from this difference to obtain an enthalpy of formation of corderoite ($Hg_3S_2Cl_2$). Solubility and field evidence related to corderoite suggest that the enthalpy value is still too high and needs further correction.

A series of papers have been published on the estimation of Gibbs free energies of formation of silicate minerals by various techniques. Slaughter (1966a,b,c) calculated the free energies of some silicate minerals based on electrostatic principles and semiempirical models. His approach is not as practical as the more empirical methods. Karpov and Kashik (1968) developed a method for estimating the free energies of silicate minerals based on oxide components by grouping them according to crystal structure classifications. Tardy and Garrels (1974) were able to give reasonable estimates of the Gibbs free energies of formation for several layer-silicate minerals by calculating fictive values for the oxide components and $Mg(OH)_2$ component. For example, by taking known free energies of talc, chrysotile, and sepiolite, we have three equations in three unknowns to be solved simultaneously:

talc $\qquad \Delta G^\circ_{f,sil,Mg(OH)_2} + 2\,\Delta G^\circ_{f,sil,MgO} + 4\,\Delta G^\circ_{f,sil,SiO_2} = -5523\ kJ\cdot mol^{-1}$

chrysotile $\qquad 2\,\Delta G^\circ_{f,sil,Mg(OH)_2} + \Delta G^\circ_{f,sil,MgO} + 2\,\Delta G^\circ_{f,sil,SiO_2} = -4038\ kJ\cdot mol^{-1}$

sepiolite $\qquad 2\,\Delta G^\circ_{f,sil,Mg(OH)_2} + 3\,\Delta G^\circ_{f,sil,SiO_2} = -4269.8\ kJ\cdot mol^{-1}$

From these equations, we can derive $\Delta G^\circ_{f,sil,Mg(OH)_2}$, $\Delta G^\circ_{f,sil,MgO}$, and $\Delta G^\circ_{f,sil,SiO_2}$. From this beginning, Tardy and Garrels were able to derive 19 fictive free energies of oxide components in layer silicates. This method is the same basic approach as that used by Robinson and Haas (1983), except that in the latter study a more consistent data base was used and a more sophisticated fitting procedure was employed. For application to orthosilicates and metasilicates, Tardy and Garrels (1976, 1977) defined the parameter

$$\Delta O^{2-} = \Delta G^\circ_{f,MO_x} - \Delta G^\circ_{f,M^{2x+}} \qquad (11\text{-}64)$$

where x equals one-half the cation valence, and ΔO^{2-} is then the difference between the fictive free energies of the metal oxide and the metal aqueous ion. A linear relation was found when the difference between the free energies of the silicate and its component oxides was correlated with ΔO^{2-}. This correlation provides a means of testing the consistency of thermodynamic data and estimating values for unmeasured minerals. Tardy and Viellard (1977), Tardy and

Gartner (1977), and Tardy (1979) have extended the ΔO^{-2} concept to phosphates, sulfates, and nitrates.

Nriagu (1975) considered hydroxides rather than oxides as the silicate components in layered silicates. An empirical difference function Q was derived and was found to correlated with the charge of the cation so that

$$\Delta G_f^\circ = \sum n_i \Delta G_{f,i,\text{hoc}}^\circ - \left(\sum n_i Z_i - 12\right) \Delta G_{f,i,\text{H}_2\text{O}}^\circ - Q \qquad (11\text{-}65)$$

where hoc represents the hydroxide component in the silicate, Z_i is the charge on the ith cation, and $Q = 0.39(\sum n_i Z_i - 12)$. Mattigod and Sposito (1978) improved Nriagu's equation for smectites by simultaneously correlating Q with the cation exchange capacity of the mineral, the radius of the exchange cation, and the charge of the exchangeable cation. In all of these calculations

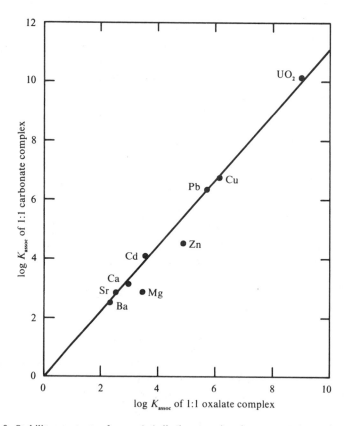

Figure 11-9. Stability constants of some 1:1 divalent metal carbonate complexes plotted against stability constants for the corresponding oxalate complexes. Data are for $I = 0$. (Langmuir, 1979)

involving smectite-like clay minerals, the assumption of stable thermodynamic equilibrium is always questionable. As Mattigod and Sposito (1978) point out, "If indeed the smectite were truly unstable solid solutions as Lippman (1977) has proposed, the thermodyamic significance of ΔG_f° for these minerals would be open to serious question."

Chen (1975) proposed a novel method of estimating free energies that does not require a "calibration" of any sort but uses an extrapolation procedure based on the correlation

$$\sum \Delta G_{f,i}^\circ = ae^{bx} + c \tag{11-66}$$

in which a, b, and c are empirical fits, and x is the integer index for a reaction involving components of the given mineral. Because silicates can be ascribed components at various levels of structural complexities (e.g., subgrouped as silicates, oxides, or elements), each level is assigned an integer index according to the complexity of the components chosen. The value for c is then found by extrapolation to $x = 0$ and differs from the ΔG_f° of the mineral by less than 0.6%.

Finally, Langmuir (1979) has reviewed several methods that have been tried to correlate and estimate the association constant of aqueous complexes. These methods can be summarized as follows.

1. *Isovalent–isostructural analogs.* If a group of complexes share the same metal or the same ligand, then there is often a good correlation of the association constants with another group in which a metal or ligand of the same charge and size is substituted. For example, Figure 11-9 shows a correlation of 1:1 oxalate and carbonate complexes, Figure 11-10 shows a correlation of Fe^{3+} and Al^{3+} complexes, and Figure 11-11 shows a correlation of OH^- and F^- complexes.

2. *Ligand valence.* Well-behaved trends of the association constant with ligand valence are useful, as shown by the phosphate ligand in Figure 11-12.

3. *Ligand number.* There is commonly a regular increase in the association constant with the number of ligands in the complex, as shown in Figure 11-13.

4. *Electrostatic models.* Linear correlations exist between $\log K_{assoc}$ and $z_+ z_-/(r_+ + r_-)$, where z_+ is the cationic charge, z_- is the anionic charge, r_+ is the cation radius, and r_- is the anion radius. Various types of electrostatic models can be derived (see Langmuir, 1979) such as the Bjerrum model or the Fuoss model, which give reasonable estimates for a few complexes. A better estimate often can be obtained from an empirical fit to the data.

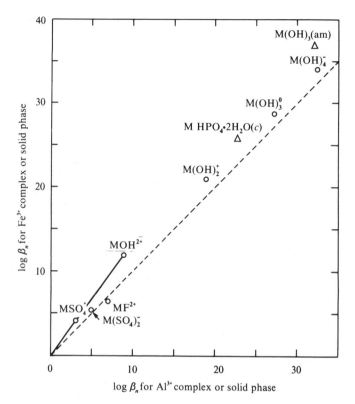

Figure 11-10. Cumulative stability constants for some Fe(III) complexes and solids at $I = 0$ plotted against the constants for corresponding Al(III) complexes and solids. The dashed line is drawn for reference through equal log β_n values for both metals. (Langmuir, 1979)

SUMMARY

- Thermodynamic data can be obtained from many different types of physicochemical measurements. The degree of both precision and accuracy of thermodynamic values retrieved from experiments is a function of (1) differences in experimental procedure and technique, (2) suitability of theoretical assumptions that may be required to calculate the values, and (3) reliability of auxiliary thermodynamic data that may be required.

- The principal experimental methods discussed in this chapter (along with the most common types of thermodynamic data that may be obtained from these experiments) are the following:

  1. calorimetry (heat capacities, third-law entropies, standard enthalpies and free energies, temperature effect on entropy, enthalpy, and free energy);

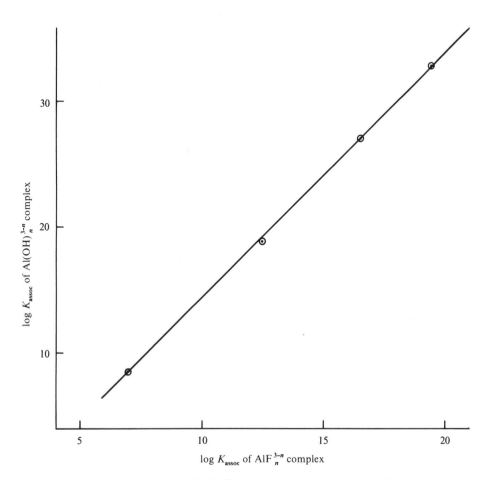

Figure 11-11. Plot of $\log K_{assoc}$ for $Al(OH)_n^{3-n}$ aqueous complexes versus $\log K_{assoc}$ for AlF_n^{3-n} aqueous complexes.

2. determination of temperature and pressure coordinates of univariant reactions (free energies of reaction);
3. volume measurements (effect of pressure on free energy, fugacities of gases and gas mixtures, partial molal volumes of electrolytes);
4. solubility measurements (solubility-product constants, hydrolysis constants, activity coefficients);
5. conductivity measurements (contributions to electrolyte theory, degree of association and/or ion pairing of electrolytes);
6. electromotive-force measurements (free energies of reaction, dissociation constants, activities of ions, activity coefficients);

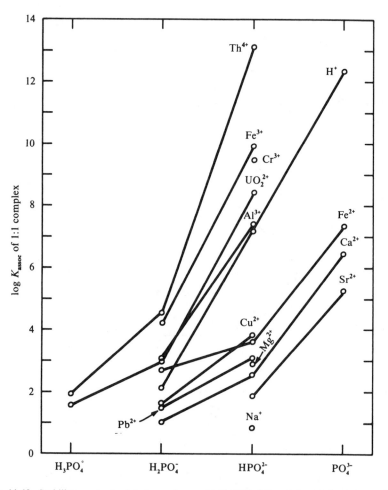

Figure 11-12. Stability constants of $1:1$ complexes with $H_n PO_4^{n-3}$ ligands ($n = 0$ through $n = 3$). All stability constants are for $I = 0$ except values for $ThH_3PO_4^{4+}$ and $UO_2H_3PO_4^{2+}$, which were obtained at $I = 2\,M$. Stability constants for $CoHPO_4^0$, $NiHPO_4^0$, and $ZnHPO_4^0$ (with log K_{assoc} values of 3.0, 2.9, and 3.2, respectively) are not plotted to avoid cluttering the figure. The constants for Al^{3+} species are estimates. (Langmuir, 1979)

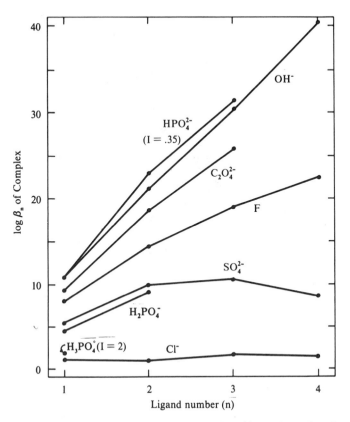

Figure 11-13. Cumulative formation constants of monomeric Th^{4+} complexes plotted against their ligand numbers. All the data are for $I = 0$ except where otherwise indicated. (Langmuir, 1979)

7. spectroscopic and spectrophotometric measurements (equilibrium constants, heat capacities).

- Many techniques are available for the estimation of thermodynamic data. Estimates are useful when thermodynamic data cannot be obtained in any other way. These methods may be empirical and/or theoretical, and they vary widely in the ability to predict a known value. Considerable success has been achieved for prediction of both finite entropies and heat capacities, but attempts to predict standard enthalpies and free energies have been less successful.

- A good measurement is better than a good estimate.

Data Acquisition

12

THERMODYNAMIC DATA: COMPILATION AND EVALUATION

No computer program can be devised that will automatically *evaluate all data supplied,* automatically *remove or correct errors, and* automatically *return* the best *set of fitted parameters for the chemical species in a given system. Evaluation of experimental data must, by its very nature, be an interactive process between the compiler and the machine.*

J. L. HAAS, JR., and J. R. FISHER (1976)

The proliferation of scientific and technical data has never been greater than it is today. The number of scientific journals, monographs, continuing series, and handbooks has grown exponentially during this century. Consequently, storage and retrieval of information are often overwhelming and tedious tasks, although the difficulties are softened by advances in computer technology. Critical evaluation of data lags seriously behind the production of data, and thermodynamic databases are no exception. Numerous incorrect values abound in the literature, and the naive acceptance of a value can have dire societal consequences (Stockmayer, 1978; Lide, 1981). Stockmayer (1978) has summarized the problem very concisely:

> Uncritical acceptance of bad scientific information can lead to social penalties.... A particularly pernicious aspect of this problem involves numerical data, which are essential in all branches of science and technology and are often needed to arrive at valid operational decisions. Unfortunately, the scientific literature contains many erroneous values. Few scientists or engineers seem to have given much thought to the magnitude of the problem, and some probably regard every numerical entry in a handbook as revealed truth. Yet anyone who has had to seek a particular number in the literature and searched

out a dozen or more reports, only to end up with a set of widely disparate values, comes to realize that a substantial intellectual effort and a considerable background in the field are needed to arrive at reliable figures.

This chapter is an introduction to the sources of compiled thermodynamic data and methods of evaluating data. A comprehensive discussion of these topics is beyond the scope of this book, but the general guidelines that you find here should provide some confidence in searching and examining thermodynamic data.

12-1 MAJOR COMPILATIONS

A large number of books, journals, and miscellaneous reports contain thermodynamic data useful for geochemical studies. We have tried to make these references more accessible by tabulating major substances or properties in Table F-1. If you are interested only in free energies of formation of silicate minerals, then you should check the references listed under "Mineral compounds" and "Inorganic compounds." The references themselves are listed in Appendix F-1. Compilations of data on a single substance or group of substances with a common element have been ignored except for comprehensive or critically evaluated studies. Data on water have been compiled separately in Appendix F-2.

These references cover a very long period of time, and many revisions have been made that have invalidated some of the older data. For example, the enthalpy of quartz was revised in the early 1960s, so that compilations published before 1965 will be inconsistent with more-recent sources. The enthalpy of quartz is tied in with the enthalpies and free energies of formation for several silicate minerals through such thermochemical cycles as those described in Chapter 11. Consequently, this revision has had a major affect on the thermodynamic data for minerals. There also have been changes in the temperature scale, energy units (all tables are being converted to joules), and reference-state pressure (from atmospheres to bars to pascals). All of these changes must be taken into account when comparing data from different sources.

Although we would prefer not to bias students toward any particular reference, we recognize the need to point out some of the more reliable and more useful ones. For thermodynamic functions of solid minerals, the compilations by Robie et al. (1978), Stull and·Prophet (1971, the JANAF Tables), Haas et al. (1981), Hemingway et al. (1982), Robinson et al. (1982), and Helgeson et al. (1978) are excellent sources; however, thermodynamic data for aluminum-bearing phases given by Helgeson et al. (1978) are inconsistent with the data from the other sources as explained at the end of this chapter. The CODATA Key Values (CODATA Task Group, 1976) are a carefully evaluated set of enthalpy and entropy data on 125 gases, solids, and aqueous ions that are primarily monatomic and diatomic substances. The National Bureau of Standards has a fairly complete table of thermodynamic properties of substances (solids, liquids, gases, aqueous species, and inorganic and organic compounds) that has recently been revised (Wagman et al., 1982). For aqueous reactions, the work of Sillén and Martell (1964), Smith and Martell (1976), and the IUPAC supplements such as Högfeldt (1982) are exceptionally good. The IUPAC is publishing a Solubility Data Series that will be the most complete and most thorough evaluation of solubilities available. In this series, the aqueous gas solubilities and the aqueous solid solubilities will be of most interest to geochemists. The work of Hultgren et al. (1973) is recommended for data on the elements.

12-2 GENERAL STRATEGIES FOR EVALUATION

There are several levels of data searching and evaluation. Sometimes a number is needed quickly to interpret a much larger problem, and the magnitude of the uncertainty is not so critical that a proper search and evaluation are needed. On the other hand, if the main purpose is to evaluate data as part of a major tabulation to be used by others, then an exhaustive search and evaluation are necessary. A whole range of complexity exists between these two extremes, and a considerable amount of personal judgment may be required to determine how much time to spend on an evaluation. For quick consultation, the major compilations cited previously should be used: the JANAF Tables, the U.S. Geological Survey Tables, the IUPAC Tables, the National Bureau of Standards Tables, the CODATA Tables, and the publications of Helgeson and coworkers. These tables have been carefully reviewed and evaluated by groups or committees of evaluators, who have removed some of the risks of choosing a number. If a more thoroughly researched retrieval is necessary, however, then the original references should be sought out and evaluative techniques such as those we outline here should be applied.

If you can find recently reviewed, compiled, and carefully evaluated numbers for species, such as the data in Hemingway et al. (1982) or Robinson et al. (1982), then there is little reason to pursue a search any further. If you cannot find such an evaluation and you have sought out original sources of data, then do not rely on any single experimental determination, but first consider the agreement or disagreement between all of the *independent* investigations that you can find. This comparison will tell you how discrepant the values are; however, when you find highly consistent data, it usually means either that the same source of data was used or that the same method of measurement was used, and it is not necessarily a measure of accuracy.

For example, suppose that you want to make some thermodynamic calculations involving the free energies and enthalpies of quartz and sillimanite. Three possible sources are the JANAF Tables (Stull and Prophet, 1971), U.S. Geological Survey Bulletin *1452* (Robie et al., 1978), and the National Bureau of Standards Tables (Wagman et al., 1968). Table 12-1 shows the free energy, enthalpy, and entropy for each of these two minerals from each source. For quartz,

TABLE 12-1
STANDARD-STATE DATA FOR QUARTZ AND SILLIMANITE

Datum	USGS	JANAF	NBS
Quartz, SiO_2			
ΔG_f° (kJ·mol^{-1})	-856.288	-856.480	-856.67
ΔH_f° (kJ·mol^{-1})	-910.700	-910.860	-910.94
S° (J·mol^{-1}·K^{-1})	41.46	41.46	41.84
Sillimanite, Al_2SiO_5			
ΔG_f° (kJ·mol^{-1})	-2438.988	-2442.45	-2625.9
ΔH_f° (kJ·mol^{-1})	-2585.760	-2589.10	-2772.3
S° (J·mol^{-1}·K^{-1})	96.11	96.18	96.13

the data are in excellent agreement, and it makes little difference which value is used. These differences may reflect different averaging techniques and different standard state pressures. For sillimanite, however, there is considerable disagreement. The three entropy values are nearly the same (undoubtedly from the same source), but the enthalpies are discrepant by as much as 187 kJ·mol^{-1}! Because the free energies of solids are commonly derived from entropy and enthalpy data, it can be safely assumed that each compiler selected or derived a different "best enthalpy" for sillimanite. More reliable measurements have been made involving sillimanite in heats of reactions, and more recent evaluations have been completed since the publication of these results. These newer data are listed in Table 12-2. Now we find that all three sources are in much better agreement, and there is a high degree of confidence in these values. The discrepancies are within experimental uncertainties with the possible exception of the entropy. In many cases, an evaluation is severely limited by the quantity of data available, and it is quite common for a significant revision within two to five years due to newly published measurements. Another example of error propagation in the thermodynamic properties of aluminosilicate minerals will be discussed at the end of this chapter.

TABLE 12-2
REVISED STANDARD-STATE PROPERTIES FOR SILLIMANITE

	USGS*	JANAF	NBS[†]
ΔG_f° (kJ·mol^{-1})	-2442.060	-2442.45	-2440.99
ΔH_f° (kJ·mol^{-1})	-2588.902	-2589.10	-2587.76
S° (J·mol^{-1}·K^{-1})	95.581	96.18	96.11

* Robinson et al. (1982).
[†] Wagman et al. (1982).

Thermodynamic data can cover a wide range of measured properties, substances, and reactions. A strategy for evaluation often depends on (1) the property of substance to be evaluated, (2) the type, quality, and amount of data available, and (3) the theoretical and empirical tools that can be used in data interpretation. An internationally accepted approach has not been agreed upon, but intensive efforts are currently underway to find such an approach. The general procedure for any evaluation can be summarized as follows.

1. *Search the literature*:
   a) search journals, books, abstracts, and proceedings;
   b) collect and store relevant papers;
   c) compile bibliographic information;
   d) prepare indexes and filing system.
2. *Extract the data*:
   a) annotate and summarize relevant data, experimental precision, and experimental details;
   b) assess the quality of data and techniques.
3. *Convert and code the data* (*if necessary*):
   a) convert to common units;
   b) apply weighting factors to all data where possible;
   c) code for computer evaluation.
4. *Initial comparison of data*:
   a) use standard thermodynamic relations with appropriate temperature, pressure, and/or composition functions to compare the data (computer programs are commonly useful at this stage);
   b) examine comparison for highly inconsistent data and for large deviations from the mean or from the best fit of correlated data;
   c) remove erroneous data;
   d) readjust weighting factors.
5. *Refine and complete data comparison*:
   repeat step 4 until the best fit can be obtained.
6. *Prepare final set of consistent data*:
   a) prepare tables of all substances and their properties, complete with measures of reliability;
   b) include in final dissemination of data fully documented references and methods of evaluation (in sufficient clarity that any reader can repeat your calculations).

A key factor in evaluating data is the determination of its "consistency." This word, unfortunately, has suffered from a variety of slightly different denotations. The phrase "internally consistent" means that a selected set of data has thermodynamic consistency but it may be inconsistent with other sets of data. We prefer to follow the clearest description of consistency, which is discussed in *CODATA Bulletin 47*: "A systematic approach to the preparation of thermodynamic tables."

Thermodynamic consistency can mean at least seven different things:

1. Data are consistent with thermodynamic relationships (the basic laws and their consequences).
2. Common scales are used for temperature, energy, atomic mass, and fundamental physical constants.
3. Conflicts among measurements have been resolved.
4. The same mathematical model is used to fit different data sets.
5. The same chemical model is used to fit different data sets.
6. Appropriate consideration has been given to starting point in applying item 1.
7. Appropriate choice of standard states has been made, and the same standard states have been used for all similar substances.

These items are not necessarily independent of each other. For example, chemical models may include a mathematical model, and both chemical models and mathematical models are constrained by the fundamental thermodynamic relationships, even though they may be extrathermodynamic in origin.

12-3 CONSISTENCY WITH THERMODYNAMIC RELATIONSHIPS

We have already seen that all thermodynamic functions are related through equations that can be derived from the basic laws of thermodynamics and the ideal gas law—e.g., $\Delta G = \Delta H - T \Delta S$ and $\Delta G_r^\circ = -RT \ln K$. Calcite solubility (see Chapter 8) provides a good example. If we choose enthalpy and entropy data for each species in the reaction

$$CaCO_{3(s)} \rightleftharpoons Ca_{(aq)}^{2+} + CO_{3(aq)}^{2-} \tag{12-1}$$

then we can calculate $\Delta G_{r(298\,K,\,1\,bar)}^\circ$ from

$$\Delta G_r^\circ = \Delta H_r^\circ - T \Delta S_r^\circ$$

and

$$\Delta H_r^\circ = \sum_i v_i \Delta H_{f_i}^\circ$$

and

$$\Delta S_r^\circ = \sum_i v_i S_i^\circ$$

Using values listed in Robie et al. (1978; see our Table 8-1), we have $\Delta G_r^\circ = 47.54\ \text{kJ} \cdot \text{mol}^{-1}$, which must be consistent with the value calculated from free energies via

$$\Delta G_r^\circ = \sum_i v_i \Delta G_{f_i}^\circ = 47.40\ \text{kJ} \cdot \text{mol}^{-1}$$

The difference of 140 $\text{J} \cdot \text{mol}^{-1}$ between these values could be due to rounding errors during conversion of the original data from calories to joules. Discrepancies much larger than this would be cause for concern, and they would indicate an inconsistency in the original tabulation of the data.

The next step is to compare ΔG_r° calculated from enthalpies and entropies with that calculated from $\Delta G_r^\circ = -RT \ln K$ and solubility data. Using the data of Plummer and Busenberg (1982), we have $\log K = -8.48$, which gives $\Delta G_r^\circ = 48.41$ kJ·mol^{-1}, a value about 1 kJ·mol^{-1} different from that obtained from the calorimetric data. This inconsistency is pertinent to both items 1 and 3 in the list at the end of Section 12-2, in that it arose by calculating the same thermodynamic property (ΔG_r°) by two independent paths and these paths began from different measurements. To determine which source of data may be more reliable, we can first look at the indicated error of precision associated with each value. From the calorimetric data listed by Robie et al. (1978), we have $\Delta G_r^\circ = 47.40 \pm 1.83$ kJ·mol^{-1}. The error is calculated from the formula

$$\text{standard deviation} = \sqrt{\delta^2} = \sqrt{\delta_a^2 + \delta_b^2}$$

The solubility data are given as $\log K = -8.48 \pm 0.02$, or $\Delta G_r^\circ = 48.41 \pm 0.12$ kJ·mol^{-1}. However, the calorimetric data are reported as two standard deviations, whereas the solubility data are reported as one standard deviation. Our comparison shows that the precision in the solubility data is about an order of magnitude better than that in the calorimetric data because the solubility value is a direct measurement of the reaction we have chosen. Had we chosen the reaction $CaCO_{3(s)} \rightleftharpoons CaO_{(s)} + CO_{2(g)}$, then the calorimetric data would have provided the more precise data. Furthermore, the data taken from Robie et al. (1978) reflect an uncertainty derived from fitting a large set of different experimental measurements. This type of uncertainty will commonly be larger than the uncertainty inherent in one carefully done investigation. The application of errors to evaluated data is not simple, and it can be misleading to compare two different types of errors, as this example shows.

One additional point should be made with reference to calcite dissolution. The interpretation of the solubility data requires a chemical model for the aqueous species—i.e., expressions for activity coefficients and stability constants. The effect of including or excluding stability constants in the interpretation of calcite dissolution has been discussed by Langmuir (1968), Jacobson and Langmuir (1974), Christ et al. (1974), and Plummer and Busenberg (1982). The evolution of thinking on this issue is interesting, and these sources should be read by those pursuing research in this area. The main conclusion to be derived is that very precise measurements are required when studying the effect of an aqueous chemical model on solubility data.

A slightly more complicated evaluation that involves the use of two different aqueous chemical calculations will help to demonstrate the effect of ion pairing. This example also demonstrates a simple iterative approach that is necessary for the ion-association-model calculation.

Example. Find the solubility product constant for gypsum, using three methods: (1) from free energy data; (2) from measured solubility data and mean activity-coefficient data; and (3) from measured solubility data and single free-ion activity coefficients, including ion-pairing effects.

1. *Calculation from free energies of formation.* For the reaction

$$CaSO_4 \cdot 2\,H_2O_{(s)} \rightleftharpoons Ca^{2+}_{(aq)} + SO^{2-}_{4(aq)} + 2\,H_2O_{(l)} \qquad (12\text{-}2)$$

we find the following free energies of formation from Robie et al. (1978):

gypsum	$\Delta G^\circ_{f(298)} =$	$-(1797.197 \pm 4.602)\ kJ \cdot mol^{-1}$
$Ca^{2+}_{(aq)}$	$\Delta G^\circ_{f(298)} =$	$-(553.540 \pm 1.200)\ kJ \cdot mol^{-1}$
$SO^{2-}_{4(aq)}$	$\Delta G^\circ_{f(298)} =$	$-(744.630 \pm 0.120)\ kJ \cdot mol^{-1}$
$H_2O_{(l)}$	$\Delta G^\circ_{f(298)} =$	$-(237.141 \pm 0.084)\ kJ \cdot mol^{-1}$

Thus we have

$$\Delta G^\circ_r = \sum_i \nu_i \Delta G^\circ_{f_i} = \Delta G^\circ_{f(Ca^{2+}_{(aq)})} + \Delta G^\circ_{f(SO^{2-}_{4(aq)})} + 2\Delta G^\circ_{f(H_2O_{(l)})} - \Delta G^\circ_{f(gypsum)}$$

$$= 24.744 \pm 2.38\ kJ \cdot mol^{-1}$$

and

$$\log K = -\frac{\Delta G^\circ_r}{2.303RT} = -\frac{24{,}744}{2.303 \times 8.314 \times 298.15}$$

$$= -4.33 \pm 0.42 \qquad (12\text{-}3)$$

where the uncertainty in each calculated value is taken as 1σ.

2. *Calculation from solubility data and mean activity coefficient.* From the compilation by Linke (1958), we find that Hewlett and Allen have measured the solubility of gypsum in pure water at 25°C to be $s = 2.080\ g \cdot L^{-1}$ with a density of 0.99911. First, we must convert this value to molality:

$$\frac{2.080}{136.142}\ mol \cdot L^{-1} \times \frac{1000}{(1000 \times 0.99911) - 2.080} = 0.01532\ mol \cdot kg^{-1}_{H_2O}$$

This value compares well with a recent, more precise value of $0.01518\ mol \cdot kg^{-1}_{H_2O}$ measured by Lilley and Briggs (1976). The mean activity coefficient has been measured by Lilley and Briggs (1976) to be $\gamma_\pm = 0.338$. The expression for the solubility product constant is

$$\log K = \log a_{Ca^{2+}}\, a_{SO^{2-}_4}\, a^2_{H_2O} \qquad (12\text{-}4)$$

The activity of water can safely be assumed equal to unity. Thus,

$$K = a_{Ca^{2+}}\, a_{SO^{2-}_4} = \gamma_{Ca^{2+}}\, \gamma_{SO^{2-}_4}\, m_{Ca^{2+}}\, m_{SO^{2-}_4} \qquad (12\text{-}5)$$

Using the definition of mean activity coefficient (Chapter 7) and the assumption of congruent dissolution—i.e.,

$$m_{Ca^{2+}} = m_{SO_4^{2-}} = m_{CaSO_4 \cdot 2H_2O_{(aq)}} \tag{12-6}$$

we have

$$K = \gamma_{\pm}^2 m^2 = (0.338)^2(0.01518)^2 = 10^{-4.58} \tag{12-7}$$

$$\log K = -4.58 \pm 0.015 \quad (1\sigma)$$

3. *Calculation from solubility data, ion pairing, and free-ion activity coefficients.* This calculation is a bit more complicated because we must solve for ionic strength and activity coefficients simultaneously with the ion-pairing effect. First, we assume that only formation of $CaSO_{4(aq)}^{\circ}$ ion pairing is important—i.e.,

$$Ca_{(aq)}^{2+} + SO_{4(aq)}^{2-} \rightleftharpoons CaSO_{4(aq)}^{\circ} \tag{12-8}$$

We must then solve simultaneously for the mass-action expression

$$K_1 = \frac{a_{CaSO_{4(aq)}^{\circ}}}{a_{Ca_{(aq)}^{2+}} a_{SO_{4(aq)}^{2-}}} = 10^{-2.30} \tag{12-9}$$

and the mass-balance expressions

$$m_{Ca} = m_{Ca^{2+}} + m_{CaSO_4^{\circ}} \tag{12-10}$$

$$m_{SO_4} = m_{SO_4^{2-}} + m_{CaSO_4^{\circ}} \tag{12-11}$$

Fortunately, in this example,

$$m_{Ca} = m_{SO_4} \tag{12-12}$$

and

$$m_{Ca^{2+}} = m_{SO_4^{2-}} \tag{12-13}$$

We can substitute for the activities in (12-9) and rearrange to solve for $m_{CaSO_4^{\circ}}$:

$$K_1 = \frac{m_{CaSO_4^{\circ}} \gamma_{CaSO_4^{\circ}}}{m_{Ca^{2+}} \gamma_{Ca^{2+}} m_{SO_4^{2-}} \gamma_{SO_4^{2-}}} = \frac{m_{CaSO_4^{\circ}} \gamma_{CaSO_4^{\circ}}}{m_{Ca^{2+}}^2 \gamma_{Ca^{2+}} \gamma_{SO_4^{2-}}} \tag{12-14}$$

$$m_{CaSO_4^{\circ}} = \frac{K_1 m_{Ca^{2+}}^2 \gamma_{Ca^{2+}} \gamma_{SO_4^{2-}}}{\gamma_{CaSO_4^{\circ}}} \tag{12-15}$$

Neutral ion pairs have activity coefficients near unity in dilute solutions, so that

$$m_{CaSO_4^{\circ}} = K_1 m_{Ca^{2+}}^2 \gamma_{Ca^{2+}} \gamma_{SO_4^{2-}} \tag{12-16}$$

Now we can substitute (12-16) into (12-10) to get

$$m_{Ca} = m_{Ca^{2+}} + m_{Ca^{2+}}^2 K_1 \gamma_{Ca^{2+}} \gamma_{SO_4^{2-}} \tag{12-17}$$

To calculate activity coefficients, we need the ionic strength, which is first calculated on the assumption that there is negligible ion association:

$$I = \tfrac{1}{2}[(0.01518) \times 4 + (0.01518) \times 4] = 0.06072$$

From the Truesdell–Jones activity-coefficient equations (Section 7-6), we find that

$$\gamma_{Ca^{2+}} = 0.450 \quad \text{and} \quad \gamma_{SO_4^{2-}} = 0.437$$

so that

$$m_{Ca} = m_{Ca^{2+}} + (0.450)(0.437)(10^{2.30})m_{Ca^{2+}}^2 = 0.01518 \qquad (12\text{-}18)$$

Rearranging to

$$39.24m_{Ca^{2+}}^2 + m_{Ca^{2+}} - 0.01518 = 0 \qquad (12\text{-}19)$$

we need only solve a quadratic equation. The only possible answer is

$$m_{Ca^{2+}} = 0.01069$$

and thus

$$m_{CaSO_4^{\circ}} = 0.00449$$

Because about one-third of the total calcium concentration is taken up by an ion pair, we must recalculate the ionic strength and the activity coefficients. These quantities now become

$$I = 0.04276 \qquad \gamma_{Ca^{2+}} = 0.493 \qquad \gamma_{SO_4^{2-}} = 0.483$$

The quadratic equation changes to

$$47.51m_{Ca^{2+}}^2 + m_{Ca^{2+}} - 0.01518 = 0 \qquad (12\text{-}20)$$

with the solution $m_{Ca^{2+}} = 0.01022$, so that $m_{CaSO_4^{\circ}} = 0.00496$. Another iteration using these values makes a negligible difference:

$$I = 0.04088 \qquad \gamma_{Ca^{2+}} = 0.499 \qquad \gamma_{SO_4^{2-}} = 0.489$$

$$48.69m_{Ca^{2+}}^2 + m_{Ca^{2+}} - 0.01518 = 0 \qquad (12\text{-}21)$$

$$m_{Ca^{2+}} = 0.01016 \qquad m_{CaSO_4^{\circ}} = 0.00502$$

Finally, we can calculate the solubility-product constant

$$a_{Ca^{2+}}a_{SO_4^{-2}} = \gamma_{Ca^{2+}}m_{Ca^{2+}}\gamma_{SO_4^{2-}}m_{SO_4^{2-}}$$

$$= 0.499 \times 0.01016 \times 0.489 \times 0.01016$$

$$= 10^{-4.60}$$

$$\log K = -4.60 \pm 0.02 \quad (1\sigma)$$

which is in excellent agreement with the result from method 2, demonstrating the reliability of the ion-association model at low ionic strength. Where there are several ion pairs present or at least several electrolytes, these calculations require a numerical-approximation procedure that is much more efficiently handled by computer programs such as those described in Appendix E. A comparison of the uncertainties in these three calculations shows excellent agreement for each method. However, the preferred choice would be to determine $\log K_{sp}$ of gypsum from method 2 or 3 because these methods provide more direct measures of the reaction.

Certain thermodynamic expressions are used so much in evaluating data that they have acquired specific names. The "second-law method" is based on the relation

$$\frac{d(\ln K)}{d(1/T)} = \frac{-\Delta H_r^\circ}{R} \tag{12-22}$$

This method calculates heats of reaction from equilibrium constants measured over a temperature range. An example of the second-law method in reverse—i.e., calculating the temperature dependence for an equilibrium constant from the enthalpy—was given in Chapter 8.

The "third-law method" is a relation based on the absolute entropy of the reacting species and uses the "Gibbs-free-energy function," defined as

$$\frac{G_T^\circ - H_{ref}^\circ}{T} \tag{12-23}$$

The reference temperature may be 0 K or 298.15 K. From equation (3-70), we have

$$S_T^\circ = -\frac{G_T^\circ - H_T^\circ}{T}$$

$$= \frac{H_T^\circ - H_{ref}^\circ}{T} - \frac{G_T^\circ - H_{ref}^\circ}{T}$$

and

$$\frac{G_T^\circ - H_{ref}^\circ}{T} = -S_T^\circ + \frac{H_T^\circ - H_{ref}^\circ}{T} \tag{12-24}$$

For solids, the third-law equation is

$$\Delta G_{P,T} = 0 = \Delta H_{298}^\circ + T\,\Delta \left[\frac{G_T^\circ - H_{298}^\circ}{T}\right] + (P - 1)\Delta V_T^\circ$$

The Gibbs-free-energy function is a slowly increasing function of temperature, and it has the advantage over the second-law method that an independent calculation can be made for each data point. With the third-law method, individual data points can be examined for consistency; if they do not fit a simple, consistently increasing function with temperature, then they are erroneous.

Both second-law and third-law methods have been used extensively in the preparation of the JANAF Tables and in the evaluations of the U.S. Geological

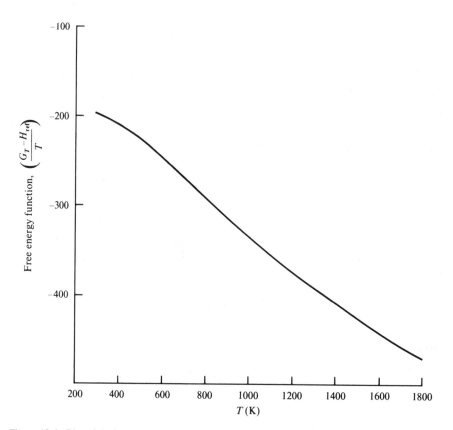

Figure 12-1. Plot of the free-energy function for anorthite versus temperature. (Data from Haas et al., 1981)

Survey, where the reference temperature is 298.15 K. To illustrate the behavior of the Gibbs-free-energy function, we show a plot for anorthite from the compilation of Haas et al. (1981) in Figure 12-1. Note that the function behaves almost linearly above 500 K. This plot is for a smoothed function but, if the actual data points from a series of measurements are plotted in such a manner, erroneous values are often quite visible. Recent examples of the application of the third-law method include Robie et al. (1982) and Krupka et al. (1979).

12-4 CONFLICTS AMONG MEASUREMENTS DUE TO TECHNIQUE PROBLEMS

When serious discrepancies occur between different investigations, there must be a reason for them. Techniques should be evaluated at the start of the evaluation procedure, after all the papers have been compiled. Several aspects of technique should be carefully examined. Some of the more important questions that can be asked are the following.

1. Was the analytical or measuring procedure adequate? Are there known interferences between or misinterpretations of the method(s) used?
2. Were the substances adequately characterized (by spectroscopic, X-ray, optical, or chemical techniques)?
3. Were the raw data reduced adequately, and was the reduction procedure stated?
4. Was enough information given to know just what the experimenter did to arrive at the reported measurements?
5. Was the reaction reversed? Was reversible equilibrium demonstrated?
6. Was particle size measured? Was the particle-size effect known?

Technique problems are best deduced by someone who has worked in the same field of research with the same kind of equipment reported in the reference paper. Consequently, if you are dealing with an evaluation in which a certain paper (or papers) is critical to your choice of thermodynamic properties but you not sure how reliable the technique is, then you had best find someone with *at least* 5 years of experience working with it; you may find a drastic change in your opinions after your consultation.

12-5 EVALUATION OF CONSISTENCY BY COMPUTER PROGRAMS

There are two main advantages of using a computer in the evaluation process: (1) the computer permits the evaluator to compile, store, execute, and retrieve large data sets with comparative ease, *and* (2) it can do simultaneous-regression analysis on large data sets. The difficulties reside in doing simultaneous calculations on large data sets containing several different types of thermochemical measurements and then revising the results when new data are published. Computer calculations are routinely used in many compilations—e.g., data in the JANAF Tables (Stull and Prophet, 1971), the CATCH Tables (Pedley and Rylance, 1977), and the U.S. Geological Survey Tables (Robie et al., 1978).

Garvin et al. (1977) at the National Bureau of Standards have described a least-sums and least-squares computer evaluation of standard-state enthalpies, entropies, and free energies at constant temperature (reference temperature of 298 K)—an evaluation that evolved from the work on the JANAF Tables and the CATCH Tables. In their procedure, there are five basic steps. First, measurements are selected by a data analyst who assigns uncertainties. Second, the data are manually extrapolated to 298.15 K and 1 atm. Third, a least-sums computation is made by finding a simultaneous solution of an equally weighted data set. Fourth, the residual on each measurement is averaged with the preassigned uncertainty to give an "average fit" for each measurement. Finally, a second simultaneous solution is made using a weighted linear least-squares computation where the weight is the reciprocal of the "average fit." The

TABLE 12-3
TYPES OF THERMODYNAMIC MEASUREMENTS FOR
NBS LEAST-SUMS/LEAST-SQUARES PROCEDURE

1. Enthalpy changes

Direct formation of a compound from the elements
Combustion (oxidation) of a compound
Reaction between two (or more) compounds
Decomposition reactions
Solution of a compound
Dilution of a solution
Phase changes
Ionization and appearance potentials

2. Gibbs-free-energy changes (equilibrium data)

Solubility
Dilution
Reaction
EMF cell data

3. Entropy measurements

Absolute entropies (from heat capacities)
Temperature coefficients from equilibria and EMF data

procedure gives the same results as sequential calculations done by hand in which each measurement is considered in sequence without simultaneous least-sums/least-squares solution. The types of measurements that can be handled by this method are listed in Table 12-3.

Another approach developed at the U.S. Geological Survey by Haas and Fisher (1976) simultaneously evaluates and correlates thermochemical data as a function of temperature and pressure with a computer program called PHAS20. The program evaluates input data in the form of thermochemical data as a function of temperature and pressure. In addition to those measurements listed in Table 12-3, it accepts heat capacities, relative heat contents ($H_T - H_{ref}$), and individual measurements of EMF, activity, entropy, etc. at any temperature. This possibility helps to remove from smoothed temperature-dependent functions bias that might occur in the NBS procedure. A general empirical equation for the heat capacity (see Section 12-6) and derived functional relations for S, H, G, and E provide a mathematical model for simultaneous multiple-regression analysis via least squares. Measurements are weighted according to the reciprocal of the square of the standard error for each. A user's manual is available (Haas, 1974), and the program is an essential feature of the data evaluation of minerals.

Excellent examples of its use are provided by the evaluations of 24 mineral phases in the four-component system $CaO-Al_2O_3-SiO_3-H_2O$ at 1 atm by Haas et al. (1981) and at 1 bar by Hemingway et al. (1982). Perusal of these papers demonstrates the extensive amount of effort that must go into any data evaluation and the importance of interaction between the evaluator and the data both before and during the computations. This point cannot be stressed too much. A computer program can store and process data—but only according to the limitations and criteria entered by the programmer. Data evaluation is the responsibility of the evaluator, and a computer is only one tool (along with a sequence of procedures) that can aid in the evaluation process.

Computer programs also can be used to evaluate thermodynamic data by applying aqueous chemical equilibrium computations to solubility data.

12-6   THE MATHEMATICAL MODEL AND THERMODYNAMIC INCONSISTENCIES

There is an empirical or semi-empirical aspect of fitting and evaluating the thermodynamic measurements. Although a convenient set of quantitative expressions exists for relating all of the fundamental thermodynamic functions, there is no universal function for their pressure, temperature, or compositional dependence. For example, we have seen in Chapter 7 how activity coefficients can be expressed by several different semi-empirical models. If two different investigators make exactly the same type of measurements but each uses a different form of activity-coefficient equation to fit the data, then the resulting equilibrium constants will be different. The same conclusion is true when fitting the temperature dependence of the enthalpy, the heat capacity, etc. Temperature-dependent solubility data must be compared with independent calorimetric values of the solid and aqueous species for evaluation but, if the temperature functions used do not share the same common form, then a direct comparison is not possible. This problem raises an important question: *Is there a rational basis for the formulation of a single mathematical model that could reliably be applied to any set of data?* Because the same question pertains to the fitting of temperature, pressure, *and* compositional dependences, we shall discuss in detail only one of these, the temperature dependence.

If we consider the exact thermodynamic relationships between heat capacity C_P, entropy S, enthalpy H, Gibbs free energy G, equilibrium constant K, and electrochemical potential E, we find that all of these functions can be derived from standard-state conventions and the heat capacity by integration. If heat capacities are precisely known as a function of temperature, then the other properties can be calculated very precisely. The reverse process (calculation of heat capacity from another property by differentiation) cannot be done as precisely. Therefore, the temperature dependence of the heat capacity is fundamental to the development of a rational model for fitting thermodynamic properties.

The heat capacity is a fairly well-behaved function that was first fitted by an empirical power-series function:

$$C_P = a + bT + cT^2 + dT^3 + \cdots \tag{12-25}$$

King and Queen (1979) criticized the power-series expansion for its poor fitting and poor extrapolation properties and suggested that a rational function would be a better fitting function for representing data. Maier and Kelley (1932) pointed out the inadequacies of the power-series model, based on their attempts to fit highly precise data on gases, and recommended instead

$$C_P = a + bT - \frac{c}{T^2} \tag{12-26}$$

The Maier–Kelley equation proved very useful until relatively recently. Continued progress in making highly precise measurements over a large temperature range had led to the formulation of two additional terms for equation (12-26) by Haas and Fisher (1976):

$$C_P = a + 2bT + \frac{c}{T^2} + fT^2 + \frac{g}{T^{1/2}} \tag{12-27}$$

Because the temperature dependence of aqueous species and water is a function of the dielectric constant of water, Haas and Fisher (1976) proposed a modified heat-capacity equation

$$C_P = a + 2bT + \frac{c}{T^2} + fT^2 - \frac{gf(T)}{\varepsilon} \tag{12-28}$$

where

$$f(T) = \alpha^2 \exp^2(\beta + \alpha T) + \alpha^2 \exp(\beta + \alpha T) + \frac{2\alpha}{\theta} \exp(\beta + \alpha T) + \frac{1}{\theta^2}$$

ε is the dielectric constant of vapor-saturated liquid H_2O, and α, β, and θ are fitted constants. Equation (12-28) is based on the equation of Helgeson (1967).

Other forms for the temperature dependence have been used. Harned and Robinson (1940) proposed

$$pK_a = \frac{A}{T} - B + CT \tag{12-29}$$

for the acid-ionization constants, whereas Everett and Wynne-Jones (1939) proposed

$$pK_a = \frac{A}{T} - B + C \log T \tag{12-30}$$

Equation (12-29) is consistent with $\Delta C_P = -4.606RCT$, and equation (12-30) is consistent with $\Delta C_P = -2.303RC$. Highly precise gas-solubility data have been fitted to an equation of the form

$$\ln K = a + \frac{b}{T} + \frac{c}{T^2} + \frac{d}{T^3} + \cdots \tag{12-31}$$

by Benson and Krause (1976, 1980) and Benson et al. (1979); equation (12-31) is consistent with

$$\Delta C_P = -\frac{2Rc}{T^2} - \frac{3Rd}{T^3} + \cdots$$

The relationships between C_P, H, S, G, and $\ln K$ are conveniently summarized in Table 12-4. For several given assumptions regarding the temperature dependence of C_P, the resultant $\ln K$ is given. The choice of function depends on over how large a temperature interval the measurements are made, the precision of the measurements, and the type of substances measured. Notice, however, that the Benson–Krause equation does not have a constant, nor does it have a linear term in T for the heat-capacity function. Therefore, it would be valid only for a restricted range of temperature, and it is inconsistent with known functions (such as the Maier–Kelley equation) that are valid for a wider temperature range.

The choice of mathematical function must meet the criterion that it *fit the data points to within the experimental error using the smallest number of terms* (or the smallest number of fitting parameters). If this is the only criterion, then the fit is empirically based, and there is not a unique solution. In addition, using a different function will result in different smoothed data and may lead to a different value at the reference temperature. An example for the equilibrium

$$HSO_4^- \rightleftharpoons H^+ + SO_4^{2-} \tag{12-32}$$

should make this clear. Lietzke et al. (1961) reported experimental values for the log K of reaction (12-32) from 25°C to 225°C. We have fitted this data using a variety of functions and three different least-squares fitting programs: PHAS20 (Haas and Fisher, 1976), POLYNOMIAL APPROXIMATION (Hewlett-Packard Manual), and FITIT (Head, 1970). The results are shown in Table 12-5. Several points should be noted:

1. Different fitting parameters will be obtained when using the same function on the same data if different algorithms are used;
2. Even though the fitting parameters may be different and the smoothed value at any given point may be different, the average value of the absolute magnitude of the deviation is nearly the same (regardless of algorithm) for the same function;
3. The smoothed value at the reference temperature (298.15 K) depends on both the algorithm used and the mathematical function;
4. The data are not precise enough to determine whether a regular power series in log K or a Maier–Kelley-type log K is preferable;
5. The sign on each parameter may be the same or different from function to function or from program to program.

TABLE 12-4

THERMODYNAMICALLY CONSISTENT TEMPERATURE FUNCTIONS

1. $C_P = 0$

   $H = d$ (constant)

   $S = e$ (constant)

   $G = d - eT$

   $\ln K = -\dfrac{D}{RT} + \dfrac{E}{R}$ where $D = \sum_i v_i d_i$ and $E = \sum_i v_i e_i$

2. $C_P = a$ (constant)

   $H = d + aT$

   $S = e + a \ln T$

   $G = (a - e)T - aT \ln T + d$

   $\ln K = \dfrac{E - A}{R} - \dfrac{D}{RT} + \dfrac{A \ln T}{R}$ where $A = \sum_i v_i a_i$

   or

   $\ln K = \bar{A} - \dfrac{\bar{B}}{T} + \bar{C} \ln T$ where $\bar{A} = \dfrac{E - A}{R}$ and $\bar{B} = \dfrac{D}{R}$ and $\bar{C} = \dfrac{A}{R}$

3. $C_P = a + 2bT$

   $H = d + aT + bT^2$

   $S = e + a \ln T + 2bT$

   $G = (a - e)T - AT \ln T + d - bT^2$

   $\ln K = \dfrac{E - A}{R} - \dfrac{D}{RT} + \dfrac{A \ln T}{R} + \dfrac{BT}{R}$

   or

   $\ln K = \bar{A} - \dfrac{\bar{B}}{T} + \bar{C} \ln T + \bar{D}T$ where $\bar{D} = \dfrac{B}{R}$ and $B = \sum_i v_i b_i$

4. $C_P = a + bT + cT^2 + \cdots$ (traditional)

   $H = d + aT + \dfrac{bT^2}{2} + \dfrac{cT^3}{3} + \cdots$

   $S = e + a \ln T + bT + \dfrac{cT^2}{2} + \cdots$

   $G = (a - e)T + d\left(aT \ln T - \dfrac{bT^2}{2} - \dfrac{cT^3}{6} - \cdots \right)$

   $\ln K = \dfrac{E - A}{R} - \dfrac{D}{RT} + \dfrac{A \ln T}{R} + \dfrac{BT}{2R} + \dfrac{CT^2}{6R} + \cdots$

   or

   $\ln K = \bar{A} - \dfrac{\bar{B}}{T} + \bar{C} \ln T + \bar{D}T + \bar{E}T^2 + \cdots$

   where $\bar{D} = \dfrac{B}{R}$ and $\bar{E} = \dfrac{C}{6R}$ and $C = \sum_i v_i c_i$

TABLE 12-4 (Continued)

5. $C_p = a + bT - \dfrac{c}{T^2}$ (Maier–Kelley)

$H = d + aT + \dfrac{bT^2}{2} + \dfrac{c}{T}$

$S = e + a \ln T + bT + \dfrac{c}{2T^2}$

$G = (a - e)T - aT \ln T + d - \dfrac{bT^2}{2} + \dfrac{c}{2T}$

$\ln K = \dfrac{E - A}{R} - \dfrac{D}{RT} + \dfrac{A \ln T}{R} + \dfrac{BT}{2R} - \dfrac{C}{2RT^2}$

or

$\ln K = \bar{A} - \dfrac{\bar{B}}{T} + \bar{C} \ln T + \bar{D}T - \dfrac{\bar{E}}{T^2}$ where $\bar{E} = \dfrac{C}{2R}$

6. $C_p = a + 2bT + \dfrac{c}{T^2} + fT^2 + \dfrac{g}{T^{1/2}}$ (Haas–Fisher)

$H = d + aT + bT^2 - \dfrac{c}{T} + \dfrac{fT^3}{3} + 2gT^{1/2}$

$S = e + a \ln T + 2bT - \dfrac{c}{2T^2} + \dfrac{fT^2}{2} - \dfrac{2g}{T^{1/2}}$

$G = (a - e)T - aT \ln T + d - \dfrac{c}{2T} - bT^2 - \dfrac{fT^3}{6} + 4gT^{1/2}$

$\ln K = \dfrac{E - A}{R} + A \ln T + \dfrac{BT}{R} - \dfrac{D}{RT} + \dfrac{FT^2}{6R} + \dfrac{C}{2RT^2} - \dfrac{4G}{RT^{1/2}}$

or

$\ln K = \bar{A} - \dfrac{\bar{B}}{T} + \bar{C} \ln T + \bar{D}T + \dfrac{\bar{E}}{T^2} + \bar{F}T^2 - \dfrac{\bar{G}}{T^{1/2}}$

where $\bar{D} = \dfrac{B}{R}$ $\bar{E} = \dfrac{C}{2R}$ $F = \sum_i v_i f_i$ $\bar{F} = \dfrac{F}{6R}$

$G = \sum_i v_i g_i$ $\bar{G} = \dfrac{4G}{R}$

Although FITIT, PHAS20, and the calculation of Lietzke et al. (1961) all used the least-squares method for obtaining a best fit, there are obvious differences when using the same function. These differences may be attributable to different numerical approximations in each program and different criteria for convergence (e.g., the deviations may be constant, or they may be a function of the temperature). Also, different weighting procedures can be used, although

TABLE 12-5
VARIOUS FITTED EQUATIONS TO DATA OF LIETZKE et al. (1961) for $HSO_4^- \rightleftharpoons H^+ + SO_4^{2-}$

	Parameters*							Average deviation[†]	$-\log K_{(298)}$
	a	b	c	d	e	f	g		
Lietzke et al.	-5.3505	0.018341	557.246	0	0	0	0	0.044	1.987
PHAS20 (a)	4.3214	-0.017041	-359.002	0	0	0	0	0.044	1.964
PHAS20 (b)	81.8161	0	-2,815.780	-30.0436	0	0	0	0.050	1.969
PHAS20 (c)	-500.904	-0.126127	15,869.85	195.348	0	0	0	0.013	1.904
HP (a)	1.2454	-0.008457	0	0	-7.797×10^{-6}	0	0	0.040	1.969
HP (b)	13.3996	-0.10302	0	0	-2.3344×10^{-4}	0	2.0197×10^{-7}	0.020	1.916
FITIT (a)	1.6476	-0.010524	0	0	-5.1913×10^{-6}	0	0	0.043	1.952
FITIT (b)	15.5399	-0.11991	0	0	2.7717×10^{-4}	0	-2.3908×10^{-7}	0.022	1.908
FITIT (c)	-14.5650	0	6,490.673	0	0	-813,899.6	0	0.064	1.951
FITIT (d)	-561.971	-0.139565	17,803.34	219.028	0	0	0	0.015	1.899

* General form of equation: $-\log K = a + bT + \dfrac{c}{T} + d \log T + eT^2 + \dfrac{f}{T^2} + gT^3$

[†] Average deviation $= \dfrac{\Sigma |d_i|}{n}$, where d_i = calculated − observed

these procedures were kept constant here. Fitting data is not a simple computer exercise, and caution should be employed when using any nonlinear least-squares fitting method.

Returning to the question of what rational basis there might be for a single, universal mathematical model, we find that it has been addressed by Clarke and Glew (1966) with supportive comments by several authors including Alexander et al. (1971), Bolton (1970), and Ramette (1977). Clarke and Glew (1966) derived a "best unbiased" and "general, nonempirical" method of calculating temperature-dependent thermodynamic functions from equilibrium data. Unfortunately, their method is neither nonempirical nor unbiased because it assumes a regularly increasing power series via a Taylor-series expansion of the heat content as a function of temperature. This type of an equation was an arbitrary starting point in their derivation, which naturally leads to a Maier–Kelley basis for the temperature dependence of the thermodynamic functions. Several other functions could have been chosen, and these would also have been mathematically and thermodynamically valid. There are some sound theoretical bases for a preferred form of the heat-capacity function at low temperatures—e.g., the Debye and Einstein functions and expanded models (Debye, 1912; Einstein, 1907; Kieffer, 1979a,b,c, 1980). Statistical-mechanical arguments combined with spectroscopic data may ultimately provide the most appropriate mathematical function for the temperature dependence. Until these relationships have been determined, the Haas–Fisher approach is more than adequate and strongly recommended for the purposes of fitting, correlating, and evaluating data.

12-7 THE CHEMICAL MODEL AND THERMODYNAMIC INCONSISTENCIES

We have already seen that the evaluation of aqueous species and reactions is model dependent, and for these cases the choice is either the ion-association (IA) model or the specific-ion-interaction (SI) model. The gypsum solubility-product calculation showed that the IA model gives the same answer as that obtained from mean activity-coefficient data. Pitzer (1979) and Rogers (1981) have shown that the same answer can be calculated from the SI method. At low ionic strengths, the two methods are (and should be) in excellent agreement. For ionic strengths greater than 0.1, especially in mixed electrolyte solutions, the agreement can be quite poor. At moderate to high ionic strengths, the SI method has been shown to be more reliable and less ambiguous. Although it has not been utilized very much, it would seem highly desirable to use the SI method for routine evaluation of equilibrium constants. Baes and Mesmer (1976) have used a simplified SI equation to evaluate cation-hydrolysis data. Their book is a highly commendable example of removing thermodynamic inconsistencies by keeping the same aqueous chemical model throughout the evaluation. It is not enough to compare the equilibrium constants from different reports because each investigator may have used a different chemical model (choice of ion-pairing constants and choice of activity-coefficient expression) to arrive at the reported constant. Instead, it is necessary to take the raw data and use the same model to fit all results as did Baes and Mesmer (1976).

12-8   APPROPRIATE CHOICE OF STARTING POINT

State functions have the property that allows thermodynamic cycles to be used to calculate one property from another. In some cases these cycles are numerous and constitute large networks of interconnecting data (e.g., Garvin et al., 1977). It is easy in such cases to get lost in these networks and not know where the most reliable starting point is. For example, take the properties of gibbsite. The free energy of formation of gibbsite is commonly calculated from solubility measurements and the free energies of aqueous aluminum and hydroxide ions. Because gibbsite solubility measurements are difficult to make and variable from one investigation to another, so is the free energy. However, the assumption that the free energy of aqueous aluminum is well established proved to be an error. In fact, the free energies of many aqueous ions are not well known. Hemingway and Robie (1977b) showed that choosing a carefully determined solubility and combining that with the free energy of gibbsite from carefully made calorimetric measurements can give a reliable value of the aqueous ion. Starting points in thermodynamic-cycle calculations must always be made from the most precise and most reliable data. Another well-known story involving aluminum brings together several of the points made in this chapter and demonstrates how easy it is to perpetrate incorrect data through thermodynamic cycles. We devote the final section of this chapter to that story.

12-9   THERMODYNAMIC INCONSISTENCIES AMONG ALUMINOUS MINERALS

During the 1970s, considerable discussion erupted in the geochemical literature regarding the most appropriate free-energy values for common aluminous minerals. The factors that underlie this debate provide an instructive review of many of the principles discussed in this chapter. We may begin with Zen (1969), who attempted to retrieve a standard free energy of formation for pyrophyllite based on several sets of previously reported phase-equilibrium experiments. Using techniques similar to those discussed at length in Appendix D, Zen found two different standard free-energy values for pyrophyllite, depending on the reaction studied. Specifically, for

$$\text{pyrophyllite} = \text{andalusite} + 3\text{ quartz} + H_2O \qquad (12\text{-}33)$$

a value of $-5263.5 \text{ kJ} \cdot \text{mol}^{-1}$ was calculated for $\Delta G_{f(\text{pyrophyllite})}^{\circ}$, whereas for

$$\text{kaolinite} + 2\text{ quartz} = \text{pyrophyllite} + H_2O \qquad (12\text{-}34)$$

Zen obtained $-5234.2 \text{ kJ} \cdot \text{mole}^{-1}$. How can this difference of $29.3 \text{ kJ} \cdot \text{mole}^{-1}$ be explained? The retrieval method described in Appendix D returns only a standard free energy of *reaction*. To obtain $\Delta G^{\circ}_{f(\text{pyrophyllite})}$, we have from (12-33)

$$\Delta G^{\circ}_{f(\text{pyrophyllite})} = -\Delta G^{\circ}_{r} + \Delta G^{\circ}_{f(\text{andalusite})} + 3\,\Delta G^{\circ}_{f(\text{quartz})} + \Delta G^{\circ}_{f(\text{steam})} \quad (12\text{-}35)$$

and from (12-34) we have

$$\Delta G^{\circ}_{f(\text{pyrophyllite})} = \Delta G^{\circ}_{r} + \Delta G^{\circ}_{f(\text{kaolinite})} + 2\,\Delta G^{\circ}_{f(\text{quartz})} - \Delta G^{\circ}_{f(\text{steam})}. \quad (12\text{-}36)$$

There are two possibilites to account for the discrepancy: (1) either one or both of the sets of phase-equilibrium data could be in error, or (2) the standard free-energy values added to ΔG°_{f} in (12-35) and (12-36) may be incompatible.

The first possibility can be discarded because temperature-reversal brackets ranging between $\pm 5°C$ and $\pm 20°C$ were reported for both reactions by the experimentalists. Figure 12-2 is a schematic isobaric ΔG_{r}–T diagram that illustrates the relationship between the width of a reversal bracket ($\pm \Delta T$) and the experimental uncertainty in $\Delta G_{r}(\pm \varepsilon)$. The slope of the line is $-\Delta S_{r}$, so the

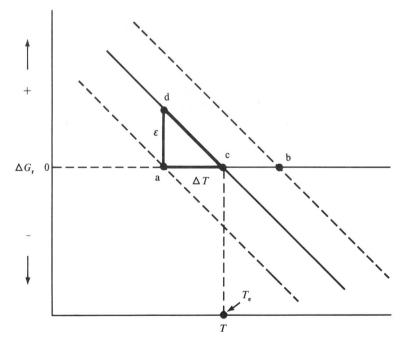

Figure 12-2. Isobaric plot of the free energy ΔG_{r} of a reaction as a function of temperature T, showing the effect of an uncertainty of $\varepsilon \text{ kJ} \cdot \text{mol}^{-1}$ in the free energy of reaction on the calculated equilibrium temperature T_{e}. Because the slope of the reaction line is $-\Delta S_{r}$, the temperature uncertainty ΔT is $\pm \varepsilon/\Delta S_{r}$.

TABLE 12-6

CALORIMETRIC SCHEMES USED BY BARANY AND KELLEY (1961)

Gibbsite, 2 Al(OH)₃

$$2(\text{AlCl}_3 \cdot 6\,\text{H}_2\text{O})_{(c,25)} = 2\,\text{Al}^{+3}_{(s,73.7)} + 6\,\text{Cl}^-_{(s,73.7)} + 12\,\text{H}_2\text{O}_{(s,73.7)} \tag{1}$$

$$70.386\,\text{H}_2\text{O}_{(l,25)} = 70.386\,\text{H}_2\text{O}_{(s,73.7)} \tag{2}$$

$$6(\text{HCl} \cdot 12.731\,\text{H}_2\text{O})_{(l,25)} = 6\,\text{H}^+_{(s,73.7)} + 6\,\text{Cl}^-_{(s,73.7)} + 73.386\,\text{H}_2\text{O}_{(s,73.7)} \tag{3}$$

$$2\,\text{Al(OH)}_{3(c,25)} + 6\,\text{H}^+_{(s,73.7)} = 2\,\text{Al}^{+3}_{(s,73.7)} + 6\,\text{H}_2\text{O}_{(s,73.7)} \tag{4}$$

$$2\,\text{Al}_{(c,25)} + 6(\text{HCl} \cdot 12.731\,\text{H}_2\text{O})_{(s,25)} = 2(\text{AlCl}_3 \cdot 6\,\text{H}_2\text{O})_{(c,25)} + 64.386\,\text{H}_2\text{O}_{(s,25)} + 3\,\text{H}_{2(s,25)} \tag{6}$$

$$\text{H}_{2(g,25)} + \tfrac{1}{2}\,\text{O}_{2(g,25)} = \text{H}_2\text{O}_{(l,25)} \tag{18}$$

$$2\,\text{Al}_{(c,25)} + 3\,\text{O}_{2(g,25)} + 3\,\text{H}_{2(g,25)} = 2\,\text{Al(OH)}_{3(c,25)} \atop \text{gibbsite} \tag{8}$$

$$(\Delta H_8 = \Delta H_1 + \Delta H_2 - \Delta H_3 - \Delta H_4 + \Delta H_6 + 6\Delta H_{18})$$

Kaolinite, Al₂Si₂O₅(OH)₄

$$2(\text{AlCl}_3 \cdot 6\,\text{H}_2\text{O})_{(c,25)} = 2\,\text{Al}^{+3}_{(s,73.7)} + 6\,\text{Cl}^-_{(s,73.7)} + 12\,\text{H}_2\text{O}_{(s,73.7)} \tag{1}$$

$$69.386\,\text{H}_2\text{O}_{(l,25)} = 69.386\,\text{H}_2\text{O}_{(s,73.7)} \tag{12}$$

$$6(\text{HCl} \cdot 12.731\,\text{H}_2\text{O})_{(l,25)} = 6\,\text{H}^+_{(s,73.7)} + 6\,\text{Cl}^-_{(s,73.7)} + 76.386\,\text{H}_2\text{O}_{(s,73.7)} \tag{14}$$

$$\text{Al}_2\text{Si}_2\text{O}_5(\text{OH})_{4(c,25)} + 6\,\text{H}^+_{(s,73.7)} + 12\,\text{HF}_{(s,73.7)} = 2\,\text{Al}^{+3}_{(s,73.7)} + 2\,\text{H}_2\text{SiF}_{6(s,73.7)} + 9\,\text{H}_2\text{O}_{(s,73.7)} \tag{13}$$

$$2\,\text{Al}_{(c,25)} + 6(\text{HCl} \cdot 12.731\,\text{H}_2\text{O})_{(s,25)} = 2(\text{AlCl}_3 \cdot 6\,\text{H}_2\text{O})_{(c,25)} + 64.386\,\text{H}_2\text{O}_{(s,25)} + 3\,\text{H}_{2(g,25)} \tag{6}$$

$$\text{H}_{2(g,25)} + \tfrac{1}{2}\,\text{O}_{2(g,25)} = \text{H}_2\text{O}_{(l,25)} \tag{18}$$

$$\text{Si}_{(c,25)} + \text{O}_{2(g,25)} = \text{SiO}_{2(c,25)} \tag{19}$$

$$2\,\text{SiO}_{2(c,25)} + 12\,\text{HF}_{(s,73.7)} = 2\,\text{H}_2\text{SiF}_{6(s,73.7)} + 4\,\text{H}_2\text{O}_{(s,73.7)} \tag{10}$$

$$2\,\text{Al}_{(c,25)} + 2\,\text{Si}_{(c,25)} + 2\,\text{H}_{2(g)} + \tfrac{9}{2}\,\text{O}_2 = \text{Al}_2\text{Si}_2\text{O}_5(\text{OH})_{4(c,25)} \atop \text{kaolinite} \tag{17}$$

$$(\Delta H_{17} = \Delta H_1 + \Delta H_{12} - \Delta H_{14} - \Delta H_{13} + \Delta H_6 + 5\Delta H_{18} + \Delta H_{19} + \Delta H_{10})$$

Subscripts: $c, 25$ = crystalline phase at 25°C
$g, 25$ = gaseous phases at 25°C
$l, 25$ = liquid phase at 25°C
$s, 73.7$ = subscripted species dissolved in the calorimeter solution at 73.7°C

uncertainty in the free energy of reaction is $\pm\varepsilon = \Delta T\,\Delta S_r$. The entropies of reaction for (12-33) and (12-34) at 298.15 K are 166.9 $J \cdot K^{-1}$ and 142.2 $J \cdot K^{-1}$, respectively (Robie et al., 1978). Assuming the largest possible temperature uncertainty ($\Delta T = 20°$), we find $\pm\varepsilon \approx 3\, kJ \cdot mol^{-1}$. It is clear that the hydrothermal experiments cannot be the principal source of error. Thus the problem must lie with the auxiliary free-energy values. Moreover, because steam and quartz are common to both reactions, it is easy to infer that the standard free energies of andalusite and kaolinite taken by Zen were incompatible. Because both values were based on calorimetric measurements, it is important to unravel their respective calorimetric schemes if we want to uncover the reason for the discrepancy.

The standard enthalpy of formation of andalusite taken by Zen was calculated by dissolving corundum, quartz, and andalusite in a molten salt solvent at 968 K, according to

$$\text{corundum} + \text{quartz} = \text{andalusite} \qquad (12\text{-}37)$$

with a heat of reaction of $-1.99\, kcal \cdot mol^{-1}$ (Holm and Kleppa, 1966). Thus, the standard enthalpy of formation of andalusite at 968 K is

$$\Delta H^{\circ}_{f,968(\text{andalusite})} = -1.99\, \text{kcal} + \Delta H^{\circ}_{f,968(\text{corundum})} + \Delta H^{\circ}_{f,968(\text{quartz})} \qquad (12\text{-}38)$$

All values were corrected to 298.15 K with heat-capacity data. Apparently, the standard calorimetric enthalpy of andalusite depends ultimately on the standard enthalpy of *corundum*, a value that was determined by bomb calorimetry and has subsequently been criticized by Helgeson et al. (1978).

In contrast, the enthalpy value for kaolinite that was adopted by Zen had a much more complicated history. Table 12-6 lists the calorimetric schemes used by Barany and Kelley (1961) to calculate the standard enthalpies of both gibbsite and kaolinite. Note that the Al in both phases is based on measurements involving the phase $AlCl_3 \cdot 6H_2O$—namely, the heat of solution of $AlCl_3 \cdot 6H_2O$ in the calorimeter solvent (reaction (1)) and the enthalpy of formation of $AlCl_3 \cdot 6H_2O$, taken from a previous measurement by Coughlin (1958) (reaction (6)). The subsequent calorimetric schemes for many aluminous minerals (e.g., anorthite, gehlenite, muscovite) were then based either on this gibbsite scheme or on $AlCl_3 \cdot 6H_2O$ itself (e.g., leucite, analcime, halloysite). For this reason, these data have been subject to careful scrutiny in past years.

Hemingway and Robie (1977b) redetermined the standard enthalpy of formation of gibbsite by means of a path that did not involve $AlCl_3 \cdot 6H_2O$. Their value ($-1293.13 \pm 1.19\, J \cdot mol^{-1}$) is more than 11 $kJ \cdot mol^{-2}$ more negative than the value of Barany and Kelley (1961). The following explanations have been offered to account for these differences.

1. The standard enthalpy of formation of $AlCl_3 \cdot 6H_2O$ reported by Coughlin (1958) could be in error. However, Helgeson et al. (1978) assert that this value has been corroborated by more-recent measurements.

2. The enthalpy of solution of $AlCl_3 \cdot 6\,H_2O$ (reaction (1), Table 12-6) could be in error. Hemingway and Robie (1977*b*) considered this possibility, emphasizing that the phase is hygroscopic and could have absorbed water prior to being dropped into the calorimeter. It seems unlikely that this could account for the entire 11 $kJ \cdot mol^{-1}$ difference, however.

3. The heats of solution of liquid H_2O in the calorimeter solution (reaction (2) and (12), Table 12-6) could be in error. Hemingway and Robie (1977*b*) suggest that the value reported by Barany and Kelley of 3507 $J \cdot (mol\ H_2O)^{-1}$ is too low, and they cite a more recent value of 3598 $J \cdot mol^{-1}$. Although this seems like a very small difference, the stoichiometric coefficient of $70.386\,H_2O$ magnifies the discrepancy to more than $6\,kJ \cdot mol^{-1}$.

4. Hemingway and Robie (1977*b*) also note that the enthalpy of solution of $HCl \cdot 12.731\,H_2O$ was in error. This result combined with the differences in the enthalpy of solution of H_2O accounts for most of the difference.

This brings us to the end of our detective story. Unfortunately, however, many conflicting standard enthalpy values for aluminous minerals are still in circulation. For example, Helgeson et al. (1978) published a set of standard thermodynamic data based primarily on a careful analysis of many reversed-phase-equilibrium experiments. The standard enthalpies of most aluminous minerals in that compilation differ substantially from those in other very careful compilations, such as that of Haas et al. (1981). The latter authors used the PHAS20 computer program to perform simultaneous regression analysis on many types of carefully evaluated thermodynamic data. Hemingway et al. (1982) reviewed these discrepancies and concluded that the values presented by Helgeson et al. (1978) for standard enthalpies and free energies of aluminous minerals contain a systematic error of about 6.5 $kJ \cdot (mol\ Al)^{-1}$—gibbsite being the only exception. Hemingway et al. (1982) contend that the source of the error can be traced to an incorrect value adopted for the reaction kaolinite + H_2O = 2 gibbsite + $2\,SiO_{2(aq)}$, which was subsequently used by Helgeson et al. (1978) to calculate the enthalpy of formation of kaolinite. Kaolinite was used as secondary reference phase to tie together the free-energy values obtained from mineral reactions in different sets of experiments. Thus, the standard enthalpy and free-energy values of other aluminous minerals were adjusted to conform to the kaolinite reference state.

The preceding discussion shows how experimental errors and incompatibilities in reference state can combine to produce widely differing standard thermodynamic values for phases. Can we ever know which values are really "correct," or at least which values are most accurate? Because arbitrary reference states are required by thermodynamics, the answer to this question is not as simple as it first appears. However, as we have tried to stress in Chapters 11 and 12, the best data are likely to emerge from careful *simultaneous* evaluation of all pertinent physicochemical data—calorimetric, phase-equilibrium, solubility, electrochemical, etc.—for as many related phases as possible. Because great effort is being expended in this direction, it seems likely that future compilations of thermodynamic data for substances of geochemical interest should be of very high quality.

SUMMARY

- Compilation and evaluation of thermodynamic data require a considerable sacrifice of time and effort.

- A reliable thermodynamic data set must be internally consistent with respect to
  1. fundamental thermodynamic relations;
  2. common scales for temperature, energy, and the fundamental constants;
  3. resolution of experimental conflicts or inconsistencies;
  4. common mathematical and chemical models;
  5. common standard states, and resolution of differences in reference states.

- The same concerns apply when comparing thermodynamic data sets that may be "consistent internally" but are inconsistent with respect to each other. If large discrepancies exist between the two sets, then different reference states may have been used.

Appendix A

UNITS, SYMBOLS, AND FUNDAMENTAL CONSTANTS

The proper use of units and symbols is essential to the application of thermodynamic equations. Symbols are the carriers of meaning for any language, and the language of thermodynamics uses a variety of subscripts, superscripts, and letters. This book follows the recommendations of the International Union of Pure and Applied Chemistry (IUPAC) for most, but not all, of its symbols and units. The IUPAC has recommended SI units for all scientific communications. One example where this book departs from SI units is the unit for pressure. Pressure is force per unit area. In SI units, force is measured in $kg \cdot m \cdot s^{-2}$, which is defined as the newton (N). Thus, the SI unit for pressure is newtons per square meter ($N \cdot m^{-2}$), or *pascals* (Pa). The advantage of pascal units is that they are consistent with all other SI units because they are derived from them. The disadvantage is that pressures commonly encountered in geochemistry are cumbersome when reported in pascals. For example, a great many thermodynamic measurements have been done at ambient pressure, usually 1 atmosphere (equivalent to the pressure exerted by the atmosphere at sea level). One atmosphere (atm) is equal to 101,325 pascals. The bar is a more convenient unit for geochemists to use than either the pascal or the atmosphere. One bar = 10^5 Pa, which is nearly identical to 1 atm; geochemists now prefer the bar over the atmosphere as a unit because the bar is related to the pascal by a simple order-of-magnitude transformation. The bar is found to be the most popular pressure unit today in both meteorology and geochemistry. For atmospheric processes, millibars (mbar) are commonly used, and for deep-seated geologic processes kilobars (kbar) are appropriate. Although the bar will probably become obsolete, its convenience for geochemical problems strongly justifies its use at this time.

Another example of a common thermodynamic unit that is not compatible with SI units is the calorie. The calorie (cal) was originally defined as the quantity of heat required to raise one gram (1 g) of water from 14.5°C to 15.5°C. This definition has a clear physical significance, but it cannot be easily related to other units. In contrast, the SI energy unit is the joule (1 J = 1 N·m), and the conversion between the two scales is 1 cal = 4.184 J. Unfortunately, both energy units are very common in the thermodynamic literature, and this inconsistency is a potential source of serious confusion, especially for students. When seeking any standard thermodynamic data from the literature, it is important to ascertain whether the reported values are based on the joule or the calorie. We use the joule

TABLE A-1
THERMODYNAMIC UNITS AND SYMBOLS

Quantity	Unit	Symbol
	SI base units	
Length	meter	m
Mass	kilogram	kg
Time	second	s
Electric current	ampere	A
Temperature	kelvin	K
Amount of substance	mole	mol
	Derived units	
Volume	liter	$1\,L = 10^{-3}\,m^3$
Force	newton	$N = kg \cdot m \cdot s^{-2}$
Energy (work, heat)	joule	$J = N \cdot m$
	kilojoule	$1\,kJ = 10^3\,J$
	calorie	$1\,cal = 4.184\,J$
	kilocalorie	$1\,kcal = 10^3\,cal$
Pressure	pascal	$Pa = N \cdot m^{-2}$
	bar	$1\,bar = 10^5\,Pa$
	atmosphere	$1\,atm = 101{,}325\,Pa$
Power	watt	$W = J \cdot s^{-1}$
Electric charge	coulomb	$C = A \cdot s$
Electric potential	volt	$V = W \cdot A^{-1}$

Additional energy conversion factors

$$1\,J = 1\,V \cdot C = 1\,W \cdot s$$
$$= 10^7\,erg$$
$$= 9.869 \times 10^{-3}\,L \cdot atm$$
$$= 0.2390\,cal$$
$$= 9.484 \times 10^{-4}\,BTU\ (\text{British thermal unit})$$

(and kilojoule, kJ) in this book, except when citing data that were originally published in calories.

An enlightening discourse on the development and justification for SI units is given by McGlashan (1973). An equally enlightening criticism of SI units has been offered by Adamson (1978). Both of these papers are also recommended as good examples of creative scientific writing from opposite sides of the same issue. The source for IUPAC symbols and units is the *Manual of Symbols and Terminology for Physicochemical Quantities and Units* (Pergamon Press, 1979). Continuing appendixes to this manual are published by *Pure and Applied Chemistry*, the official journal of the IUPAC. A valuable addition to the manual is "Appendix IV: Notation for states and processes, significance of the word *standard* in chemical thermodynamics, and remarks on functions used in thermodynamic tables" (*Pure Appl. Chem.* 51:393–403).

Table A-1 summarizes the important units and conversion factors you should know. Table A-2 lists generalized thermodynamic quantities and their symbols as used in this book. The fundamental physical constants reproduced in Table A-3 have been evaluated by Cohen and Taylor (1973). Since their evaluation was published, there has been a chemical redetermination of the Faraday constant that has caused physicists to reexamine the elementary charge and Avogadro's number (Diehl, 1979). In this text, we use Cohen and Taylor's values for the elementary charge and the molar gas constant. We use Diehl's value for the Faraday constant, and we then derive the Avogadro constant and the Boltzmann constant.

TABLE A-2
THERMODYNAMIC SYMBOLS

A	Helmholtz free energy
a_i^n	activity of the ith component in the nth phase
$a_\pm$	mean ionic activity
$°C$	degrees Celsius (temperature unit)
C_P	heat capacity per mole at constant pressure
C_V	heat capacity per mole at constant volume
c	number of components
E	electromotive force
E_H	electromotive force measured relative to standard hydrogen electrode
F	Faraday constant
f	number of degrees of freedom
f_i^n	fugacity of the ith component in the nth phase
G	Gibbs free energy
$\bar{G}$	Gibbs free energy per mole

TABLE A-2 (Continued)

ΔG_f°	standard Gibbs free energy of formation
ΔG_r	Gibbs free energy of reaction
$_f\Delta G$	Gibbs free energy of fusion
H	enthalpy
$\bar{H}$	enthalpy per mole
ΔH_f°	standard enthalpy of formation
ΔH_r	enthalpy of reaction
$_f\Delta H$	enthalpy of fusion
$\bar{h}_i$	partial molal enthalpy of the ith component
h	Henry's-law constant
I	ionic strength
IAP	ionic activity product
K	equilibrium constant
k	Boltzmann constant
k	thermal conductivity
M	molarity
m	molality
$m_\pm$	mean ionic molality
N_A	Avogadro's constant
n	number of moles
P	pressure
P_c	critical pressure
P_e	equilibrium pressure
P_r	reduced pressure
p	number of phases
pa_i	$-\log a_i$
pH	$-\log a_{H^+}$
pε	$-\log a_{e^-}$
Q	heat
R	gas constant per mole
S	entropy
$\bar{S}$	entropy per mole
S_0	zero-point or residual entropy
ΔS_e	entropy exchanged between system and surroundings
ΔS_f°	standard entropy of formation
ΔS_i	entropy produced within a system during an irreversible reaction
ΔS_r	entropy of reaction
$_f\Delta S$	entropy of fusion

TABLE A-2 (Continued)

s	solubility
$\bar{s}_i$	partial molal entropy of the ith component
T	thermodynamic temperature
T_c	critical temperature
T_e	equilibrium temperature
T_r	reduced temperature
U	internal energy
V	volume
$\bar{V}$	volume per mole
$\bar{V}_c$	critical volume per mole
ΔV_r	volume change for reaction
ΔV_s	volume change for reaction involving solids only
$\bar{v}_i$	partial molal volume of the ith component
W	work
W_{G_1}, W_{G_2}	interchange parameters in Margules equations
X_i	mole fraction of the ith component
Y	any extensive thermodynamic property
$\bar{Y}$	any extensive thermodynamic property per mole
$\bar{y}_i$	any partial molal property of the ith component
Z	compressibility factor
Z	formula units per unit cell
Z_c	critical compressibility factor
α	coefficient of thermal expansion, $V^{-1}(\partial V/\partial T)_P$
β	isothermal compressibility, $-V^{-1}(\partial V/\partial P)_T$
Γ_i	fugacity coefficient of the ith substance
γ_i	activity coefficient of the ith substance, molality basis
$\gamma_{\pm}$	mean activity coefficient, molality basis
θ	constant in gas-law equation
θ	fraction of ion pairing
Λ	equivalent conductance
Λ_e	equivalent conductance for fully dissociated electrolyte
λ_i	rational activity coefficient, mole-fraction basis
μ_i^n	chemical potential of the ith substance in the nth phase
$\mu_{\pm}$	mean chemical potential of an aqueous species
ν_i	stoichiometric coefficient of the ith substance
ρ	density
ϕ	apparent molal volume
Ω	saturation index (IAP/K)

TABLE A-3
PHYSICAL CONSTANTS

	Fundamental constants	
e	elementary charge	$1.6021892(46) \times 10^{-19}$ C
F	Faraday constant	$96{,}484.56(33)$ C$\cdot$mol^{-1}
R	molar gas constant	$8.31441(26)$ J$\cdot$mol$^{-1}\cdot$K^{-1}
N_A	Avogadro constant	$6.022045(33) \times 10^{23}$ mol^{-1}
k	Boltzmann constant	$1.38063(4) \times 10^{-23}$ J$\cdot$K^{-1}

Commonly used constants	
RT (at 298.15 K)	2478.9 J$\cdot$mol^{-1}
$2.30259R$	19.1447 J$\cdot$K$^{-1}\cdot$mol^{-1}
$\dfrac{RT}{F} \ln 10$ (at 298.15 K)	59.159 mV

Appendix B

EXACT DIFFERENTIALS AND PARTIAL DIFFERENTIATION

B-1 EXACT DIFFERENTIALS

Any expression that may be written as

$$M(x,y)\,dx + N(x,y)\,dy$$

is an *exact differential* if there exists a function $z = f(x,y)$ such that

$$df(x,y) = M(x,y)\,dx + N(x,y)\,dy$$

For example, the total differential of the function $z(x,y)$ is written

$$dz = \left(\frac{\partial z}{\partial x}\right)_y dx + \left(\frac{\partial z}{\partial y}\right)_x dy = M\,dx + N\,dy$$

If dz is an exact differential, then

$$\frac{\partial M}{\partial y} = \frac{\partial N}{\partial x}$$

which is equivalent to

$$\left(\frac{dM}{dy}\right)_x = \left(\frac{dN}{dx}\right)_y$$

This rule of *cross-differentiation* is a necessary and sufficient test for exactness. If dz is exact, then z is a continuous function of x and y, and z is uniquely defined for every value of x and y.

B-2   PARTIAL DIFFERENTIATION

Partial differentials retain operations similar to those in algebra. For instance, if z is constant, then we may write

$$\left(\frac{\partial x}{\partial y}\right)_z \left(\frac{\partial y}{\partial x}\right)_z = 1$$

and

$$\left(\frac{\partial z}{\partial x}\right)_y dx = -\left(\frac{\partial z}{\partial y}\right)_x dy$$

and

$$\left(\frac{\partial z}{\partial x}\right)_y \left(\frac{\partial y}{\partial z}\right)_x = -\left(\frac{\partial y}{\partial x}\right)_z$$

and

$$\left(\frac{\partial x}{\partial y}\right)_z \left(\frac{\partial y}{\partial z}\right)_x \left(\frac{\partial z}{\partial x}\right)_y = -1$$

Appendix C

DERIVATION OF ACTIVITY RELATIONSHIP FOR IDEAL MIXING OF IONS ON CRYSTALLOGRAPHIC SITES

The contribution to the entropy of mixing that arises from exchanging ions on crystallographically and energetically equivalent sites can be derived from statistical thermodynamics. The fundamental equation (see Lewis and Randall, 1961, p. 130) is

$$S_{mix} = \frac{R}{N_A} \ln W \qquad (C-1)$$

where R is the gas constant, N_A is Avogadro's number, and W (the thermodynamic multiplicity) is equal to the total number of quantum states accessible to the system for a given P and T. It can be shown for crystals that W can be evaluated by calculating the number of ways in which a fixed number of different ions can be mixed on a fixed number of crystallographic sites. Because we will be counting individual ions and because entropy is an extensive property, it is important to keep track of the number of sites per formula unit on which mixing occurs. We shall evaluate W in terms of one site per mole but, if there are n sites in the mineral formula, then the total entropy of mixing will be

$$S_{mix} = \frac{nR}{N_A} \ln W \qquad (C-2)$$

Consider now the case of mixing two different ions (labelled 1 and 2—e.g., Mg^{2+} and Fe^{2+}) on the same site. If the numbers of the ions mixed are N_1 and N_2, respectively, and if no other ions occupy the site, then the evaluation of W is a simple problem in permutation: how many ways can N_1 and N_2 ions be distributed over $N_1 + N_2$ distinct positions? The answer from probability theory is

$$W = \frac{(N_1 + N_2)!}{N_1! \, N_2!} \tag{C-3}$$

Substituting back into equation (C-1), we obtain

$$S_{\text{mix}} = \frac{nR}{N_A} [\ln (N_1 + N_2)! - \ln N_1! - \ln N_2!] \tag{C-4}$$

Because N_1 and N_2 are very large numbers (on the order of N_A), the factorial terms can be expanded according to *Stirling's approximation**:

$$S_{\text{mix}} = \frac{nR}{N_A} [(N_1 + N_2)\ln (N_1 + N_2) - (N_1 + N_2) - (N_1 \ln N_1 - N_1)$$

$$- (N_2 \ln N_2 - N_2)] \tag{C-5}$$

By expanding the first term, collecting like terms, and rearranging, we obtain

$$S_{\text{mix}} = \frac{nR}{N_A} \{N_1 [\ln (N_1 + N_2) - \ln N_1] + N_2 [\ln (N_1 + N_2) - \ln N_2]\} \tag{C-6}$$

which is equal to

$$S_{\text{mix}} = -nR \left(\frac{N_1}{N_A} \ln \frac{N_1}{N_1 + N_2} + \frac{N_2}{N_A} \ln \frac{N_2}{N_1 + N_2} \right) \tag{C-7}$$

Because we have calculated W in terms of 1 mol of exchangeable sites, we have $N_1 + N_2 = N_A$, and thus

$$S_{\text{mix}} = -nR(X_1 \ln X_1 + X_2 \ln X_2) \tag{C-8}$$

The free energy of the ideal solution is then (cf 5-50)

$$G_s = X_1 \mu_1^\circ + X_2 \mu_2^\circ + nRT(X_1 \ln X_1 + X_2 \ln X_2) \tag{C-9}$$

Differentiating with respect to X_1, and substituting $X_2 = 1 - X_1$, we obtain

$$\left(\frac{\partial G_s}{\partial X_1} \right)_{P,T} = \mu_1^\circ - \mu_2^\circ + nRT[\ln X_1 - \ln (1 - X_1)] \tag{C-10}$$

Multiplying both sides by $(1 - X_1)$, we obtain

$$(1 - X_1) \left(\frac{\partial G_s}{\partial X_1} \right)_{P,T} = \mu_1^\circ (1 - X_1) - \mu_2^\circ (1 - X_1)$$

$$+ nRT(1 - X_1)[\ln X_1 - \ln (1 - X_1)] \tag{C-11}$$

Turn back to Chapter 5 and review the derivation of equation (5-27) and the

---

* Stirling's approximation: $\ln x! = x \ln x - x$ for very large values of x.

related equations. Substituting G_s for $\bar{V}$ and substituting chemical potentials for partial molal volumes, we have

$$G_s = X_1 \mu_1 + X_2 \mu_2 \tag{C-12}$$

Replacing X_2 with $(1 - X_1)$, differentiating with respect to X_1, and substituting the result back into (C-12), we find that

$$G_s = \mu_1 - \left(\frac{\partial G}{\partial X_1}\right)_{P,T} (1 - X_1) \tag{C-13}$$

which may be rearranged in the form

$$\left(\frac{\partial G_s}{\partial X_1}\right)_{P,T} (1 - X_1) = \mu_1 - G_s \tag{C-14}$$

Substituting (C-14) into (C-11), and substituting for G_s according to (C-9), we are left with

$$\mu_1 - X_1 \mu_1^\circ - (1 - X_1)\mu_2^\circ - nRT[X_1 \ln X_1 + (1 - X_1)\ln(1 - X_1)]$$
$$= \mu_1^\circ(1 - X_1) - \mu_2^\circ(1 - X_1) + nRT(1 - X_1)[\ln X_1 - \ln(1 - X_1)] \tag{C-15}$$

which reduces to

$$\mu_1 = \mu_1^\circ + nRT \ln X_1 \tag{C-16}$$

By comparison with equation (5-62),

$$\mu_1 = \mu_1^\circ + RT \ln a_1 \tag{5-62}$$

we see that

$$\boxed{a_1 = X_1^n} \tag{C-17}$$

Appendix D

RETRIEVAL OF STANDARD THERMODYNAMIC DATA FROM HYDROTHERMAL PHASE EQUILIBRIUM EXPERIMENTS

The equations that are used to extract standard thermodynamic data from reversed phase equilibrium studies (see Chapter 11) are no more than rearrangements of the fundamental relations encountered in earlier chapters. Many different approaches to the problem have been suggested in the literature, but they differ only in the style of rearrangement, the choice of standard states, and the number of simplifying assumptions. At every point on a univariant curve, the fundamental equation is

$$\Delta G_{r(T_e, P_e)} = 0 = \Delta G_r^\circ + RT \ln K_{(T_e, P_e)} \tag{D-1}$$

Taking standard states as pure phases at P_e and T_e, we have

$$\Delta G_{r(T_e, P_e)} = 0 = \Delta G_{r(298,1)}^\circ - \int_{298}^{T_e} \Delta S(T)\, dT + \int_{1}^{P_e} \Delta V_r\, dP + RT \ln K_{(T_e, P_e)} \tag{D-2}$$

D-1 REACTIONS INVOLVING SOLIDS ONLY

The most straightforward case is a reaction involving only solid phases of fixed composition. In this case, the last term in (D-2) equals zero at every temperature and pressure. To illustrate the method, we return one last time to jadeite + quartz = analbite as an example of a solid-state reaction with some interesting complications. This reaction has been directly measured by a large number of investigators, and there is general agreement as to its location in P–T space, which makes it a reasonable candidate for a retrieval exercise. Because such agreement is rare in high-pressure research, it has been suggested that the reaction be used as a calibrant for high-pressure apparatus. Based on reversal brackets at 600°C that were determined at six independent laboratories, Johannes et al. (1971) proposed a "consensus" pressure of 16.3 kbar that is consistent with all six reversal brackets. Although similar careful comparisons have not been made at higher temperatures, Holland (1980) obtained reversals over a very wide temperature range, and his bracket at 600°C (16.0 to 16.5 kbar) includes the consensus pressure. We shall use these data (Table D-1) as an example of how to retrieve a standard free energy of reaction from the coordinates of a univariant curve.

Equation (D-2) cannot strictly apply to *any* of the experimental points in Table D-1 (because a reaction was observed in every case, the pressures and temperatures in the table cannot be equilibrium pressures and temperatures) but must apply exactly to some unknown pressure between the extreme values given for each bracket. Thus, we may use (D-2) to calculate maximum and minimum

TABLE D-1

RETRIEVAL OF A STANDARD FREE ENERGY OF REACTION FOR
JADEITE + QUARTZ = ALBITE

P (kbar)	T (K)	Observation	$(P-1)\Delta V_s$ (kJ)	$\int_{298}^{T} \Delta S(T)\,dT$ (kJ·K^{-1})	$\Delta G_{r(1,298)}$ (kJ)
16.0	873	J + Q → A	27.74	28.87	1.13
16.5	873	A → J + Q	28.61		0.26
21.0	1073	J + Q → A	36.41	38.41	2.00
22.0	1073	A → J + Q	38.15		0.26
23.5	1173	J + Q → A	40.75	42.96	2.21
25.0	1173	A → J + Q	43.35		−0.39
26.0	1273	J + Q → A	45.08	47.34	2.26
27.5	1273	A → J + Q	47.68		−0.34
29.0	1373	J + Q → A	50.29	51.52	1.23
30.0	1373	A → J + Q	52.02		−0.50
31.0	1473	J + Q → A	53.75	55.49	1.74
33.0	1473	A → J + Q	57.22		−1.73
					Mean = 0.68 ± 1.22

Notes: Reversal brackets from Holland (1980, p. 130). J = jadeite; Q = quartz; A = albite. Entropy, volume, and heat-capacity data from Robie et al. (1978). Integrals evaluated by trapezoidal approximation.

values of $\Delta G_{r(298,1)}$. Equation (D-2) can be rearranged to obtain

$$\Delta G_{r(298,1)}^{\circ} \gtrless \int_{298}^{T} \Delta S_r(T)\, dT - \int_{1}^{P} \Delta V_r\, dP \qquad \text{(D-3)}$$

where P and T are the *experimental* pressure and temperature, in contrast to the *equilibrium* pressure P_e and temperature T_e. To solve (D-3), we need entropy and volume data for the reaction. Because of the high pressures involved, better accuracy is assured by making use of compressibility and thermal expansion data for all phases to calculate ΔV_r as a function of P and T. Although some data are available for all phases, large extrapolations from the existing database would be required to cover the observed range in both P and T in Table D-1. To simplify the example, we assume that ΔV_r is constant, although this assumption introduces some error. A more significant problem involves the entropy integral. The degree of Al/Si disorder in albite must be known to calculate the entropy of reaction. The simplest possibilities are complete order (low albite) and complete disorder (analbite). Because of the high temperatures involved, we shall initially assume that analbite was present and proceed with the calculations.

Very accurate heat-capacity data are available for analbite (Hemingway et al., 1981) as well as for quartz and jadeite, so ΔS_r may be calculated exactly as a function of temperature (Figure D-1). It would be a serious error to assume $\Delta C_P = 0$ and use the 298.15 K value for ΔS_r, because ΔS_r falls from $51.47 \text{ J} \cdot \text{K}^{-1}$ at 298.15 K to $40.22 \text{ J} \cdot \text{K}^{-1}$ at 1400 K. Furthermore, it would be

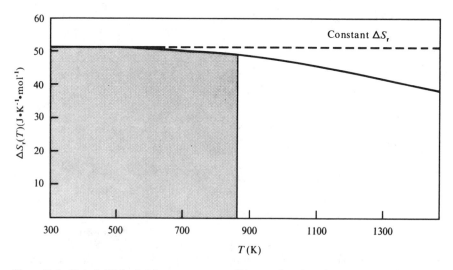

Figure D-1. Plot of ΔS_r for jadeite + quartz = analbite as a function of temperature. The shaded area is equal to $\int_{298}^{873} \Delta S_r(T)\, dT$.

foolish to make this assumption in the light of the available heat-capacity data. The integrand in Figure D-1 can be determined analytically in the usual way—for instance, the shaded area under the curve is equal to the integrand evaluated between 298.15 K and 873 K, which is equal to 28.87 kJ $\cdot$ K^{-1}.

The standard free energy of the reaction can now be determined by subtracting the $P \, \Delta V$ term from the entropy integral (last column in Table D-1). The "true" value of ΔG_r must lie at some intermediate value between those at the high and low pressures for each bracket. Note that the experimental uncertainty in ΔG_r is independent of any assumptions regarding the entropy of the reaction and is proportional to the difference between the minimum and maximum pressures for each bracket. The total experimental uncertainty must include the uncertainties in control and measurement of both pressure and temperature and must be larger than the "bracket" uncertainty. The results in Table D-1 are an example of a "point-by-point" retrieval, because a standard free energy value is obtained from every experimental point. If all six reversal brackets are consistent with the entropy data chosen, then the mean value of 0.68 $\pm$ 1.22 kJ for $\Delta G_{r(1,298)}$ should pass through every bracket. Figure D-2(a) shows that this is indeed the case. This test is very sensitive for judging the internal consistency of a retrieval exercise, as we illustrate in Figure D-2(b,c) by adopting two less desirable assumptions. For instance, if no heat-capacity data had been available, the values of entropy integrals in Table D-1 would have become progressively larger than those shown in the table, because ΔS_r would have been assumed to be equal to $\Delta S_{r(298)}$, or 51.47 J $\cdot$ K^{-1}. This assumption must result in a progressive *increase* in $\Delta G_{r(1,298)}$ as a function of temperature, as shown in Figure D-2(b). The mean value of 2.09 kJ misses one-half of the reversal brackets by a wide mark, so that the result obtained from ignoring heat-capacity effects is unsatisfactory. An even more outrageous example is shown by using both entropy and heat-capacity data for low albite instead of analbite [Figure D-2(c)]. Here, the retrieved standard free energy of reaction shows a 15 kJ range—compelling evidence that ordered albite was not present in any of the experimental runs. Although this conclusion should have been obvious because of the high temperatures involved, this exercise is not as foolish as it might appear, because most of these experiments were of very short duration and because disequilibrium with respect to Al/Si order should always be tested.

The case is not yet closed, however, because other ordering configurations for albite are possible. The most noteworthy additional model is based on aluminum avoidance, which states that Al–O–Al tetrahedral linkages are preferentially avoided in silicates. When applied to analbite, this partial ordering reduces the configurational entropy correction from

$$4R(0.25 \ln 0.25 + 0.75 \ln 0.75) = 18.7 \text{ J} \cdot \text{K}^{-1} \cdot \text{mol}^{-1}$$

representing complete disorder, to something less than 13.1 J $\cdot$ K$^{-1} \cdot$ mol^{-1} (Mazo, 1977). Assuming the Al-avoidance model will reduce the value of the entropy integral in Table D-1 proportionally, the result is a mean value for

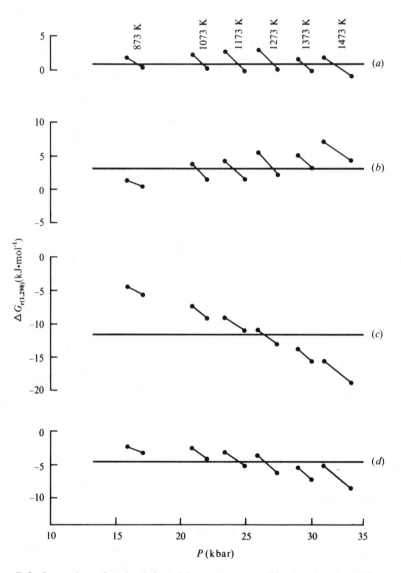

Figure D-2. Comparison of retrievals for jadeite + quartz = analbite, based on four different sets of assumptions: (*a*) analbite, $\Delta C_P \neq 0$; (*b*) analbite, $\Delta C_P = 0$; (*c*) low albite, $\Delta C_P \neq 0$; (*d*) Al-avoidance model for albite, $\Delta C_P \neq 0$. The horizontal lines show the mean $\Delta G_{r(298)}$ value for each set of reversal experiments, and the diagonal lines connect extreme values from reversal pairs.

$\Delta G_{r(1,298)}$ of -4.81 kJ. This mean does not exactly pass through every reversal bracket [Figure D-2(d)] but, in light of the experimental uncertainties and the assumption of constant ΔV_r, this model cannot be excluded. The same conclusion was reached by Hemingway et al. (1981), who calculated the univariant curves based on the completely disordered albite and aluminum-avoidance analbite and compared these curves with all experimental data on the reaction (Figure D-3). Like Figure D-2(d), Figure D-3 suggests that the experiments cannot discriminate between the two models.

Both curves in Figure D-3 were constrained to pass through the "consensus" point of Johannes et al. (1971). This calculation is achieved by integrating the Clapeyron equation

$$\int_{P_r}^{P} dP = \int_{T_r}^{T} \frac{\Delta S_r}{\Delta V_r} dT \tag{D-4}$$

Assuming ΔV_r is constant, we have

$$P - P_r = \frac{1}{\Delta V_r} \int_{T_r}^{T} \Delta S_r(T) dT \tag{D-5}$$

Setting $P_r = 16.3$ kbar and $T_r = 873$ K, we obtain

$$P = 16.3 + \frac{1}{\Delta V_r} \int_{873}^{T} \Delta S_r(T) dT \tag{D-6}$$

Equation (D-6) will generate a family of curves passing through 16.3 kbar and 873 K, depending on the entropy, volume, and heat-capacity data chosen.

This exercise has resulted in retrieval of a standard free energy for the *reaction* jadeite + quartz = analbite at 1 bar and 25°C, but it has not given any information regarding the thermodynamic properties of the individual phases. It is possible, however, to combine the free energy of reaction with any two standard free energies of formation to retrieve an unknown $\Delta G_{f(1,298)}^{\circ}$ for the third phase. Because the standard free energy of quartz is very well known, and because we were obliged to make specific assumptions regarding the structural state of the albite, it makes sense to attempt a calculation for jadeite. The answer clearly will depend on the source of the data for quartz and albite. Taking values from the compilation of Robie et al. (1978) for

$$\Delta G_{f(1,298),\text{jadeite}}^{\circ} = \Delta G_{f(1,298),\text{analbite}}^{\circ} - \Delta G_{f(1,298),\text{quartz}}^{\circ} - \Delta G_{r(1,298)}^{\circ}$$

we obtain a result of -2850.90 kJ·mol^{-1}, which is virtually identical to the value of -2850.83 ± 4.23 kJ·mol^{-1} reported by Robie et al. (1978). The latter value also was obtained by combining calorimetric and phase-equilibrium data, but was based on a much more comprehensive compilation.

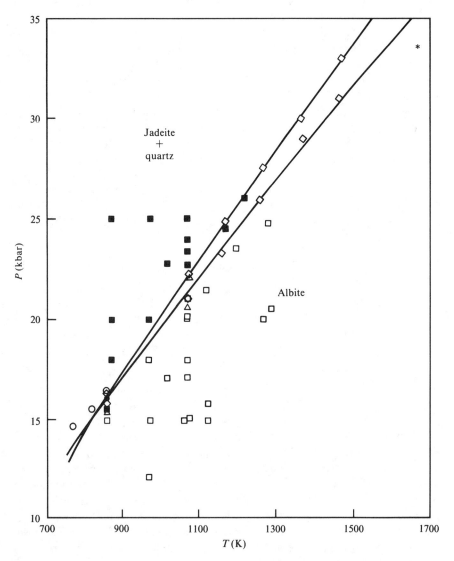

Figure D-3. Experimental data and calculated equilibrium curves for the reaction jadeite + quartz = analbite. The upper curve is calculated on the basis of $S_{0,\text{analbite}} = 18.7 \text{ J} \cdot \text{K}^{-1} \cdot \text{mol}^{-1}$; the lower curve is calculated on the basis of $S_{0,\text{analbite}} = 12.6 \text{ J} \cdot \text{K}^{-1} \cdot \text{mol}^{-1}$ (the Al-avoidance model). Both lines pass through the consensus value of Johannes et al. (1971). Diamonds represent data points of Holland (1980) referred to in the text; other symbols represent data from earlier investigations. (Hemingway et al., 1981)

D-2 REACTIONS INVOLVING SOLIDS AND GASEOUS H₂O

Although many different and equally effective approaches have been used for incorporating the steam phase into retrieval equations, we shall follow Fisher and Zen (1971). Because all dehydration reactions can be written in the form

$$A = B + H_2O \tag{D-7}$$

where A and B represent any number of solid reactant and product phases, it is useful to separate the energetic contributions of the mineral phases from that of the steam phase. At equilibrium,

$$\Delta G_{r(T_e,P_e)} = 0 = \Delta G_{f,s(T_e,P_e)} + \Delta G_{f,H_2O(T_e,P_e)} \tag{D-8}$$

where $\Delta G_{f,s}$ is the free energy of formation for the *solids only*—i.e., $\Delta G_{f,s} = \Delta G_{f,B} - \Delta G_{f,A}$. Introducing standard free energies, we have

$$\Delta G^\circ_{f,s(298,1)} + \Delta G^\circ_{f,H_2O(298,1)} - \int_{298}^{T_e} \Delta S_{f,r}(T)\,dT + \int_{1}^{P_e} \Delta V_{f,r}(P)\,dP = 0 \tag{D-9}$$

where subscripts r refer to the reaction as a whole. We define the entropies of formation ($\Delta S_{f,s}$) and the volumes of formation ($\Delta V_{f,s}$) of the solid phases to be

$$\Delta S_{f,s} = S_{f,B} - S_{f,A} = S_B - S_A + S_{H_2} + \tfrac{1}{2}S_{O_2} \tag{D-10}$$

$$\Delta V_{f,s} = V_{f,B} - V_{f,A} \tag{D-11}$$

Equations (D-9) and (D-10) assume that 1 mol of H_2O is released by dehydration, in accordance with (D-7). The entropy and volume terms can be expanded to

$$\Delta S_{f,r}(T) = \Delta S_{f,s}(T) + \Delta S_{f,H_2O}(T) \tag{D-12}$$

$$\Delta V_{f,r}(P) = \Delta V_{f,s}(P) + \Delta V_{f,H_2O}(P) \tag{D-13}$$

Because the elemental H_2 and O_2 terms that appear in ΔV_f for 1 mol of hydrous phase will exactly cancel identical terms in ΔV_f for 1 mol of H_2O, the righthand side of (D-13) can be simplified to

$$\Delta V_{f,s} + \Delta V_{f,H_2O} = \Delta V_s + V_{H_2O} \tag{D-14}$$

Substituting (D-12), (D-13), and (D-14) into (D-9), we obtain

$$\Delta G^\circ_{f,s(298,1)} + \Delta G^\circ_{f,H_2O(298,1)} - \int_{298}^{T_e} \Delta S_{f,s}(T)\,dT - \int_{298}^{T_e} S_{f,H_2O}(T)\,dT$$

$$+ \int_{1}^{P_e} \Delta V_s(P)\,dP + \int_{1}^{P_e} V_{H_2O}(P)\,dP = 0 \tag{D-15}$$

The integrals involving H_2O may be evaluated independently:

$$-\int_{298}^{T_e} S_{f,H_2O}(T)\,dT = \Delta G_{f,H_2O(T_e,1)}^\circ - \Delta G_{f,H_2O(298,1)}^\circ \qquad \text{(D-16)}$$

$$\int_{1}^{P_e} V_{H_2O}(P)\,dP = G_{H_2O(T_e,P_e)} - G_{H_2O(T_e,1)} \qquad \text{(D-17)}$$

To simplify the solution of these integrals, Fisher and Zen (1971) define $G_{H_2O(T_e,P_e)}^*$ as

$$G_{H_2O(T_e,P_e)}^* \equiv \Delta G_{f,H_2O(T_e,1)}^\circ + G_{H_2O(T_e,P_e)} - G_{H_2O(T_e,1)} \qquad \text{(D-18)}$$

This function can readily be calculated from compilations such as those by Robie et al. (1978), and Burnham et al. (1969a,b); Fisher and Zen (1971, pp. 302–312) have simplified the task by tabulating $G_{H_2O}^*$ (in calories) from 100°C to 1000°C and from 100 to 10,000 bar. Allowing for the customary assumption that ΔV_s is independent of pressure and substituting $G_{H_2O}^*$ into (D-15), we obtain

$$\Delta G_{f,s(298,1)} - \int_{298}^{T_e} \Delta S_{f,s}(T)\,dT + (P_e - 1)\Delta V_s + G_{H_2O(T_e,P_e)}^* = 0 \quad \text{(D-19)}$$

Equation (D-19) has the same form as (D-2). Thus, a pair of extreme values for $\Delta G_{f,s(298,1)}^\circ$ can be retrieved from bracketing experiments on a point-by-point basis so long as entropy, volume, and $G_{H_2O}^*$ data are supplied. As before, you must decide whether to evaluate the entropy integral explicitly. As a general rule, ΔS_r for a dehydration reaction is more sensitive to changes in temperature than $\Delta S_{f,s}$ for the same reaction. Thus $\Delta S_{f,s}$ for a dehydration reaction changes more slowly with temperature than does ΔS_r, and many have advocated replacing the integral in (D-19) by $\Delta S_{f,s}(T - 298.15)$, where T is the experimental temperature in kelvins. Certainly there is no alternative if the heat capacity for one or more phases is not known or cannot be reliably estimated. Figure D-4 contrasts the variation in ΔS_r and $\Delta S_{f,s}$ with temperature for a number of dehydration reactions, and Figure D-5 shows the error in $\Delta G_{f,s(298,1)}$ that would result for each reaction if the assumption $\Delta C_P = 0$ were adopted. Note that at 800 K the error is only about 1 kJ·mol^{-1} for pyrophyllite dehydration but that it is about three times greater for the reaction of muscovite and quartz. Another point to consider is the shape of the ΔS curve. If $\Delta S_{f,s}$ goes through a maximum or minimum as temperature increases (e.g., reaction 4 in Figure D-4), then heat-capacity effects will either tend to cancel out or be minimized. On the other hand, if $\Delta S_{f,s}$ increases or decreases consistently with temperature (e.g., reactions 1 and 2), then the higher the observed dehydration temperature, the more important the

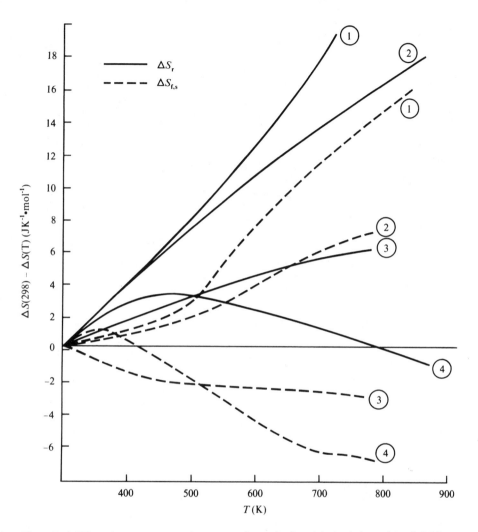

Figure D-4. Effect of temperature on the entropy change for four dehydration reactions. Solid lines represent ΔS_r for A = B + H₂O; dashed lines represent $\Delta S_{f,s}$ for A = B + H₂ + $\frac{1}{2}$O₂. The reactions are (1) muscovite + quartz = sanidine + andalusite + H₂O; (2) brucite = periclase + H₂O; (3) pyrophyllite = andalusite + 3 quartz + H₂O; (4) diaspore = corundum + H₂O. All entropy data from Robie et al. (1978).

heat-capacity correction will be, because of the necessity to extrapolate from the experimental temperature back to 298.15 K.

As shown in the Section D-1, every point-by-point retrieval requires prior knowledge of the standard entropy change. If this is unknown, Chatterjee (1977) has recommended a graphical method to obtain independent measurements of both $\Delta S_{f,s}$ and $\Delta H_{f,s}$. Because this method neglects heat-capacity effects,

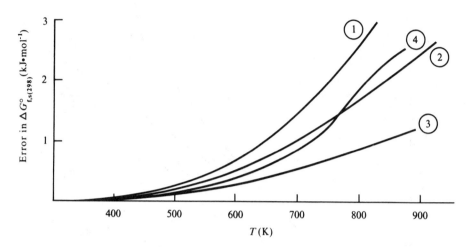

Figure D-5. Error in $\Delta G^\circ_{f,s(298)}$ that results from extrapolating point-by-point data from experimental T back to 298 K for four dehydration reactions (see legend of Figure D-4 for identification of reactions).

we introduce the approximation

$$\Delta G^\circ_{f,s(298,1)} - \int_{298}^{T_e} \Delta S_{f,s}(T)\,dT = \Delta G^\circ_{f,s(T_e)} \approx \Delta H^\circ_{f,s(298,1)} - T_e\,\Delta S_{f,s(298,1)} \quad \text{(D-20)}$$

Substituting into (D-19), we obtain

$$\Delta H^\circ_{f,s(298,1)} - T_e\,\Delta S_{f,s(298,1)} + (P_e - 1)\Delta V_s + G^*_{H_2O(T_e,P_e)} = 0 \quad \text{(D-21)}$$

which can be rearranged as

$$(P_e - 1)\Delta V_s + G^*_{H_2O(T_e,P_e)} = T_e\,\Delta S_{f,s(298,1)} - \Delta H^\circ_{f,s(298,1)} \quad \text{(D-22)}$$

The form of equation (D-22) implies that, if the lefthand side is plotted against temperature, then a straight line fitted through all reversal brackets will have a slope equal to $\Delta S_{f,s(298,1)}$ and an intercept equal to $-\Delta H^\circ_{f,s(298,1)}$. The accuracy of both these numbers will depend to a great extent on the width of the reversal brackets and the range of experimental temperatures.

Figure D-4 suggests that the reaction

$$\text{pyrophyllite} = \text{andalusite} + 3\text{ quartz} + H_2O$$

would be a good candidate for this graphical method, because $\Delta S_{f,s}$ for that reaction is not very sensitive to temperature. Haas and Holdaway (1973) report four reversals for the equilibrium ranging from extreme values of 370°C at

2.4 kbar to 463°C at 7.0 kbar. Taking

$$\Delta V_s = V_{\text{andalusite}} + 3V_{\text{quartz}} - V_{\text{pyrophyllite}} = -0.8226 \text{ J} \cdot \text{bar}^{-1}$$

the quantity $(P - 1)\Delta V_s + G^*_{\text{H}_2\text{O}(T_e, P_e)}$ is calculated from each experimental data point, and the results are plotted against experimental temperature (Figure D-6). A least-squares straight line fitted through the points has a slope of 212.09 J·K^{-1}·mol^{-1} and an intercept of -318.92 kJ·mol^{-1}. The entropy retrieved in this way is in satisfactory agreement with the calorimetric value for $\Delta S_{\text{f,s}(298,1)}$ of 211.46 J·K^{-1}·mol^{-1}.

We may also calculate $\Delta G^\circ_{\text{f,s}(298,1)}$ for one end of a reversal bracket as further illustration of the point-by-point method. The following data of Haas and Holdaway (1973, p. 456) pertain to the low-temperature end of the 2.4 kbar

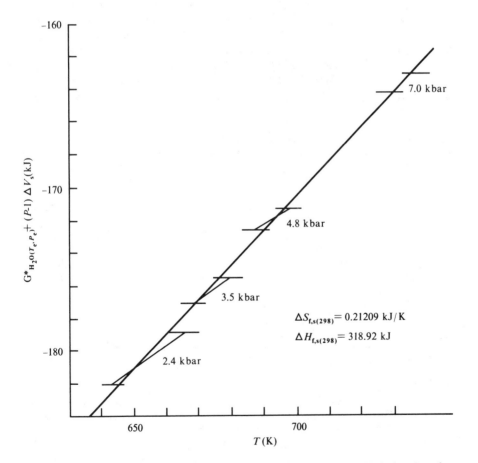

Figure D-6. Graphical retrieval for pyrophyllite = andalusite + 3 quartz + H$_2$O, based on four reversals of Haas and Holdaway (1973). See text for explanation of method.

bracket:

$$P_e = 2.4 \text{ kbar}$$

$$T_e = 370°C = 643 \text{ K}$$

$$\Delta S_{f,s(298)} = 0.21146 \text{ kJ} \cdot \text{K}^{-1} \cdot \text{mol}^{-1} \quad \text{(calorimetric value)}$$

$$G^*_{H_2O(2.4,643)} = -180.067 \text{ kJ} \cdot \text{mol}^{-1} \quad \text{(from Fisher and Zen, 1971, p. 304, converted to kJ from kcal)}$$

Solving for $\Delta G^°_{f,s(298,1)}$ in (D-19) gives, for ΔS independent of T,

$$\Delta G^°_{f,s(298,1)} = 180.07 + (2400 - 1) \times 0.8226 \times 10^{-3} + (643 - 298) \times 0.21146$$

$$= 254.99 \text{ kJ} \cdot \text{mol}^{-1}$$

Solving similarly for the high-temperature end of the 2.4 kbar bracket ($T = 392°C$) gives $\Delta G^°_{f,s(298,1)} = 256.47 \text{ kJ} \cdot \text{mol}^{-1}$, amounting to an uncertainty of $\pm 1.48 \text{ kJ} \cdot \text{mol}^{-1}$ in $\Delta G^°_{f,s}$, taking into account only the width of the reversal bracket. In contrast, if we evaluate the entropy integral in (D-19) explicitly, then the corresponding values for $\Delta G^°_{f,s(298,1)}$ are $255.61 \text{ kJ} \cdot \text{mol}^{-1}$ at $370°C$ and $257.15 \text{ kJ} \cdot \text{mol}^{-1}$ at $394°C$. Table D-2 contrasts values obtained from the means of the high and low ends of all brackets—both with and without heat-capacity

TABLE D-2

COMPARISON OF VALUES OBTAINED FROM VARIOUS METHODS
FOR STANDARD THERMODYNAMIC PROPERTIES OF PYROPHYLLITE

| | | Point-by-point | | |
	Graphical	$\Delta C_P = 0$	$\Delta C_P \neq 0$	Calorimetric
Reaction:	pyrophyllite = andalusite + 3 quartz + H_2O			
$\Delta H_{f,s(1,298)}$ (kJ·mol^{-1})	318.92	318.56	319.29	320.20
$\Delta S_{f,s(298)}$ (J·K^{-1}·mol^{-1})	212.09	——	——	211.50
$\Delta G_{f,s(298)}$ (kJ·mol^{-1})	255.68	255.50	256.24	257.18
	Pyrophyllite formation			
$\Delta G^°_{f(1,298)}$ (kJ·mol^{-1})	−5264.41	−5264.25	−5265.00	−5265.92
$\Delta S^°_{f(1,298)}$ (J·K^{-1}·mol^{-1})	−1254.76	——	——	−1254.12

Notes: For the point-by-point values of $\Delta G_{f,s(1,298)}$ assuming $\Delta C_P = 0$, the mean of four low-temperature points was $255.09 \pm 0.14 \text{ kJ} \cdot \text{mol}^{-1}$, and the mean of four high-temperature points was $255.92 \pm 0.40 \text{ kJ} \cdot \text{mol}^{-1}$; assuming $\Delta C_P \neq 0$, the mean of four low-temperature points was $255.80 \pm 0.19 \text{ kJ} \cdot \text{mol}^{-1}$, and the mean of four high-temperature points was $256.68 \pm 0.34 \text{ kJ} \cdot \text{mol}^{-1}$. The standard Gibbs free energy of formation of pyrophyllite is calculated for the point-by-point method as

$$\Delta G^°_{f(1,298),\text{pyrophyllite}} = \Delta G^°_{f(1,298),\text{andalusite}} + 3\Delta G^°_{f(1,298),\text{quartz}} - \Delta G_{f,s(1,298)}$$

using standard Gibbs free energies of formation of andalusite and quartz given by Robie et al. (1978). Calorimetric values are from Robie et al. (1978).

data—with the values obtained from the graphical method and from calorimetry.

D-3 SIMULTANEOUS RETRIEVAL

The preceding examples show how standard thermodynamic data can be retrieved from experiments performed on a single univariant reaction. This procedure has one serious drawback. We have seen that it is possible to obtain accurate measurements of the free energy differences between product and reactant assemblages at 1 bar and 298.15 K, but standard data for individual phases (such as $\Delta G_{f,s}^{\circ}$ for pyrophyllite calculated in the last example) depend on prior knowledge of standard data for all the other phases in the reaction. These data are most commonly obtained from calorimetric sources and thus are independent of the phase-equilibrium experiments. It should be obvious then that the value of the standard property you seek will depend ultimately on the calorimetric data you choose.

Retrieval experiments of the kind described in this appendix can never be totally independent of thermodynamic data from other sources, but this dependence can be reduced considerably by considering more than one reaction simultaneously. For instance, consider the following univariant reactions in the system $CaO-Al_2O_3-SiO_2-H_2O$:

$$\text{grossular} + \text{quartz} = \text{anorthite} + 2\,\text{wollastonite} \qquad \text{(D-23)}$$

$$4\,\text{zoisite} + \text{quartz} = \text{grossular} + 5\,\text{anorthite} + 2\,H_2O \qquad \text{(D-24)}$$

$$2\,\text{grossular} = \text{anorthite} + 3\,\text{wollastonite} + \text{gehlenite} \qquad \text{(D-25)}$$

$$\text{grossular} + 2\,\text{kyanite} + \text{quartz} = 3\,\text{anorthite} \qquad \text{(D-26)}$$

$$\text{grossular} + 2\,\text{corundum} = 3\,\text{Ca-Al-pyroxene} \qquad \text{(D-27)}$$

Each of the above reactions could be analyzed individually using either point-by-point or graphical techniques to retrieve standard free energies of reaction from the five different sets of phase-equilibrium data. These five reactions contain nine mineral phases. Thus, it is possible to think of the combined system in algebraic terms as nine unknown standard free energies of formation linked by five known standard free energies of reaction. The system has a unique solution if four standard free energies are supplied. For instance, if standard free energies of quartz, wollastonite, kyanite, and corundum are supplied from calorimetric or other sources, then the standard free energies of grossular, anorthite, zoisite, and Ca-Al-pyroxene can be calculated from the retrieved $\Delta G_{f,s(298,1)}^{\circ}$ values. The resulting set of linear simultaneous equations can be conveniently treated using linear algebra methods (e.g., Spear et al., 1982). For instance, equations (D-23) through (D-27) can be written in the form $\mathbf{A} \cdot \mathbf{X} = \mathbf{Y}$, where $\mathbf{A}$ is the coefficient matrix (equal to the stoichiometric coefficients of all phases in all five reactions), $\mathbf{X}$ is the vector of unknown $\Delta G_{f,s}^{\circ}$, and $\mathbf{Y}$ is the vector of all known free energies. When the calorimetric standard free energies of formation are combined with the retrieved reaction free energies, the matrix equation becomes

$$
\begin{vmatrix}
-1 & 1 & 0 & 0 & 0 \\
1 & 5 & -4 & 0 & 0 \\
-2 & 1 & 0 & 1 & 0 \\
-1 & 3 & 0 & 0 & 0 \\
-1 & 0 & 0 & 0 & 3
\end{vmatrix}
\cdot
\begin{vmatrix}
\Delta G^{\circ}_{f(gr)} \\
\Delta G^{\circ}_{f(an)} \\
\Delta G^{\circ}_{f(zo)} \\
\Delta G^{\circ}_{f(ge)} \\
\Delta G^{\circ}_{f(ca)}
\end{vmatrix}
$$

$$
=
\begin{vmatrix}
\Delta G^{\circ}_{r(23)} - \Delta G^{\circ}_{f(qz)} - 2\,\Delta G^{\circ}_{f(wo)} \\
\Delta G^{\circ}_{r(24)} + \Delta G^{\circ}_{f(qz)} + 2\,\Delta G^{\circ}_{f(steam)} \\
\Delta G^{\circ}_{r(25)} - 3\,\Delta G^{\circ}_{f(wo)} \\
\Delta G^{\circ}_{r(26)} + 2\,\Delta G^{\circ}_{f(ky)} + \Delta G^{\circ}_{f(qz)} \\
\Delta G^{\circ}_{r(27)} + 2\,\Delta G^{\circ}_{f(co)}
\end{vmatrix}
\qquad \text{(D-28)}
$$

where gr = grossular, an = anorthite, zo = zoisite, ge = gehlenite, Ca = Ca-Al-pyroxene, qz = quartz, wo = wollastonite, ky = kyanite, co = corundum, and the numbers in parentheses are the reaction numbers. All free energies are at 1 bar and 298.15 K. The vector of unknown free energies can be solved by matrix inversion, by Cramer's rule, or by numerical approximation.

The benefit of this broader approach is that it produces a set of thermodynamic data that is consistent with both a number of reversed phase-equilibrium studies and the chosen calorimetric base. Any additional calculated phase boundaries involving these nine phases must also be consistent with the measured reactions, because they are not independent. The key to success is to choose the reactions used in the retrieval scheme with great care--all phases involved must be very well characterized, and reversal of the observed phase boundaries must be proved.

Helgeson et al. (1978) have compiled a very comprehensive set of thermodynamic data for rock-forming minerals that is based primarily on regression analysis of multiple sets of phase-equilibrium data. Their method relies on the fundamental equilibrium-constant equation (D-1) and is similar to the approach just described for the $CaO–Al_2O_3–SiO_2–H_2O$ system, although it is substantially more comprehensive and more elegant. The clear advantages of this approach are that thermodynamic data for several reactions must be satisfied simultaneously, and the precision of the standard data tends to increase as more reactions are added to the network. A clear disadvantage is that the networks are interrelated by arbitrary assignment of standard properties (especially enthalpies of formation) to secondary reference phases. If one of the reference phases is shown to be in error, then all values based on that reference phase will be shifted by a proportional amount. Thermodynamic data for aluminous minerals in the data set of Helgeson et al. (1978) have been criticized by Hemingway et al. (1982) for exactly this reason. See Chapter 12 for a discussion.

Appendix E

THE COMPUTATION OF GEOCHEMICAL EQUILIBRIUM

Now that you are equipped with the principles and data of thermodynamics, we shall introduce the subject of computerized geochemical modeling. Although geochemical modeling can apply to solid phases only as well as to combined water–rock reactions, this section will deal with the latter type, especially as it applies to natural waters. Water–mineral equilibrium computations have become widespread and popular in teaching, research, and water-quality management. These computations are a valuable tool in interpreting ground-water and surface-water chemistry, understanding aqueous hazardous waste problems, and simulating water–rock interactions. There are currently more than 50 computer programs mentioned in the literature that can calculate chemical equilibrium for a multicomponent electrolyte solution such as a natural water.

The state of the art in geochemical modeling has been reviewed by Nordstrom et al. (1979), Wolery (1979), Potter (1979), Jenne (1981), and Nordstrom and Ball (1984). This appendix draws largely from the information found in Nordstrom et al. (1979) and Nordstrom and Ball (1984), which should be consulted for more details. Nordstrom et al. (1979) made direct comparisons of several different computer programs, and Nordstrom and Ball (1984) give an extensive list of computer programs that do speciation computations. Both papers describe the major strengths and weaknesses of the existing chemical models used in the programs.

The following terms are useful when discussing equilibrium computations:

speciation, the equilibrium distribution of aqueous species between free ions and ion pairs or complexes

phase distribution, the equilibrium distribution of components between two or more phases

chemical model, a theoretical construction that allows us to predict the thermodynamic properties of chemical substances *mass transfer*, simple transfer of mass between two or more phases, such as the precipitation or dissolution of soluble minerals

mass transport, solute movement during fluid flow (a time-dependent quantity)

These definitions help to distinguish mass "transfer" from mass "transport," terms that sometimes are used interchangeably. We also point out that there are "chemical models" and "computer programs" but that there is no such thing as a "computer model."

It has been about 22 years since the first publication of a paper on the speciation of ions in a natural water (Garrels and Thompson, 1962) and about 15 years since the pioneering work of Helgeson et al. (1969, 1970) in which computers were used to model the geochemistry of water–mineral reactions. These classic papers really signaled the beginning of computerized geochemical modeling. The use of computers to solve problems involving aqueous systems has expanded enormously, and "chemical modeling" is now a common phrase that describes the application of physicochemical principles to the interpretation of hydrogeochemical systems (Jenne, 1979). Chemical equilibrium computations can be broken down into four main aspects: (1) the mathematical formalism, (2) the thermodynamic formalism, (3) the numerical-approximation algorithm, and (4) the computer coding.

E-1   THE CHEMICAL EQUILIBRIUM PROBLEM AND ITS MATHEMATICAL AND THERMODYNAMICAL BASIS

The chemical-equilibrium problem can be simply stated as the attempt to find the most stable state of a system for a given set of pressure, temperature, and compositional constraints. The variables P, T, and X need not be kept fixed at constant values, but they must be defined. For example, the temperature of a system undergoing retrograde metamorphism may decrease with time according to a well-defined cooling curve. The chemical equilibria for such a process can still be modeled, as long as the temperature variations can be described by well-defined functions. In thermodynamic terms, a chemical equilibrium calculation attempts to find the minimum value for the total Gibbs free energy of a system. In mathematical terms, we want to solve the partial differential

$$\left(\frac{\partial G}{\partial \xi}\right)_{P,T} = 0 \qquad \text{(E-1)}$$

where ξ represents the progress variable—i.e., the variable that measures the extent of the reaction (see Prigogine and Defay, 1954). During the progress of the reaction, the free energy goes through its lowest value somewhere between the state of pure reactants and that of pure products; this minimum condition is the goal of the calculation. At this state, the equilibrium distribution of mineral phases and/or aqueous species is established.

The chemical equilibrium problem consists of two parts: the constraint of chemical equilibrium, and the constraint of conservation of mass. The equilibrium constraint can be considered in terms of free energies of the reacting species or in terms of the equilibrium constants (or "mass action" expressions) for the reacting species. Obviously, these expressions are equivalent because they are related by the familiar equation $\Delta G_r^\circ = -RT \ln K$. The conservation-of-mass constraint is expressed by a set of mass-balance equations, one equation for each component in the system. When the mass-balance equations are substituted into the equilibrium conditions, the result is a set of nonlinear equations. A set of nonlinear equations for a natural water containing hundreds of possible aqueous species and a dozen or more components can be practically impossible to solve without using a numerical approximation procedure and a computer.

Although more could be said about the mathematical nature of the equilibrium problem formulation, it tends to go beyond the scope of this appendix. Further information can be found in excellent books by Van Zeggeren and Storey (1970) and by Smith and Missen (1982). Additional reviews by Zeleznik and Gordon (1968) and Wolery (1979) are also recommended. An outstanding review by Rubin (1983) outlines the mathematical formalism necessary to solve the chemistry of reacting systems during fluid flow for both equilibrium and nonequilibrium conditions. These five references cover most of the mathematical, thermodynamical, and computational aspects of the chemical equilibrium problem.

E-2 NUMERICAL APPROXIMATION ALGORITHMS

Generalized algorithms for solving the chemical equilibrium problem began with the papers of Brinkley (1946, 1947). Although he intended the calculations for desk calculators, he represented them in a form well suited for computers. He showed the importance of organizing mass-balance equations around component species and showed how these can be substituted into equilibrium-constant expressions. Most of the early programs were developed for solving equilibrium problems associated with high-temperature gaseous systems for use in rocket propellant research and in the space program. White et al. (1958) viewed the problem as one involving free energy minimization and pointed out the advantage of using alternative techniques such as linear programming. These different viewpoints led to the classification of algorithms as either the "free energy minimization" method or the "equilibrium-constant" method. This distinction has some mathematical support in that "minimization" or "optimization" techniques are used in the free energy minimization method, whereas numerical approximations suited for solving sets of nonlinear equations are used in the equilibrium-constant method. Another approach, introduced by Johansen (1967), classified algorithms according to the manner in which the mass-balance constraint and the equilibrium constraint are solved. Some algorithms satisfy the mass-balance constraint at every iteration and converge on the equilibrium constraint, whereas others satisfy the equilibrium constraint and converge on the mass-balance constraint. Algorithms of a third group satisfy neither condition and converge on both of them simultaneously. Finally, Smith and Missen (1982) have suggested a division between stoichiometric and nonstoichiometric algorithms, depending on whether or not stoichiometric reactions are used explicitly. All of these classification schemes have drawbacks; each fails to represent fairly a difference in structure of the programming strategy. For example, equilibrium constants can be used in a minimization technique, whereas free energy minimization, usually a nonstoichiometric algorithm, can be employed in a stoichiometric algorithm. The only difference worth underscoring here is that a program that accepts only free energy values for the operating data base makes it far more difficult to obtain reliable values for aqueous species than it is with a program that uses equilibrium constants. Some free energy values of aqueous species do not exist because the measurements have not been made, whereas the equilibrium constants typically are available.

For aqueous geochemical calculations, it is convenient to use the categories of Van Zeggeren and Storey (1970) whereby algorithms are characterized as (1) pure iteration, (2) Newton–Raphson iteration, or (3) integration of ordinary differential equations. Wigley (1977) has described the method of successive approximation (pure iteration) in which either the "brute force" or the "continued fraction" method can be used. The brute force method is simple back-substitution of the mass-action expressions into the mass-balance equations directly on each iteration. In the continued fraction method, the equation is rearranged for faster convergence. The Newton–Raphson method is one of the oldest and still most widely used numerical methods of approximation. The general procedure is described in many textbooks, and a good description applied to the aqueous chemical equilibrium problem is given by Westall et al. (1976) and Westall (1979). The Newton–Raphson and continued fraction methods can best be understood in a simple example involving the gypsum solubility problem from Section 12-3.

Example. Use the continued fraction method and the Newton–Raphson method to find the distribution of species in a gypsum-saturated solution of 0.01518 m at 25°C.

First, we state the problem chemically and set it up. As shown in Chapter 12, there are two mass-balance equations,

$$m_{Ca} = m_{Ca^{2+}} + m_{CaSO_4^\circ}$$

$$m_{SO_4} = m_{SO_4^{2-}} + m_{CaSO_4^\circ}$$

and the mass-action equation,

$$K_1 = \frac{a_{CaSO_4^\circ}}{a_{Ca^{2+}} a_{SO_4^{2-}}} = 10^{2.3}$$

leading to the master equation

$$m_{Ca} = m_{Ca^{2+}} + m_{Ca^{2+}}^2 K_1 \gamma_{Ca^{2+}} \gamma_{SO_4^{2-}} \qquad (12\text{-}17)$$

In the continued fraction method, we begin by rearranging equation (12-17) in terms of the component Ca^{2+} to obtain

$$m_{Ca^{2+}}(1 + m_{Ca^{2+}} K_1 \gamma_{Ca^{2+}} \gamma_{SO_4^{2-}}) = m_{Ca}$$

and

$$m_{Ca^{2+}} = \frac{m_{Ca}}{1 + m_{Ca^{2+}} K_1 \gamma_{Ca^{2+}} \gamma_{SO_4^{2-}}} \qquad (E\text{-}2)$$

For the first iteration, we assume that the total concentration of calcium is equal to the free-ion concentration. For the second iteration, we calculate the free-calcium concentration from equation (E-2). This cycle is then repeated until a reasonable convergence is obtained. In this example, the convergence criterion will be to obtain an answer identical to that found in Chapter 12 when the problem was solved analytically. The results are shown in Table E-1, where five iterations were required for convergence.

The Newton–Raphson method works if the derivative of a function can be easily obtained. First we rearrange equation (12-17) so that it is of the form $f(x) = 0$, and then we can obtain an improved value of $x \, (=m_{Ca^{2+}})$ from

$$x = x_0 + \Delta x$$

where $\Delta x = -f(x)/f'(x)$, and $f'(x)$ is the first derivative. Thus, equation (12-17) becomes

$$f(x) = m_{Ca^{2+}} + (m_{Ca^{2+}})^2 K_1 \gamma_{Ca^{2+}} \gamma_{SO_4^{2-}} - 0.01518 = 0 \qquad (E\text{-}3)$$

and

$$f'(x) = 1 + 2m_{Ca^{2+}} K_1 \gamma_{Ca^{2+}} \gamma_{SO_4^{2-}} \qquad (E\text{-}4)$$

The results of the iterative calculations are shown in Table E-2, where the

TABLE E-1
SOLUTION OF EXAMPLE BY CONTINUED FRACTION METHOD

Iteration	$m_{Ca^{2+}}$	I	$\gamma_{Ca^{2+}}$	$\gamma_{SO_4^{2-}}$
1	0.01518	0.06072	0.450	0.437
2	0.00951	0.03805	0.507	0.498
3	0.01026	0.04106	0.497	0.487
4	0.01015	0.04060	0.499	0.489
5	0.01016	0.04065	0.499	0.489

TABLE E-2
SOLUTION OF EXAMPLE BY NEWTON–RAPHSON METHOD

Iteration	$m_{Ca^{2+}}$	I	$\gamma_{Ca^{2+}}$	$\gamma_{SO_4^{2-}}$	Δx
1	0.01518	0.06072	0.450	0.437	-0.0041
2	0.01105	0.04422	0.488	0.477	-0.00076
3	0.01029	0.04116	0.497	0.487	-0.00011
4	0.01018	0.04071	0.499	0.489	-0.00002
5	0.01016	0.04065	0.499	0.489	-0.000003

number of iterations required is the same as that required with the continued fraction method (Table E-1).

E-3 COMPUTERIZED CHEMICAL MODELS FOR WATER–MINERAL REACTIONS

Soon after the appearance of the classic paper by Garrels and Thompson (1962) on the chemical speciation of seawater, several computer programs were developed for chemical equilibrium calculations. Some of these programs are described briefly here.

HALTAFALL program

Ingri et al. (1967) developed a successive-approximation algorithm (in ALGOL) that was originally intended for laboratory mixing operations such as titrations, precipitation separations, and solvent extractions. More recently, this HALTAFALL algorithm has been used for the speciation of seawater (Dyrssen and Wedborg, 1974) and the speciation of estuarine waters (Dyrssen and Wedborg, 1980).

SOLGASWATER program

In an effort to improve the efficiency of HALTAFALL computations for high-temperature equilibria, Eriksson (1971, 1979) produced a new program, SOLGASWATER (originally SOLGAS), that utilizes a free energy minimization technique with Gaussian elimination. This program is gaining popularity in Sweden, where it has been used to model laboratory solutions and natural waters (Öhman, 1983; Lidén, 1983).

REDEQL series

The FORTRAN program called REDEQL was first described by Morel and Morgan (1972) as a speciation and phase-distribution program that uses the

Newton–Raphson iteration technique. This program has undergone numerous improvements and modifications, including REDEQL2 (McDuff and Morel, 1973), MINEQL (Westall et al., 1976), MICROQL (Westall, 1979), GEOCHEM (Mattigod and Sposito, 1979; Sposito and Mattigod, 1980) and REDEQL.EPAK (Ingle et al., 1979). These programs all perform the same type of computations, but significant differences exist between their data bases and in some cases between the algorithms. A noteworthy feature of these programs is the capability to do adsorption calculations. The only other program noted for adsorption calculations is ADSORP, written in BASIC (Bourg, 1982).

WATEQ series

Truesdell and Jones (1974) produced a speciation program for natural waters based on a continued fraction algorithm and written in PL/I. This program also evolved through several generations of modification, including WATEQF (FORTRAN version, Plummer et al., 1976), WATEQ2 (trace element version in PL/I, Ball et al., 1979, 1980), WATEQ3 (uranium added, Ball et al., 1981), and WATSPEC (Wigley, 1977). The SOLMNEQ program (Kharaka and Barnes, 1973) developed in parallel to the WATEQ series and performs the same type of calculations, but its data base and algorithm also permit calculations for geothermal waters at high temperature and pressure.

Brine computations

Progress in making equilibrium calculations with brine solutions was made by Wood (1975, 1976), who developed a program that uses Harned's rule and the SI model. Alternatively, Van Luik and Jurinak (1978, 1979) applied an algorithm based on statistical theory to compute the equilibria among major and trace components in the Great Salt Lake, Utah. The most successful approach has been obtained by Harvie and Weare (1980) through the application of the Pitzer equations.

Reaction-path programs

An additional degree of computational complexity arises when consecutive stages of partial equilibrium are considered en route to complete equilibrium. Such approaches are called reaction-path or reaction-progress calculations. For example, during the dissolution of feldspar in water, the sequential precipitation and dissolution of gibbsite, kaolinite, and montmorillonite may occur before feldspar saturation is reached. The first reaction-path model came from the pioneering efforts of Helgeson and his colleagues, who developed the PATHI program (Helgeson et al., 1970). This program has been used to simulate weathering processes, diagenetic processes, metasomatic processes, and ore deposition. The PATHI program is based on the method of integration of ordinary differential equations. A set of linear equations is established by differentiating the mass-balance and mass-action equations with respect to the progress variable ξ. Further modifications of this approach have been used by Fritz (1975, 1981) and Droubi (1976) to simulate granite weathering, evaporation processes, hydrothermal reactions, and diagenesis with the programs EQUIL (for species distribution), DISSOL, EVAPOR, and THERMAL.

Wolery (1979) greatly improved the reaction-path computations with EQ3/6, which utilizes a dual algorithm. EQ3 calculates the speciation of an aqueous phase, and EQ6 calculates mass transfer and reaction progress. EQ3 has been replaced by EQ3NR, which uses a Newton–Raphson iteration to compute speciation (Wolery, 1983), which then becomes input to EQ6 (Wolery, 1984). The algorithm in EQ6 combines Newton–Raphson iteration to solve the governing mass-balance and mass-action equations for mass transfer with a finite-difference method to allow the reaction to progress to the next step. The SOLVEQ program (Reed, 1982) also solves reaction-path calculations based on a Newton–Raphson iteration. A simplified version of the reaction-path computation is available with PHREEQE (Parkhurst et al., 1980), a dual algorithm approach that calculates speciation with the continued fraction method for everything except pH, pε, and mass transfer, which are done by Newton–Raphson iteration. PHREEQE was developed especially for the interpretation of groundwater chemistry and, when used in conjunction with its companion program BALANCE (Parkhurst et al., 1982), provides a valuable tool in the evaluation of groundwater geochemistry (Plummer et al., 1983).

Mass transport

Initial attempts have been made to combine mass transport with chemical equilibrium computations. Chapman et al. (1982) have combined a physical model for fluid flow of a river with the chemical model MINEQL to calculate chemical changes during short-term neutralization of an acid mine drainage. The resulting marriage, called RIVEQL, was reasonably successful in predicting the precipitation of trace metals compared to actual field data. Ion-exchange equilibria have been combined with groundwater-flow equations to simulate exchange equilibria in a groundwater system by Vallochi et al. (1981). Miller and Benson (1983) have combined simple equilibrium reactions with fluid flow in a porous media by the program CHEMTRN. They used Newton–Raphson iteration to solve a set of differential equations for fluid transport and algebraic equations for chemical equilibria.

The best program?

You may well wonder whether there is any one best program among the many available. The answer is unambiguous: No. Three important parts of a program are the algorithm, the application, and the data base. Zeleznik and Gordon (1968) have pointed out that the algorithm chosen is not important as long as it is reliable. As we have seen, several different algorithms have been equally successful in solving the chemical equilibrium problem. Each program has different objectives and applications—i.e., some do adsorption calculations, some do brine calculations, and some do reaction-path calculations—so one should be properly chosen for the particular problem. The data base can vary enormously from one program to another, and this aspect must remain the responsibility of the user. The development and use of these programs must be integrated as much as possible with the compilation and evaluation of thermodynamic data.

Limitations to chemical equilibrium computations

The following limitations are very important in applying computerized chemical models to the interpretation of water–mineral reactions.

1. The equilibrium assumption may not be valid. Equilibrium is not always achieved between natural waters and minerals because of kinetic barriers.
2. The thermodynamic data base may be incomplete or in error. Sensitivity analyses should be carried out to determine the effect of errors on any geochemical computation.
3. Chemical models for nonideality in aqueous systems must be appropriate for the ionic strength of the solution. SI models should be used for ionic strengths greater than that of seawater. At lower ionic strengths, the IA or SI models should be equally valid.
4. Redox conditions probably cannot be quantitatively determined at this time. More information on reaction rates and the validity of the equilibrium assumption for redox reactions in natural waters is needed. Individual redox species in natural waters must be analyzed and computed separately from any other measure of redox potential.

E-4 EXAMPLES OF GEOCHEMICAL MODELING

Plummer et al. (1983) have concisely summarized the purpose of chemical modeling:

> In chemical reaction modeling, we use the available data to determine (1) what chemical reactions have occurred, (2) the extent to which the reactions have proceeded, (3) the conditions under which the reactions have occurred (such as open *vs* closed, equilibrium *vs* disequilibrium, constant or variable temperature), and (4) how the water quality and mineralogy will change in response to natural processes and perturbations to the system. The available data may

include chemical analyses of the aqueous phase, hydrology, mineralogy, gas compositions, isotopic data, and other relevant information. Chemical reaction modeling is largely a mental exercise, but it is facilitated by calculations of (1) equilibrium speciation, (2) mass balance, and (3) reaction-path.

Equilibrium speciation calculations are an estimation of the distribution of aqueous ionic forms (free ions and ion pairs) and an estimation of the saturation state of the water with respect to the solubility of various minerals and gases. This type of computation is probably the most common type found in geochemical investigations. One example is provided by the iron(II/III) speciation in acid mine waters calculated by Nordstrom et al. (1979). The distribution of dissolved iron between free ions and hydrolysis-paired and sulfate-paired species was shown to change significantly with oxidation.

Mass-balance calculations can determine the mass amounts of minerals that must have dissolved or precipitated to account for the observed changes in water chemistry between two points, such as the input and output points of a watershed or the recharge and discharge points of an aquifer. Mass-balance calculations are essential in interpreting water quality changes upon mixing of two or more bodies of water, such as the confluence of two rivers whose chemistries are markedly different. Hydrologic data such as discharge are necessary for these calculations, and useful quantitative results can be obtained when combined with speciation calculations.

Reaction-path calculations are true exercises in simulation. They can provide information about the resultant mineral composition and aqueous-phase speciation for a given set of initial conditions and a given set of hypothetical reactions. For example, Helgeson et al. (1969) modeled the hydrolysis of K-feldspar with the PATHI program. During feldspar dissolution, a series of chemical changes takes place, beginning with the precipitation of gibbsite and a rise in pH, followed by the dissolution of gibbsite and the precipitation of kaolinite, followed by the dissolution of kaolinite and the precipitation of K-mica, before final equilibrium with K-feldspar is reached. These reactions are generally consistent with experimental data on the dissolution kinetics of feldspar (Busenberg and Clemency, 1976).

Another valuable role for reaction-path calculations is to test the thermodynamic validity of mass-balance calculations. An excellent example of testing mass-balance calculations for a groundwater aquifer is described by Plummer et al. (1983). In this study, the major water chemistry involves reactions within a carbonate aquifer and is therefore dominated by carbonate reactions and redox reactions (owing to the presence of sulfate, pyrite, and organic matter). Three wells along the aquifer flow path were chosen for mass-balance and reaction-path computations, along with ten plausible phases: gypsum, calcite, Mg-calcite, dolomite, CO_2, CH_4, CH_2O (organic matter), FeOOH, FeS, and pyrite. From a total of twelve possible mineral reaction sets, seven could be eliminated from mass-balance calculations using BALANCE and reaction-path calculations using PHREEQE. The remaining five models were similar, suggesting that the net reaction is the incongruent dissolution of dolomite (i.e., dolomite dissolution with calcite precipitation) driven irreversibly by gypsum dissolution, with accompanying sulfate reduction and conservation of iron as ferric hydroxide and pyrite. The modeling suggests that the groundwater system is initially open to CO_2 and reacts with organic carbon along the flow path.

Appendix F

SOURCES OF THERMODYNAMIC DATA

Section F-1 lists sources of thermodynamic data for a variety of systems, excluding data on H_2O. Table F-1 is a guide to specific sources for various substances and properties, identifying references by the numbers that precede them in the listing. Section F-2 lists sources of thermodynamic data on H_2O.

F-1 SOURCES OF THERMODYNAMIC DATA, EXCLUDING DATA ON H_2O

1. Abas-Zade, A. K., and Shpil'rain, E. E., eds. 1970. *Thermophysical properties of liquid.* Academy of Sciences, USSR. 224 pp.

2. Ahrland, S. 1973. Thermodynamics of the stepwise formation of metal-ion complexes in aqueous solution. *Struc. Bonding,* 15:167–188.

3. Albert, A., and Serjeant, E. P. 1962. *Ionization constants of acids and bases.* Wiley. 179 pp.

4. Anderegg, G. 1977. *Critical survey of stability constants of EDTA complexes.* IUPAC Chemical Data Series No. 15. Pergamon Press. 42 pp.

5. Anderson, P. D. 1966. *Thermodynamic properties of metals and alloys.* UCRL-11821. Clearinghouse. 136 pp.

6. Angus, S., and Armstrong, B., eds. 1972. *International thermodynamic tables of the fluid state, argon.* IUPAC Chemical Data Series No. 5. Butterworths. 102 pp.

7. Angus, S., Armstrong, B., and de Rueck, K. M., eds. 1974. *International thermodynamic tables of the fluid state, ethylene.* IUPAC Chemical Data Series No. 6. Butterworths. 204 pp.

8. Angus, S., Armstrong, B., and de Rueck, K. M., eds. 1976. *International thermodynamic tables of the fluid state, carbon dioxide.* IUPAC Chemical Data Series No. 7. Pergamon Press. 385 pp.

9. Angus, S., and de Rueck, K. M., eds. 1977. *International thermodynamic tables of the fluid state—4 helium.* IUPAC Chemical Data Series No. 8. Pergamon Press. 265 pp.

10. Angus, S., Armstrong, B., and de Rueck, K. M., eds. 1978. *International thermodynamic tables of the fluid state—5 methane.* IUPAC Chemical Data Series No. 16. Pergamon Press. 251 pp.

11. Angus, S., de Rueck, K. M., and Armstrong, B., eds. 1979. *International thermodynamic tables of the fluid state—6 nitrogen.* IUPAC Chemical Data Series No. 20. Pergamon Press. 241 pp.

12. Angus, S., Armstrong, B., and de Rueck, K. M., eds. 1980. *International thermodynamic tables of the fluid state—7 propylene.* IUPAC Chemical Data Series No. 25. Pergamon Press. 400 pp.

13. Armstrong, G. T., and Goldberg, R. N. 1976. An annotated bibliography of sources of compiled thermodynamic data relevant to biochemical and aqueous systems (1930–1975). *N.B.S. Spec. Pub.* 454. 60 pp.

14. Ashcroft, S. J., and Mortimer, C. T. 1970. *Thermochemistry of transition metal complexes.* Academic Press. 478 pp.

15. Astakhov, K. V., ed. 1970. *Thermodynamic and thermochemical constants.* Science Pub. House. 480 pp.

16. Aylward, G. H., and Findlay, T. J. H. 1974. *SI chemical data.* Wiley. 127 pp.

17. Baehr, H. D., Hartmann, H. Z., and Schomacker, H. 1968. *Thermodynamic functions for ideal gases at temperatures up to 6000°K.* Springer-Verlag. 73 pp.

18. Baes, C. F., Jr., and Mesmer, R. E. 1976. *The hydrolysis of cations.* Wiley-Interscience. 489 pp.

19. Baes, C. F., Jr., and Mesmer, R. E. 1981. The thermodynamics of cation hydrolysis. *Am. J. Sci.* 281:935–962.

20. Bard, A. J., ed. 1973–1982. *Encyclopedia of electrochemistry of the elements.* 14 vols. Marcel Dekker.

21. Barin, I., and Knacke, O. 1973. *Thermochemical properties of inorganic substances.* Springer-Verlag. 921 pp.

22. Barin, I., Knacke, O., and Kubaschewski, O. 1977. *Thermochemical properties of inorganic substances—supplement.* Springer-Verlag. 861 pp.

23. Barner, H. E., and Scheurman, R. V. 1978. *Handbook of thermochemical data for compounds and aqueous species.* Wiley. 176 pp.

24. Barnes, H. L., ed. 1979. *Geochemistry of hydrothermal ore deposits.* 2d ed. Wiley-Interscience. 798 pp.

25. Battino, R., and Clever, H. L. 1966. The solubility of gases in liquids. *Chem. Rev.* 66:395–463.

26. Benson, S. W. 1976. *Thermochemical kinetics.* 2d ed. Wiley-Interscience. 320 pp.

27. Benson, S. W., Cruickshank, F. R., Golden, D. M., Hangen, G. R., O'Neil, H. E., Rogers, A. S., Shaw, R., and Walsh, R. 1969. Additivity rules for the estimation of thermochemical properties. *Chem. Rev.* 69:279–324.

28. Bethune, A. J. de, and Loud, N. A. S. 1964. *Standard aqueous electrode potentials and temperature coefficients at 250°C.* C. A. Hampel. 19 pp.

29. Bichowsky, F. R., and Rossini, F. D. 1936. *The thermochemistry of the chemical substances.* Reinhold. 460 pp.

30. Bond, A. M., and Hefter, G. T. 1980. *Critical survey of stability constants and related thermodynamic data of fluoride complexes in aqueous solution.* IUPAC Chemical Data Series No. 27. Pergamon Press. 71 pp.

31. Boublik, T., Fried, V., and Hala, E. 1973. *The vapor pressures of pure substances.* Elsevier. 626 pp.

32. Brandrup, J., and Immergut, E. H. 1966. *Polymer handbook.* Several vols. (2d ed., 1975.) Wiley-Interscience.

33. Britske, E. V., Kapustinskii, A. F., Veselovskii, B. K., Shamovskii, L. M.,Chentsova, L. G., and Anvaer, B. I. 1949. *Thermal constants of inorganic substances.* Academy of Sciences, USSR. 1014 pp.

34. Bulakh, A. G. 1968. *Methods in thermodynamics for mineralogy.* Nidra. 175 pp.

35. Bulakh, A. G., and Bulakh, K. G. 1978. *Physiochemical data for minerals and the chief components of hydrothermal fluids.* Nidra. 167 pp.

36. Canjar, L. E., and Manning, F. S. 1967. *Thermodynamic properties and reduced correlations for gases.* Gulf Pub. Co. 212 pp.

37. Charlot, G., Collumeau, A., and Marchon, M. J. 1971. *Selected constants, oxidation-reduction potentials of inorganic substances in aqueous solution.* IUPAC Chemical Data Series No. 13. Butterworths. 78 pp.

38. Christensen, J. J., Eatough, D. J., and Izatt, R. M. 1975. *Handbook of metal ligand heats and related thermodynamic quantities.* 2d ed. Marcel Dekker. 495 pp.

39. Christensen, J. J., Hansen, L. D., and Izatt, R. M. 1975. *Handbook of proton ionization heats and related thermodynamic quantities.* Wiley-Interscience. 320 pp.

40. Clark, S. M. 1960. *Oxidation-reduction potentials of organic systems.* Williams and Wilkins. 584 pp.

41. Clark, S. P., ed. 1966. Handbook of physical constants. *Geol. Soc. Am. Mem.* 97. 587 pp.

42. Clever, H. L., ed. 1979. *Helium and neon.* IUPAC Solubility Data Series, vol. 1. Pergamon Press. 393 pp.

43. Clever, H. L., ed. 1979. *Krypton, xenon and radon.* IUPAC Solubility Data Series, vol. 2. Pergamon Press. 357 pp.

44. Clever, H. L. 1980. *Argon.* IUPAC Solubility Data Series, vol. 4. Pergamon Press. 331 pp.

45. Cobble, J. W. 1953. Empirical considerations of entropy. I. The entropies of the oxyanions and related species. II. The entropies of inorganic complex ions. III. A structural approach to the entropies of aqueous organic solutes and complex ions. *J. Chem. Phys.* 21:1443–1456.

46. CODATA. 1978. CODATA recommended key values for thermodynamics 1977. *CODATA Bull.* 28. 17 pp.

47. Conway, B. E. 1952. *Electrochemical data.* Elsevier. 374 pp.

48. Cottrell, T. L. 1958. *The strengths of chemical bonds.* 2d ed. Butterworths. 310 pp.

49. Cox, J. D., and Pilcher, G. 1970. *Thermochemistry of organic and organometallic chemistry.* Academic Press. 643 pp.

50. Criss, C. M., and Cobble, J. W. 1964. The thermodynamic properties of high temperature aqueous solutions. IV. Entropies of the ions up to 200° and the correspondence principle. V. The calculation of ionic heat capacities up to 200°. Entropies and heat capacities above 200°. *J. Am. Chem. Soc.* 86:5385–5393.

51. Davies, C. W. 1962. *Ion association.* Butterworths. 190 pp.

52. Din, F., ed. 1961. *Thermodynamic functions of gases.* Butterworths. 218 pp.

53. Dobos, D. 1975. *Electrochemical data.* Elsevier. 339 pp.

54. Domalski, E. S. 1972. Selected values of heats of combustion and heats of formation of organic compounds containing the elements C, H, N, O, P and S. *J. Phys. Chem. Ref. Data* 1:221–278.

55. Dreisbach, R. R. 1952. *PVT relationships of organic compounds.* Handbook Publ. 303 pp.

56. Dymond, J. H., and Smith, E. B. 1969. *The virial coefficients of gases: a critical compilation.* Oxford Clarendon Press. 231 pp.

57. Ellis, A. J., and McFadden, I. M. 1972. Partial molal volumes of ions in hydrothermal solutions. *Geochim. Cosmochim. Acta* 36:413–426.

58. Feitknecht, W., and Schindler, P. W. 1963. *Solubility constants of metal oxides, metal hydroxides and metal hydroxide salts in aqueous solution.* Butterworths. 199 pp. (Also in *Pure Appl. Chem.* 6:130–199.)

59. Franklin, J. L. 1969. Ionization potentials, appearance potentials and heats of formation of gaseous positive ions. *NSRDS–NBS* 26. 285 pp.

60. Frantz, J. D., Popp, R. K., and Boctor, N. Z. 1981. Mineral-solution equilibria. V. Solubilities of rock-forming minerals in supercritical fluids. *Geochim. Cosmochim. Acta* 45:69–77.

61. Freeman, R. D. 1962. Thermodynamic properties of binary sulfides. *Research Foundation Report* 60 (Oklahoma State Univ.) pp. 28–41.

62. Freier, R. K. 1977. *Aqueous solutions: data for inorganic and organic compounds.* 2 vols. Walter de Gruyter. 917 pp.

63. Fuger, J., and Oetting, F. L. 1976. *The chemical thermodynamics of actinide elements and compounds. Part 2: The actinide aqueous ions.* IAEA. 65 pp.

64. Furukawa, G. T., Reilly, M. L., and Gallagher, J. S. 1974. Critical analysis of heat-capacity data and evaluation of thermodynamic properties of ruthenium, rhodium, palladium, iridium and platinum from 0 to 300 K. A survey of the literature on osmium. *J. Phys. Chem. Ref. Data* 3:163–209.

65. Furukawa, G. T., Saba, W. G., and Reilly, M. L. 1968. Critical analysis of the heat-capacity data of the literature and evaluation of thermodynamic properties of copper, silver and gold from 0 to 300 K. *NSRDS–NBS* 18. 49 pp.

66. Galkin, N. P., Tumanov, Y. N., and Butylkin, Y. P. 1972. *Thermodynamic properties of inorganic fluorides.* Atomizdat. 144 pp.

67. Gedansky, L. M., and Hepler, L. G. 1969. Thermochemistry of gold and its compounds. *Engelhard Ind. Tech. Bull.* 10:5–9.

68. Gedansky, L. M., Woolley, E. M., and Hepler, L. G. 1970. Thermochemistry of compounds and aqueous ions of copper. *J. Chem. Therm.* 2:561–576.

69. Glushko, V. P., ed. 1962. *Thermodynamic properties of chemical substances. Vol. I, Calculations of thermodynamic properties. Vol. II, Tables of thermodynamic properties.* Academy of Sciences, USSR. 2080 pp.

70. Glushko, V. P., ed. 1965–1971. *Handbook of thermal constants.* Vol. 1 (1965): O, H, D, T, F, Cl, Br, I, At, 3He, Ne, Ar, Kr, Xe, Rn. Vol. 2 (1966): S, Se, Te, Po. Vol. 3 (1968): N, P, As, Sb, Bi. Vol. 4 1969: C, Si, Ge, Sn, Pb. Vol. 5 (1971): B, Al, Ga, In, Tl. Nauka. 1422 pp.

71. Glushko, V. P., ed. 1971–1973. *Thermodynamic and thermophysical properties of combustion products. Vol. I, Methods of calculation. Vol. II, Oxygen-based propellants. Vol. III, Oxygen and air-based propellants. Vol. IV, Nitrogen tetroxide-based propellants.* Academy of Sciences, USSR. 1913 pp.

72. Glushko, V. P., Alemasov, V. E., Gurvich, L. V., and Medvedev, V. A. 1974. Systems of reference books of the Academy of Sciences of the USSR on the thermodynamic properties of substances. *Teplofiz. Vys. Temp.* 12:970–977.

73. Gmelin, L., ed. 1788–1975. *Handbuch der anorganischen Chemie.* Many vols. Verlag Chemie.

74. Goldberg, R. N. 1979. Evaluated activity and osmotic coefficients for aqueous solutions: bi-univalent compounds of lead, copper, manganese and uranium. *J. Phys. Chem. Ref. Data* 8:1005–1050.

75. Goldberg, R. N. 1981. Evaluated activity and osmotic coefficients for aqueous solutions: thirty-six uni-bivalent electrolytes. *J. Phys. Chem. Ref. Data* 10:671–764.

76. Goldberg, R. N., and Hepler, L. G. 1968. Thermochemistry and oxidation potentials of the platinum group metals and their compounds. *Chem. Rev.* 68:229–252.

77. Goldberg, R. N., and Nuttall, R. L. 1978. The alkaline earth metal halides. *J. Phys. Chem. Ref. Data* 7:263–310.

78. Haas, J. L., Jr. 1976. Physical properties of the coexisting phases and thermochemical properties of the H_2O component in boiling NaCl solutions. *U.S. Geol. Surv. Bull.* 1421-A. 73 pp.

79. Haas, J. L., Jr. 1976. Thermodynamic properties of the coexisting phases and thermochemical properties of the NaCl component in boiling NaCl solutions. *U.S. Geol. Surv. Bull.* 1421-B. 71 pp.

80. Haas, J. L., Jr., Robinson, G. R., Jr., and Hemingway, B. S. 1981. Thermodynamic tabulations for selected phases in the system $CaO–Al_2O_3–SiO_2–H_2O$ at 101.325 kPa (1 atm) between 273.15 and 1800 K. *J. Phys. Chem. Ref. Data* 10:575–669.

81. Hala, E., et al. 1968. *Vapor-liquid equilibrium data at normal pressures.* Pergamon Press. 541 pp.

82. Hamann, S. D. 1957. *The physico-chemical effects of pressure.* Butterworths. 246 pp.

83. Hamer, W. J., ed. 1959. *The structure of electrolytic solutions*. Wiley. 453 pp.

84. Hamer, W. J. 1968. Theoretical mean activity coefficients of strong electrolytes in aqueous solutions from 0 to 100°C. *NSRDS–NBS* 24. 271 pp.

85. Hamer, W. J., and Wu, Y. C. 1972. Osmotic coefficients and mean activity coefficients of uni-univalent electrolytes in water at 25°C. *J. Phys. Chem. Ref. Data* 1:1047–1099.

86. Harned, H. S., and Owen, B. B. 1958. *The physical chemistry of electrolytic solutions*. 3d ed. Reinhold. 354 pp.

87. Helgeson, H. C. 1967. Thermodynamics of complex dissociation in aqueous solution at elevated temperatures. *J. Phys. Chem.* 71:3121–3136.

88. Helgeson, H. C. 1969. Thermodynamics of hydrothermal systems at elevated temperatures and pressure. *Am. J. Sci.* 267:729–804.

89. Helgeson, H. C. 1982. Errata: thermodynamics of minerals, reactions, and aqueous solutions at high pressures and temperatures. *Am. J. Sci.* 282:1144–1149.

90. Helgeson, H. C., and Kirkham, D. H. 1974. Theoretical prediction of the thermodynamic behavior of aqueous electrolytes at high pressures and temperatures. II. Debye-Hückel parameters for activity coefficients and relative partial molal properties. *Am. J. Sci.* 274:1199–1261.

91. Helgeson, H. C., and Kirkham, D. H. 1976. Theoretical prediction of the thermodynamic properties of aqueous electrolytes at high pressures and temperatures. III. Equation of state for aqueous species at infinite dilution. *Am. J. Sci.* 276:97–240.

92. Helgeson, H. C., Kirkham, D. H., and Flowers, G. C. 1982. Theoretical prediction of the thermodynamic behavior of aqueous electrolytes at high pressures and temperatures. IV. Calculation of activity coefficients, osmotic coefficients, and apparent molal and standard and relative partial molal properties to 600°C and 5 kb. *Am. J. Sci.* 281:1249–1516.

93. Helgeson, H. C., Delany, J. M., Nesbitt, H. W., and Bird, D. K. 1978. Summary and critique of the thermodynamic properties of rock-forming minerals. *Am. J. Sci.* 278-A:1–229.

94. Hemingway, B. S., and Robie, R. A. 1977. Enthalpies of formation of low albite (NaAlSi₃O₈), gibbsite (Al(OH)₃), and NaAlO₂: revised values for ΔH_f°, 298 and ΔG_f°, 298 of some aluminosilicate minerals. *J. Res. U.S. Geol. Surv.* 5:413–429.

95. Hepler, L. G., and Olafsson, G. 1975. Mercury: thermodynamic properties, chemical equilibria, and standard potentials. *Chem. Rev.* 75:585–602.

96. Hill, J. O., Worsley, I. G., and Hepler, L. G. 1971. Thermochemistry and oxidation potentials of vanadium, niobium, and tantalum. *Chem. Rev.* 71:127–137.

97. Högfeldt, E. 1982. *Stability constants of metal ion complexes: Part A. Inorganic ligands*. IUPAC Chemical Data Series No. 21. Pergamon Press. 310 pp.

98. Hultgren, R., and Desai, P. D. 1973. *Selected thermodynamic values and phase diagrams for copper and some of its binary alloys*. Monograph I. Intern. Copper Res. Assoc. 204 pp.

99. Hultgren, R., Desai, P. D., Hawkins, D. T., Gleiser, M., and Kelly, K. K. 1973. *Selected values of the thermodynamic properties of binary alloys.* American Society for Metals. 1435 pp.

100. Hultgren, R., Desai, P. D., Hawkins, D. T., Gleiser, M., Kelly, K. K., and Wagman, D. D. 1973. *Selected values of the thermodynamic properties of the elements.* American Society for Metals. 636 pp.

101. Inczedy, J. 1976. *Analytical applications of complex equilibria.* Wiley. 415 pp.

102. Ives, D. J. G., and Janz, G. J., eds. 1961. *Reference electrodes: theory and practice.* Academic Press. 651 pp.

103. Janz, G. J. 1967. *Thermodynamic properties of organic compounds.* 2d ed. Academic Press, 249 pp.

104. Janz, G. J., and Dijkhuis, C. G. M. 1969. Molten salts: vol. 2, sec. 1. Electrochemistry of molten salts: Gibbs free energies and excess free energies from equilibrium type cells. *NSRDS–NBS* 28:1–48.

105. Janz, G. J., and Tomkins, R. P. T. 1972. *Nonaqueous electrolytes handbook.* 2 vols. Academic Press. 2041 pp.

106. Jordan, T. E. 1954. *Vapor pressure of organic compounds.* Interscience. 266 pp.

107. Karapet'yants, M. K., and Karapet'yants, M. L. 1970. *Thermodynamic constants of inorganic and organic compounds.* Humphrey Science Publ. 461 pp.

108. Karpov, I. K., Kashik, S. A., and Pampura, V. D. 1968. *Thermodynamic constants for calculations in geochemistry and petrology.* Nauka. 143 pp.

109. Karpov, I. K., Kiselev, A. I. and Letnikov, F. A. 1971. *Chemical thermodynamics in petrology and geochemistry.* Irkutsk. Akad. Nank. SSSR Sibr. Otdel, Inst. Geochem. Inst. Earth's Crust. 385 pp.

110. Kaufmann, D. W., ed. 1960. *Sodium chloride.* Reinhold.

111. Kelly, K. K. 1969. Contributions to the data on theoretical metallurgy. XIII. High-temperature heat content, heat capacity and entropy data for the elements and inorganic compounds. *U.S. Bur. of Mines Bull.* 584. 232 pp.

112. Kelly, K. K., and King, E. G. 1961. Contributions to the data on theoretical metallurgy. XIV. Entropies of the elements and inorganic compounds. *U.S. Bur. Mines Bull.* 592. 149 pp.

113. Khodakovskiy, I. L., Ryzhenko, B. N., and Naumov, G. B. 1968. Thermodynamics of aqueous electrolyte solutions at elevated temperature (temperature dependence of the heat capacities of ions in aqueous solution). *Geochem. Int.* 5:1200–1219.

114. King, E. G., Mah, A. D., and Prankratz, L. B. 1973. *Thermodynamic properties of copper and its inorganic compounds.* Monograph II. Inter. Copper Res. Assoc. 257 pp.

115. Kirillina, V. A., ed. 1970. *Thermophysical properties of the alkali metals.* Izdatel'stvo Standartov. 487 pp.

116. Kortum, G., Vogel, W., and Andrussow, K. 1961. *Dissociation constants of organic acids in aqueous solutions.* Butterworths. 558 pp. [Also in *Pure App. Chem.* 1:187–536 (1960).]

117. Kracek, F. C. 1963. Melting and transformation temperatures of minerals and allied substances. *U.S. Geol. Surv. Bull.* 1144-D. 81 pp.

118. Kragten, J. 1978. *Atlas of metal-ligand equilibria in aqueous solution.* Wiley. 781 pp.

119. Kubaschewski, O., and Alcock, C. B. 1977. *Metallurgical thermochemistry.* 5th ed. Pergamon Press. 500 pp.

120. Kudchadker, A. P., Alani, G. H., and Zwolinski, B. J. 1968. Critical constants of organic substances. *Chem. Rev.* 68:659–735.

121. Langmuir, D. 1969. The Gibbs free energies of substances in the system $Fe-O_2-H_2O-CO_2$ at 25°C. *U.S. Geol. Surv. Prof. Paper* 650-B:180–184.

122. Larson, J. W., and Hepler, L. G. 1969. Heats and entropies of ionization. In *Solute-Solvent Interactions*, ed. J. F. Coetzee and C. D. Ritchie, pp. 1–44. Marcel Dekker.

123. Latimer, W. M. 1952. *The oxidation states of the elements and their potentials in aqueous solutions.* Prentice-Hall. 392 pp.

124. Linke, W. F., and Seidell, A. 1958. *Solubilities of inorganic and metal organic compounds.* Vol. I. 4th ed. Am. Chem. Soc. 1487 pp.

125. Linke, W. F., and Seidell, A. 1966. *Solubilities of inorganic and metal organic compounds.* Vol. II. 4th ed. Am. Chem. Soc. 1941 pp.

126. Liu, C. T., and Lindsay, W. T., Jr. 1971. Thermodynamic properties of aqueous solutions at high temperatures. *U.S. Office of Saline Water, Research and Devel. Prog. Rept.* 722. 124 pp.

127. Lobo, V. M. M. 1975. *Electrolyte solutions: literature data on thermodynamic and transport properties.* Coimbra Editora. 448 pp.

128. Mah, A. D., and Pankratz, L. B. 1976. Contributions to the data on theoretical metallurgy. XVI. Thermodynamic properties of nickel and its inorganic compounds. *U.S. Bur. Mines Bull.* 668. 125 pp.

129. Marcus, Y., and Howery, D. G. 1975. *Ion exchange equilibrium constants.* IUPAC Chemical Data Series No. 10. Butterworths. 41 pp.

130. Martell, A. E., and Smith, R. M. 1974–1976. *Critical stability constants, Vol. 1, amino acids. Vol. 2, amines. Vol. 3, other organic ligands. Vol. 4, inorganic complexes.* Plenum Press. 1636 pp.

131. Meites, L. 1980–1981. *CRC handbook series in inorganic electrochemistry.* 2 vo s. CRC Press. 1042 pp.

132. Meites, L., and Zuman, P. 1974. *Electrochemical data. Part I: organic, organometallic and biochemical substances.* Vol. A. Krieger. 742 pp.

133. Mel'nik, Y. P. 1972. *Thermodynamic constants for the analysis of conditions of formation of iron ores.* Institute of the Geochemistry and Physics of Minerals, Academy of Sciences, Ukrainian S.S.R. 193 pp.

134. Millero, F. J. 1971. The molal volumes of electrolytes. *Chem. Rev.* 71:147–176.

135. Mills, K. C. 1974. *Thermodynamic data for inorganic sulphides, selenides and tellurides.* Butterworths. 845 pp.

136. Mischenko, K. P., and Poltoratzkii, G. M. 1968. *Aspects of the thermodynamics and structure of aqueous and non-aqueous electrolyte solutions.* Khimia. 353 pp.

137. Mischenko, K. P., and Samoilav, O. Ya. 1967. Solutions. *Razv. Fiz. Khim. SSSR 1917–1967; Akad. Nauk SSSR, Inst. Istor. Estestvozn. Tekh.* 177–222.

138. Nancollas, G. H. 1966. *Interactions in electrolyte solutions.* Elsevier. 214 pp.

139. National Research Council. 1928. *International critical tables of numerical data of physics, chemistry and technology.* McGraw-Hill. 444 pp.

140. Naumov, G. B., Ryzhenko, B. N., and Khodakovskiy, I. L. 1971. *Handbook of thermodynamic data.* Atomizdat. 240 pp.

141. Novikov, I. I., and Gordon, A. I. 1969*a. Thermophysical properties of gases, liquids and plasmas at high temperatures.* State Service for Standard and Reference Data, U.S.S.R. 407 pp.

142. Novikov, I. I., and Gordon, A. I. 1969*b. Thermophysical properties of solid substances at high temperatures.* State Service for Standard and Reference Data, U.S.S.R. 496 pp.

143. Oetting, F. L., Rand, M. H., and Ackermann, R. J. 1976. *The chemical thermodynamics of actinide elements and compounds. Part 1: The actinide elements.* IAEA. 111 pp.

144. Parker, V. B. 1965. Thermal properties of aqueous uni-univalent electrolytes. *U.S. Natl. Bur. Stand. Ref. Data Ser.* 2. 66 pp.

145. Parker V. B., Wagman, D. D., and Evans, W. H. 1971. Selected values of chemical thermodynamic properties: tables for the alkaline earth elements (elements 92 through 97 in the standard order of arrangement). *NBS Tech. Note* 270-6. 106 pp.

146. Parker, V. B., Wagman, D. D., and Garvin, D. 1976. Selected thermochemical data compatible with the CODATA recommendations. NBSIR 75-968. 34 pp.

147. Parsons, R. 1959. *Handbook of electrochemical constants.* Butterworths. 113 pp.

148. Pascal, P. 1964. *Nouveau traite de chimie minerale.* 2 vols. Masson.

149. Pedley, J. B., ed. 1972–. *Computer analysis of thermochemical data* (CATCH Tables). University of Sussex.

150. Perrin, D. D. 1965. *Dissociation constants of organic bases in aqueous solution.* Butterworths. 514 pp.

151. Perrin, D. D. 1969. *Dissociation constants of inorganic acids and bases in aqueous solution.* IUPAC Chemical Data Series No. 19. Pergamon Press. 110 pp. (Also in *Pure Appl. Chem.* 20:133–236.)

152. Perrin, D. D. 1972. *Dissociation constants of organic bases in aqueous solution. Supplement.* IUPAC Chemical Data Series No. 12. Pergamon Press. 524 pp.

153. Perrin, D. D. 1979. *Stability constants of metal-ion complexes part B: Organic ligands.* IUPAC Chemical Data Series No. 22. Pergamon Press. 1280 pp.

154. Potter R. W., II, Shaw, D. R., and Haas, J. L., Jr. 1975. Annotated bibliography of studies on the density and other volumetric properties for major components in geothermal waters 1928–74. *U.S. Geol. Surv. Bull.* 1417. 78 pp.

155. Pourbaix, M. 1966. *Atlas of electrochemical equilibrium in aqueous solutions.* Pergamon Press. 644 pp.

156. Rabinovich, V. A., ed. 1970–1972. *Thermophysical properties of substances and materials.* 2d ed. State Service for Standard and Reference Data. 299 pp. (Translated, 1974–1975, Natl. Tech. Infor. Serv.)

157. Reid, R. C., and Sherwood, T. K. 1966. *The properties of gases and liquids.* McGraw-Hill. 646 pp.

158. Robie, R. A., Bethke, P. M., and Beardsley, K. M. 1967. Selected X-ray crystallographic data, molar volumes and densities of minerals and related substances. *U.S. Geol. Surv. Bull.* 1248. 87 pp.

159. Robie, R. A., Hemingway, B. S., and Fisher, J. R. 1978. Thermodynamic properties of minerals and related substances at 298.15 K and one atmosphere pressure and at higher temperatures. *U.S. Geol. Surv. Bull.* 1259. 465 pp. (Reprinted with corrections, 1979)

160. Robinson, G. R., Haas, J. L., Jr., Schafer, C. M., and Hazelton, H. T., Jr. 1982. Thermodynamic and thermophysical properties of selected phases in the MgO–SiO_2–H_2O–CO_2, CaO–Al_2O_3–SiO_2–H_2O–CO_2, and Fe–FeO–Fe_2O_3–SiO_2 chemical systems, with special emphasis on the properties of basalts and their mineral components. *U.S. Geol. Surv. Open-File Rept.* 83-79. 429 pp.

161. Robinson, R. A., and Stokes, R. H. 1959. *Electrolyte solutions.* 2d ed. Academic Press. 571 pp.

162. Rossini, F. D., Pitzer, K. S., Arnett, R. L., Brown, R. M., and Pimental, G. C. 1953. *Selected values of the thermodynamic properties of hydrocarbons and related compounds.* Carnegie Press.

163. Rossini, F. D., Wagman, D. D., Evans, W. H., Levine, S., and Jaffe, I. 1952. Selected values of chemical thermodynamic properties. *NBS Circ.* 500. 1268 pp.

164. Sadiq, M., and Lindsay, W. L. 1979. *Selection of standard free energies of formation for use in soil chemistry.* Colo. State Univ. Tech. Bull. 134. 1069 pp.

165. Samsonov, G. V. 1968. *Handbook of the physicochemical properties of the elements.* Plenum Press. 941 pp.

166. Samsonov, G. V., ed. 1973. *The oxide handbook.* Plenum Press. 941 pp.

167. Schafer, D., and Lax, E., eds. 1961. *Landolt-Börnstein.* Springer-Verlag.

168. Schick, H. L. 1966. *Thermodynamics of certain refractory compounds. Vol. I, Discussion of theoretical studies. Vol. II, Thermodynamic tables, bibliography and property file.* Academic Press. 1424 pp.

169. Schumm, R. H., Wagman, D. D., Bailey, S., Evans, W. H., and Parker, V. B. 1973. Selected values of chemical thermodynamic properties: tables for the lanthanide (rare earth) elements (elements 62 through 76 in the standard order of arrangement). *NBS Tech. Note* 270-7. 75 pp.

170. Serjeant, E. P., and Dempsey, B. 1979. *Ionization constants of organic acids in aqueous solution.* IUPAC Chemical Data Series No. 23. Pergamon Press. 987 pp.

171. Shmonov, V. M., and Shmulovich, K. I. 1978. *Tables of thermodynamic properties of gases and liquids (CO_2).* Gosdarstvennaya Sluzhba Standartnykh Dannykh. 165 pp.

172. Silcock, H. L., ed. 1979. *Solubilities of inorganic and organic compounds*, vol. 3, parts 1, 2, 3. Pergamon Press. 3321 pp.

173. Sillen, L. G., and Martell, A. E. 1964. *Stability constants of metal-ion complexes*. Spec. Pub. No. 17. The Chemical Society. 754 pp. (Supplement No. 1, 1971, 865 pp.)

174. Smith-Magowan, D., and Goldberg, R. 1979. A bibliography of sources of experimental data leading to thermal properties of binary aqueous electrolyte solutions. *NBS Spec. Pub.* 537. 85 pp.

175. Stephen, H., and Stephen, T., eds. 1963. *Solubilities of inorganic and organic compounds*, vol. 1, parts 1, 2. Pergamon Press. 1933 pp.

176. Stephen, H., and Stephen, T., eds. 1964. *Solubilities of inorganic and organic compounds*, vol. 2, parts 1, 2. Pergamon Press. 2053 pp.

177. Stull, D. R. 1947. Vapor pressure of pure substances. Inorganic compounds. *Ind. Eng. Chem.* 39:540–550.

178. Stull, D. R. 1947. Vapor pressure of pure substances. Organic compounds. *Ind. Eng. Chem.* 39:517–539.

179. Stull, D. R., and Prophet, H. 1971. *JANAF thermochemical tables*. NSRDS–NBS 37. U.S. Natl. Bur. Stand. 1141 pp. (1974 supplement by Chase, M. W., Curnutt, J. L., Hu, A. T., Prophet, H., and Walker, L. C., *J. Phys. Chem. Ref. Data* 3:311–480; 1975 supplement by Chase, M. W., Curnutt, J. L., Prophet, H., McDonald, R. A., and Syverud, A. N., *J. Phys. Chem. Ref. Data* 4:1–175; 1978 supplement by Chase, M. W., Jr., Curnutt, J. L., McDonald, R. A., and Syverud, A. N., *J. Phys. Chem. Ref. Data* 7:793–940)

180. Stull, D. R., and Sinke, G. C. 1956. *Thermodynamic properties of the elements*. Adv. Chem. Ser. No. 18. Am. Chem. Soc. 234 pp.

181. Stull, D. R., Westrum, E. F., Jr., and Sinke, G. C. 1969. *The chemical thermodynamics of organic compounds*. Wiley. 895 pp.

182. Timmermans, J. 1950, 1965. *Physical-chemical constants of pure organic compounds*. 2 vols. Elsevier. 1175 pp.

183. Timmermans, J. 1959–1960. *The physico-chemical constants of binary systems in concentrated solutions*. 4 vols. 5185 pp.

184. Touloukian, Y. S., and Buyco, E. H. 1970. *Thermophysical properties of matter. Vol. 4, specific heat, metallic elements and alloys. Vol. 5, Specific heat, non-metallic solids*. Plenum Press. 2568 pp.

185. Touloukian, Y. S., and Makita, T. 1970. *Thermophysical properties of matter. Vol. 6, Specific heat, non-metallic liquids and gases*. Plenum Press. 384 pp.

186. Touloukian, T. S., Judd, W. R., and Roy, R. F., eds. 1981. *Physical properties of rocks and minerals*. McGraw-Hill. 548 pp.

187. Travers, J. G., Dellieu, I., and Hepler, L. G. 1976. Scandium: thermodynamic properties, chemical equilibrium and standard potentials. *Thermochim. Acta* 15:89–104.

188. Truesdell, A. H., and Jones, B. F. 1974. WATEQ, a computer program for calculating chemical equilibria of natural waters. *J. Res. U.S. Geol. Surv.* 2:1233–1248.

189. Vasserman, A. A., Kazavchinskii, Ya. Z., and Rabinovich, V. A. 1966. *Thermophys-*

ical properties of air and its components. A. M. Zhuravlev, ed. Izdatel'stov Nauka. 375 pp.

190. Vargaftik, N. B. 1975. *Tables on the thermophysical properties of liquids and gases: in normal and dissociated states.* 2d ed. Halstead Press. 758 pp.

191. Wagman, D. D., Evans, W. H., Parker, V. B., Halow, I., Bailey, S. M., and Schumm, R. H. 1968. Selected values of chemical thermodynamic properties: Tables for the first thirty-four elements in the standard order of arrangement. *NBS Tech. Note* 270-3. 270 pp.

192. Wagman, D. D., Evans, W. H., Parker, V. B., Halow, I., Bailey, S. M., and Schumm, R. H. 1969. Selected values of chemical thermodynamic properties: tables for elements 35 through 53 in the standard order of arrangement. *NBS Tech. Note* 270-4. 141 pp.

193. Wagman, D. D., Evans, W. H., Parker, V. B., Halow, I., Bailey, S. M., Schumm, R. H., and Churney, K. L. 1971. Selected values of chemical thermodynamic properties: tables for elements 54 through 61 in the standard order of arrangement. *NBS Tech. Note* 270-5. 37 pp.

194. Wagman, D. D., Evans, W. H., Parker, V. B., and Schumm, R. H. 1976. Chemical thermodynamic properties of sodium, potassium and rubidium: an interim tabulation of selected material. *NBSIR* 76-1034. 76 pp.

195. Wagman, D. D., Evans, W. H., Parker, V. B., Schumm, R. H., and Nuttall, R. L. 1981. Selected values of chemical thermodynamic properties: compounds of uranium, protactinium, thorium, actinium, and the alkaline metals. *NBS Tech. Note* 270-8. 134 pp.

196. Wagman, D. D., Evans, W. H., Parker, V. B., Schumm, R. H., Halow, I., Bailey, S. M., Churney, K. L., and Nuttall, R. L. 1982. The NBS tables of chemical thermodynamic properties: selected values for inorganic and C_1 and C_2 organic substances in SI units. *J. Phys. Chem. Ref. Data* 11 (Suppl. 2):1–392.

197. Walther, J. V., and Helgeson, H. C. 1977. Calculation of the thermodynamic properties of aqueous silica and the solubility of quartz and its polymorphs at high pressures and temperatures. *Am. J. Sci.* 277:1315–1351.

198. Westrum, E. F., ed. 1958–. *Bulletin of Chemical Thermodynamics.* University of Michigan.

199. Wicks, C. E., and Block, F. E. 1963. Thermodynamic properties of 65 elements— their oxides, halides, carbides and nitrides. *U.S. Bur. Mines Bull.* 605. 146 pp.

200. Wilhelm, E., and Battino, R. 1973. Thermodynamic functions of the solubilities of gases in liquids at 25°C. *Chem. Rev.* 73:1–9.

201. Wilhelm, E., Battino, R., and Wilcock, R. J. 1977. Low pressure solubility of gases in liquid water. *Chem. Rev.* 77:219–262.

202. Wilhoit, R. C., and Zwolinski, B. J. 1971. *Handbook on vapor pressures and heats of vaporization of hydrocarbons and related compounds.* Thermodynamics Research Center, Texas A&M University. 337 pp.

203. Wu, Y. C., and Hamer, W. J. 1969. Osmotic coefficients and mean activity coefficients of a series of uni-bivalent and bi-univalent electrolytes in aqueous solutions at 25°C. *NBSIR* 10052, part XIV. 83 pp.

204. Yatsimirskii, K. B., and Vasil'ev, V. P. 1960. *Instability constants of complex compounds.* Pergamon Press. 220 pp.

205. Zordan, T. A., and Hepler, L. G. 1968. Thermochemistry and oxidation potentials of manganese and its compounds. *Chem. Rev.* 68:737–745.

206. Zwolinskii, B. J., Wilhoit, R. C., and Chao, J. 1971. *Handbook on chemical thermodynamic properties of hydrocarbons and related sulfur compounds.* Thermodynamic Research Center, Texas A&M University. 518 pp.

TABLE F-1
GUIDE TO MAJOR COMPILATIONS OF THERMODYNAMIC DATA

Substance or Property	References listed in Section F-1
Acid-base dissociation	3, 18, 19, 39, 116, 150, 151, 152, 164
Aqueous complexes (reactions)	2, 14, 18, 19, 24, 30, 34, 35, 38, 51, 81, 85, 86, 89, 90, 95, 99, 110, 115, 126, 133, 135, 140, 148, 150, 168, 198, 204
Aqueous ions (individual species)	16, 23, 33, 34, 46, 57, 68, 90, 91, 92, 113, 134, 140, 169, 191, 192, 193, 194, 195, 196, 197
Aqueous solutes	74, 75, 77, 78, 79, 83, 84, 85, 86, 88, 90, 110, 126, 127, 139, 140, 154, 169, 174, 191, 192, 193, 194, 195, 196, 203
Redox potentials	16, 20, 28, 37, 40, 53, 59, 67, 95, 96, 102, 123, 128, 131, 132, 140, 147, 155, 167, 198, 205
Elements	5, 15, 16, 20, 21, 22, 29, 33, 46, 63, 65, 67, 69, 70, 73, 95, 96, 98, 100, 107, 111, 112, 114, 115, 128, 141, 142, 143, 145, 146, 156, 158, 159, 163, 165, 167, 169, 172, 179, 180, 184, 185, 187, 191, 192, 193, 194, 195, 196, 205
Gases	6, 7, 8, 9, 10, 11, 12, 16, 17, 25, 31, 36, 41, 42, 43, 44, 52, 54, 55, 56, 59, 69, 70, 71, 81, 82, 106, 120, 140, 141, 142, 157, 162, 171, 177, 178, 179, 182, 185, 189, 190, 196, 200, 201, 202, 206
General	13, 15, 16, 23, 26, 29, 69, 70, 71, 72, 73, 107, 139, 140, 156, 167, 196, 198
H_2O	See Section F-2
Inorganic compounds	18, 19, 21, 22, 30, 33, 62, 66, 73, 107, 124, 125, 127, 196, 198
Liquids	1, 16, 81, 103, 105, 136, 137, 141, 157, 171, 185, 190, 196, 198
Mineral compounds	24, 34, 35, 41, 58, 60, 80, 87, 88, 89, 90, 91, 92, 93, 94, 108, 109, 113, 114, 117, 121, 133, 135, 140, 154, 158, 159, 160, 164, 186, 188, 196, 197, 198
Organic compounds	4, 13, 27, 32, 38, 39, 40, 49, 54, 55, 81, 103, 105, 106, 107, 116, 120, 130, 150, 152, 162, 167, 170, 181, 182, 196, 198, 202, 206

F-2 SOURCES OF THERMODYNAMIC DATA ON H_2O

Ben-Naim, A. 1974. *Water and aqueous solution: introduction to molecular theory.* Plenum Press. 480 pp.

Bezboruah, C. P., Camoes, M. F. G. F. C., Covington, A. K., and Dobson, J. V. 1973.

Enthalpy of ionization of water from electro-motive force measurements. *J. Chem. Soc. Faraday Trans.* 5:949–962.

Bignold, G. J., Brewer, A. D., and Hearn, B. 1971. Specific conductivity and ionic product of water between 50 and 271°C. *Trans. Faraday Soc.* 67:2419–2430.

Bradley, D. J., and Pitzer, K. S. 1979. Thermodynamics of electrolytes. 12. Dielectric properties of water and Debye–Hückel parameters to 350°C and 1 kbar. *J. Phys. Chem.* 83:1599–1603.

Burnham, C. W., Holloway, J. R., and Davis, N. F. 1969. Thermodynamic properties of water to 1,000°C and 10,000 bars. *Geol. Soc. Am. Spec. Paper* 132. 96 pp.

Busey, R. H., and Mesmer, R. E. 1978. Thermodynamic quantities for the ionization of water in sodium chloride media to 300°C. *J. Chem. Eng. Data.* 23:175–176.

Clever, H. L. 1968. The ion product constant of water. *J. Chem. Educ.* 45:231–235.

Covington, A. K., Ferra, M. I. A., and Robinson, R. A. 1977. Ionic product and enthalpy of ionization of water from electromotive force measurements. *J. Chem. Soc. Faraday Trans. I* 73:1721–1730.

Eisenberg, D., and Kauzmann, W. 1969. *The structure and properties of water*. Oxford Univ. Press. 300 pp.

Fine, R. A., and Millero, F. J. 1973. Compressibility of water as a function of temperature and pressure. *J. Chem. Phys.* 59:5529–5536.

Fisher, J. R., and Barnes, H. L. 1972. The ion-product constant of water to 350°. *J. Phys. Chem.* 76:90–99.

Fisher, J. R., and Zen, E. 1971. Thermochemical calculations from hydrothermal phase equilibrium data and the free energy of H₂O. *Am. J. Sci.* 270:297–314.

Franks, F., ed. 1972. *Water, a comprehensive treatise. Vol. 1, The physics and physical chemistry of water*. Plenum Press. 596 pp.

―――― ed. 1973. *Water, a comprehensive treatise. Vol. 2, Water in crystalline hydrates. Aqueous solutions of simple nonelectrolytes*. Plenum Press. 681 pp.

―――― ed. 1973. *Water, a comprehensive treatise. Vol. 3, Aqueous solutions of simple electrolytes*. Plenum Press. 472 pp.

―――― ed. 1975. *Water, a comprehensive treatise. Vol. 4, Aqueous solutions of amphiphiles and macromolecules*. Plenum Press. 839 pp.

―――― ed. 1975. *Water, a comprehensive treatise. Vol. 5, Water in disperse systems*. Plenum Press. 366 pp.

―――― ed. 1979. *Water, a comprehensive treatise. Vol. 6, Recent advances*. Plenum Press. 465 pp.

―――― ed. 1982. *Water, a comprehensive treatise. Vol. 7, Water and aqueous solutions at subzero temperature*. Plenum Press. 400 pp.

Gildseth, W., Habenschuss, A., and Spedding, F. H. 1972. Precision measurements of densities and thermal dilation of water between 5° and 80°C. *J. Chem. Eng. Data* 17:402–409.

Gregorio, P., and Merlini, C. 1969. Thermodynamic properties of water in the critical region. *Termotecnica* 23:41–54.

Guildner, L. A., Johnson, D. P., and Jones, F. E. 1976. Vapor pressure of water at its triple point: highly accurate value. *Science* 191:1261.

Halbach, H., and Chatterjee, N. D. 1982. An empirical Redlich-Kwong-type equation of state for water to 1,000°C and 200 kbar. *Contrib. Mineral. Petrol.* 79:337–345.

Hawkins, D. T. 1976. *Physical and chemical properties of water: a bibliography (1957–1974)*. Plenum Press. 556 pp.

Helgeson, H. C., and Kirkham, D. H. 1974. Theoretical prediction of the thermodynamic behavior of aqueous electrolytes at high pressures and temperatures: 1. Summary of the thermodynamic/electrostatic properties of the solvent. *Am. J. Sci.* 274:1089–1198.

Holloway, J. R., Eggler, D. H., and Davis, N. F. 1971. Analytical expression for calculating the fugacity and free energy of H_2O to 10,000 bars and 1,300°C. *Geol. Soc. Am. Bull.* 82:2639–2642.

Horne, R. A., ed. 1972. *Water and aqueous solutions: structure, thermodynamics and transport processes*. Wiley-Interscience. 837 pp.

Kavanau, J. L. 1964. *Water and solute-water interactions*. Holden-Day. 101 pp.

Keenan, J. H., Keyes, F. G., Hill, P. G., and Moore, J. G. 1969. *Steam tables: thermodynamic properties of water, including vapor, liquid and solid phases*. Wiley. 162 pp.

Luck, W. A. 1974. *Structure of water and aqueous solutions*. Verlag Chemie. 590 pp.

Marshall, W. L., and Franck, E. U. 1981. Ion product of water substance, 0–1000°C, 1–10,000 bars new international formulation and its background. *J. Phys. Chem. Ref. Data* 10:295–304.

Meyer, C. A., McClintock, R. B., Silvestri, G. J., and Spencer, R. C., Jr. 1967. *Thermodynamic and transport properties of steam*. Am. Soc. Mech. Eng. 328 pp.

Olofsson, G., and Hepler, L. G. 1975. Thermodynamics of ionization of water over wide ranges of temperature and pressure. *J. Solution Chem.* 4:127–143.

Olofsson, G., and Olofsson, I. 1977. The enthalpy of ionization of water between 273 and 323 K. *J. Chem.* 9:65–69.

———— 1981. Empirical equations for some thermodynamic quantities for the ionization of water as a function of temperature. *J. Chem. Thermodyn.* 13:437–440.

Rivikin, S. L. 1970. *Thermophysical properties of water in the critical region, a handbook*. Izdatel'stvo Standartov. 635 pp.

Rivikin, S. L., Aleksandrov, A. A., and Kremenévskaya, E. A. 1978. *Thermodynamic derivatives of water and steam*. Transl. by Joseph Kastin. Halstead Press. 264 pp.

Schmidt, E. 1969. *Properties of water and steam in SI units, 0–800°C and 0–1000 bar*. Springer-Verlag. 205 pp.

Uematsu, M., and Franck, E. U. 1980. Static dielectric constant of water and steam. *J. Phys. Chem. Ref. Data* 9:1291–1306.

Yukalovich, M. P., Rivkin, S. L., and Aleksandrov, A. A. 1969. *Tables of thermophysical properties of water and steam*. Izdatel'stvo Standartov. 408 pp.

REFERENCES

Adamson, A. W. 1978. SI units? A camel is a camel. *J. Chem. Educ.* 55:634–637.

Alexander, D. M., Hill, D. J. T., and White, L. R. 1971. The evaluation of thermodynamic functions from aqueous solubility measurements. *Aust. J. Chem.* 24:1143–1155.

Andèrsen, D. J., and Lindsley, D. H. 1981. A valid Margules formulation for an assymmetric ternary solution: Revision of the olivine-ilmenite thermometer, with applications. *Geochim. Cosmochim. Acta* 45:847–854.

Anderson, G. M. 1970*a*. Standard states at fixed and variable pressures. *J. Chem. Educ.* 47:676–679.

———. 1970*b*. Some thermodynamics of dehydration equilibria. *Am. J. Sci.* 269:392–401.

———. 1977*a*. The accuracy and precision of calculated mineral dehydration equilibria. In *Thermodynamics in geology*, ed. D. G. Fraser, pp. 115–136. D. Reidel.

———. 1977*b*. Uncertainties in calculations involving thermodynamic data. In *Application of thermodynamics to petrology and ore deposits*, ed. H. J. Greenwood, pp. 199–215. Mineralogical Association of Canada.

Asimov, I. 1964. *Asimov's biographical encyclopedia of science and technology.* Doubleday. 662 pp.

Baes, C. F., Jr., and Mesmer, R. E. 1976. *The hydrolysis of cations.* Wiley. 489 pp.

Ball, J. W., Jenne, E. A., and Cantrell, M. W. 1981. WATEQ3—a geochemical model with uranium added. *U.S. Geol. Surv. Open-File Rept. 81-1183.* 81 pp.

Ball, J. W., Jenne, E. A., and Nordstrom, D. K. 1979. WATEQ2—a computerized chemical model for trace and major element speciation and mineral equilibria of natural waters. In *Chemical modeling in aqueous systems*, ed. E. A. Jenne, pp. 815–835. American Chemical Society (Symposium Series 93).

Ball, J. W., Nordstrom, D. K., and Jenne, E. A. 1980. Additional and revised thermochemical data and computer code for WATEQ2—a computerized chemical model for trace and major element speciation and mineral equilibria of natural waters. *U.S. Geol. Surv. Water Resour. Invest. Rept. 78-116.* 109 pp.

Barany, R., and Kelley, K. K. 1961. Heats and free energies of formation of gibbsite, kaolinite, halloysite, and dickite. *U.S. Bur. Mines Rept. Invest. 5825.* 13 pp.

Bates, R. G. 1973. *Determination of pH.* 2d ed. Wiley. 479 pp.

Benson, B. B., and Krause, D., Jr. 1976. Empirical laws for dilute aqueous solutions of nonpolar gases. *J. Chem. Phys.* 64:689–709.

———. 1980. The concentration and isotopic fractionation of gases dissolved in freshwater in equilibrium with the atmosphere. I. Oxygen. *Limnol. Oceanogr.* 25:662–671.

Benson, B. B., Krause, D., Jr., and Peterson, M. A. 1979. The solubility and isotopic fractionation of gases in dilute aqueous solution. I. Oxygen. *J. Solution Chem.* 8:655–690.

Berner, R. A. 1963. Electrode studies of hydrogen sulfide in marine sediments. *Geochim. Cosmochim. Acta* 27:563–575.

———. 1978. Equilibrium, kinetics and the precipitation of magnesian calcites from seawater. *Am. J. Sci.* 278:1475–1477.

Bevington, P. R. 1969. *Data reduction and error analysis for the physical sciences.* McGraw-Hill. 336 pp.

Bockris, J. O'M., Kitchner, J. A., Ignatowics, S., and Tomlinson, J. W. 1952. Electrical conductance in liquid silicates. *Trans. Faraday Soc.* 48:75–91.

Bockris, J. O'M., and Reddy, M. M. 1970. *Modern electrochemistry.* Plenum Press. 1432 pp.

Bolton, P. D. 1970. Calculation of thermodynamic functions from equilibrium data. *J. Chem. Educ.* 47:638–641.

Bottinga, Y., and Richet, P. 1981. High pressure and temperature equation of state and calculation of the thermodynamic properties of gaseous carbon dioxide. *Am. J. Sci.* 281:567–614.

Bottinga, Y., Weill, D. F., and Richet, P. 1981. Thermodynamic modeling of silicate melts. In *Thermodynamics of minerals and melts,* ed. R. C. Newton, A. Navrotsky, and B. J. Wood, pp. 207–256. Springer-Verlag.

Bourg, A. C. M. 1982. ADSORP, a chemical equilibria computer program accounting for adsorption processes in aquatic systems. *Environ. Tech. Letters* 3:305–310.

Boyd, F. R., and England, J. L. 1960. Apparatus for phase-equilibrium measurements at pressures up to 50 kilobars and temperatures up to 1750°C. *J. Geophys. Res.* 65:741–748.

Brinkley, S. R., Jr. 1946. Note on the conditions of equilibrium for systems for many constituents. *J. Chem. Phys.* 14:563–564.

———. 1947. Calculation of the equilibrium composition of systems of many constituents. *J. Chem. Phys.* 15:107–110.

Brønsted, J. N. 1922. Studies on solubility, IV. The principle of specific interaction of ions. *J. Am. Chem. Soc.* 44:877–898.

Bruton, C. J., and Helgeson, H. C. 1983. Calculation of the chemical and thermodynamic consequences of differences between fluid and geostatic pressure in hydrothermal systems. *Am. J. Sci.* 283A:540–588.

Buddington, A. F., and Lindsley, D. H. 1964. Iron-titanium oxide minerals and synthetic equivalents. *J. Petrol.* 4:138–169.

Burnham, C. W. 1975. Water and magmas: a mixing model. *Geochim. Cosmochim. Acta* 39:1077–1084.

———. 1979. The importance of volatile constituents. In *The evolution of the igneous rocks, fiftieth anniversary perspective*, ed. H. S. Yoder, Jr., pp. 439–482. Princeton Univ. Press.

———. 1981. The nature of multicomponent aluminosilicate melts. In *Chemistry and geochemistry of solutions at high temperatures and pressures*, ed. D. T. Rickard and F. E. Wickman, pp. 197–229. Pergamon Press.

Burnham, C. W., Darken, L. S., and Lasaga, A. C. 1978. Water and magmas: application of the Gibbs-Duhem equation. *Geochim. Cosmochim. Acta* 42:277–280.

Burnham, C. W., Holloway, J. R., and Davis, N. F. 1969a. The specific volume of water in the range 1000 to 8900 bars, 20° to 900°C. *Am. J. Sci.* 267A:70–95.

———. 1969b. Thermodynamic properties of water to 1,000°C and 10,000 bars. *Geol. Soc. Am. Spec. Pap.* 132. 96 pp.

Burt, D. M. 1975. Hydrolysis equilibria in the system $K_2O-Al_2O_3-SiO_2-H_2O-Cl_2O_{-1}$; comments on topology. *Econ. Geol.* 71:665–671.

———. 1981. Acidity-salinity diagrams—application to greisen and porphyry deposits. *Econ. Geol.* 76:832–843.

Busenberg, E. and Clemency, C. V. 1976. The dissolution kinetics of feldspars at 25°C and 1 atm CO_2 partial pressure. *Geochim. Cosmochim. Acta* 40:41–49.

Butler, J. N. 1964a. *Solubility and pH calculations*. Addison-Wesley. 104 pp.

———. 1964b. *Ionic equilibrium, a mathematical approach*. Addison-Wesley. 547 pp.

———. 1982. *Carbon dioxide equilibria and their applications*. Addison-Wesley. 260 pp.

Carmichael, I. S. E., Nicholls, J., and Smith, A. L. 1970. Silica activity in igneous rocks. *Am. Mineral.* 55:246–263.

Carmichael, I. S. E., Turner, F. J., and Verhoogen, J. 1974. *Igneous petrology*. McGraw-Hill. 739 pp.

Carnahan, N. F., and Starling, K. E. 1969. Equation of state for non-attracting rigid spheres. *J. Chem. Phys.* 51:635–636.

Chapman, B. M., Jones, D. R., and Jung, R. F. 1983. Processes controlling metal ion attenuation in acid mine drainage streams. *Geochim. Cosmochim. Acta* 47:1957–1973.

Chapman, B. M., James, R. O., Jung, R. F., and Washington, H. G. 1982. Modelling the transport of reacting chemical contaminants in natural waters. *Aust. J. Mar. Freshwater Res.* 33:617–628.

Chatterjee, N. D. 1977. Thermodynamics of dehydration equilibria. In *Thermodynamics in geology*, ed. D. G. Fraser, pp. 137–160. D. Reidel.

Chatterjee, N. D., and Johannes, W. 1974. Thermal stability and standard thermodynamic properties of synthetic $2M_1$ muscovite $KAl_2AlSi_3O_{10}(OH)_2$. *Contrib. Mineral. Petrol.* 48:89–114.

Chen, C.-H. 1975. A method for estimation of standard free energies of formation of silicate minerals at 298.15 K. *Am. J. Sci.* 275:801–817.

Chinner, G. A. 1960. Pelitic gneisses with varying ferrous/ferric ratios from Glen Clova, Angus, Scotland. *J. Petrol.* 1:178–217.

Christ, C. L., Hostetler, P. B., and Siebert, R. M. 1974. Stabilities of calcite and aragonite. *J. Res. U.S. Geol. Surv.* 2:175–184.

Clarke, E. C. W., and Glew, D. N. 1966. Evaluation of thermodynamic functions from equilibrium constants. *Trans. Faraday Soc.* 62:539–547.

CODATA Task Group. 1976. CODATA recommended key values for thermodynamics, 1975. *J. Chem. Thermodyn.* 8:603–605.

Cohen, E. R., and Taylor, B. N. 1973. The 1973 least-squares adjustment of the fundamental constants. *J. Phys. Chem. Ref. Data* 2:663–734.

Coughlin, J. P. 1958. Heats of formation and hydration of anhydrous aluminum chloride. *J. Phys. Chem.* 62:419–421.

Covington, A. K., Ferra, M. I. A., and Robinson, R. A. 1977. Ionic product and enthalpy of ionization of water from electromotive force measurements. *J. Chem. Soc. Faraday Trans. I* 73:1721–1730.

Criss, C. M., and Cobble, J. W. 1964a. The thermodynamic properties of high temperature aqueous solutions. IV. Entropies of the ions up to 200° and the correspondence principle. *J. Am. Chem. Soc.* 86:5385–5390.

———. 1964b. The thermodynamic properties of high temperature aqueous solutions. V. The calculation of ionic heat capacities up to 200°. Entropies and heat capacities above 200°. *J. Am. Chem. Soc.* 86:5390–5393.

Crockett, J. W., and Winchester, J. W. 1966. Coprecipitation of zinc with calcium carbonate. *Geochim. Cosmochim. Acta* 30:1093–2009.

Davies, C. W. 1938. The extent of dissociation of salts in water. VIII. An equation for the mean ionic activity coefficient of an electrolyte in water, and a revision of the dissociation constant of some sulfates. *J. Chem. Soc.* 2093–2098.

———. 1962. *Ion association.* Butterworths. 190 pp.

Debye, P. 1912. On the theory of specific heat. *Ann. Phys. Leipzig,* 39:789–839.

Debye, P., and Hückel, E. 1923. On the theory of electrolytes. *Phys. Z.* 24:185–208, 305–325.

Deer, W. A., Howie, R. A., and Zussman, J. 1966. *An introduction to the rock-forming minerals,* Wiley, 528 pp.

Denbigh, K. G. 1981. *The principles of chemical equilibrium.* Cambridge Univ. Press, 494 pp.

de Santis, R., Breedveld, G. J. F., and Prausnitz, J. M. 1974. Thermodynamic properties of aqueous gas mixtures at advanced pressures. *Ind. Eng. Chem. Proc. Des. Dev.* 13:374–377.

Dickerson, R. E. 1969. *Molecular thermodynamics,* Benjamin/Cummings, 452 pp.

Diehl, H. 1979. High-precision coulometry and the value of the Faraday. *Anal. Chem.* 51:318A–330A.

Douglas, T. B., and King, E. G. 1968. High-temperature drop calorimetry. In *Experimental thermodynamics, Volume I: Calorimetry of non-reacting systems,* ed. J. P. McCullough and D. W. Scott, pp. 293–332. Butterworths.

Droubi, A. 1976. Geochemistry of salts and solution concentrations upon evaporation. Thermodynamic simulation model. Application to the salt deposits of Tchad. Ph.D. Thesis, Univ. of Louis Pasteur, Strasbourg. 177 pp.

Dulong, P. L., and Petit, A. T. 1819. Recherches sur quelques points importants de la theorie de la chaleur. *Ann. Chim. Phys.* 10:395–413.

Dyrssen, D., and Wedborg, M. 1974. Equilibrium calculations of the speciation of elements in seawater. In *The sea, Volume 5: Marine chemistry,* ed. E. D. Goldberg, pp. 181–195. Wiley.

———. 1980. Major and minor elements, chemical speciation in estuarine waters. In *Chemistry and biogeochemistry of estuaries,* ed. E. Olausson and I. Cato, pp. 71–119. Wiley.

Edmister, W. C. 1968. Applied hydrocarbon thermodynamics, part 32: Compressibility factors and fugacity coefficients from the Redlich-Kwong equation of state. *Hydrocarbon Prov.* 47:239–244.

Einstein, A. 1907. The Planck theory of radiation and the theory of specific heat. *Ann. Phys.* 22:180–186.

Ellis, A. J. 1966. Partial molal volumes of alkali chlorides in aqueous solution to 200° *J. Chem. Soc.*: 1579–1584.

Emerson, S. M. 1976. Early diagenesis in anaerobic lake sediments: chemical equilibria in interstitial waters. *Geochim. Cosmochim. Acta* 40:925–934.

Emerson, S. M., and Widmer, G. 1978. Early diagenesis in anaerobic lake sediments: II. Thermodynamic and kinetic factors controlling the formation of iron phosphate. *Geochim. Cosmochim. Acta* 42:1307–1316.

Eriksson, G. 1971. Thermodynamic studies of high temperature equilibria. *Acta Chem. Scand.* 25:2651–2658.

———. 1979. An algorithm for the computation of aqueous multicomponent, multiphase equilibria. *Anal. Chim. Acta* 112:375–383.

Essene, E. J. 1982. Geologic thermometry and barometry. *Rev. Mineralogy* 10:153–206.

Eugster, H. P. 1957. Heterogeneous reactions involving oxidation and reduction at high pressures and temperatures. *J. Chem. Phys.* 26:1760.

———. 1959. Reduction and oxidation in metamorphism. In *Researches in geochemistry,* ed. P. H. Abelson, pp. 397–426. Wiley.

———. 1972. Reduction and oxidation in metamorphism (II). *24th Int. Geol. Congr., Montreal, sec. 10,* pp. 3–11.

———. 1981. Metamorphic solutions and reactions. In *Chemistry and geochemistry of solutions at high temperatures and pressures,* ed. D. T. Richard and F. E. Wickman, pp. 461–508. Pergamon Press.

Eugster, H. P., Albee, A. L., Bence, A. E., Thompson, J. B., Jr., and Waldbaum, D. R. 1972. The two-phase region and excess mixing properties of paragonite-muscovite crystalline solutions. *J. Petrol.* 13:147–179.

Eugster, H. P., and Wones, D. R. 1962. Stability relations of the ferruginous biotite, annite. *J. Petrol.* 3:82–125.

Evans, R. C. 1966. *An introduction to crystal chemistry.* Cambridge Univ. Press. 410 pp.

Everett, D. H., and Wynne-Jones, W. F. K. 1939. Thermodynamics of acid-base equilibria. *Trans. Faraday Soc.* 35:1380–1401.

Ferry, J. M., and Burt, D. M. 1982. Characterization of metamorphic fluid composition through mineral equilibria. *Rev. Mineralogy* 10:207–262.

Fisher, J. R., and Zen, E-an. 1971. Thermochemical calculations from hydrothermal phase equilibrium data and the free energy of H_2O. *Am. J. Sci.* 270:297–314.

Franck, E. U. 1973. Concentrated electrolyte solutions at high temperature and pressure. *J. Solution Chem.* 2:339–356.

Frantz, J. D., and Popp, R. K. 1979. Mineral-solution equilibria—I. An experimental study of complexing and thermodynamic properties of aqueous $MgCl_2$ in the system $MgO–SiO_2–H_2O–HCl$. *Geochim. Cosmochim. Acta* 43:1223–1239.

Fraser, D. G. 1977. Thermodynamic properties of silicate melts. In *Thermodynamics in geology,* ed. D. G. Fraser, pp. 301–326. D. Reidel.

French, B. M., and Eugster, H. P. 1965. Experimental control of oxygen fugacities by graphite-gas equilibriums. *J. Geophys. Rev.* 70:1529–1539.

Fritz, B. 1975. Thermodynamic study and simulation of the reactions between minerals and solutions. Application to the geochemistry of alteration and of continental waters. Eng.D. thesis, University of Strasbourg. *Sci. Geol. Mem.* 41. 152 pp.

———. 1981. Thermodynamic study and simulation of hydrothermal and diagenetic reactions. Ph.D. thesis, University of Strasbourg. *Sci. Geol. Mem.* 65. 197 pp.

Fyfe, W. S., Turner, F. J., and Verhoogen, J. 1958. Metamorphic reactions and metamorphic facies. *Geol. Soc. Am. Mem.* 75. 253 pp.

Garrels, R. M., and Christ, C. M. 1965. *Solutions, minerals and equilibria.* Freeman and Cooper. 450 pp.

Garrels, R. M., and Thompson, M. E. 1962. A chemical model for seawater at 25°C and one atmosphere total pressure. *Am. J. Sci.* 260:57–66.

Garrels, R. M., and Wollast, R. 1978. Discussion: equilibrium criteria for two-component solids reacting with fixed composition in an aqueous phase—example: the magnesian calcite. *Am. J. Sci.* 278:1469–1474.

Garvin, D., Parker, V. B., Wagman, D. D., and Evans, W. H. 1977. A combined least sums and least squares approach to the solution of thermodynamic data networks. In *Proc. Fifth Biennial Int. CODATA Conf.,* ed. B. Dreyfus, pp. 515–518.

Ghent, E. D. 1976. Plagioclase–garnet–Al_2SiO_5–quartz: a potential geobarometer-geothermometer. *Am. Mineral.* 61:710–714.

Gibbs, J. W. 1906. On the equilibrium of heterogeneous substances. In *The scientific papers of J. Willard Gibbs, volume I: Thermodynamics*, pp. 55–371. Longmans Greens. (Reprinted by Dover, 1961).

Gillespie, L. J., and Coe, J. R., Jr. 1933. The heat of expansion of a gas of varying mass. *J. Chem. Phys.* 1:103–113.

Gordon, P. 1968. *Principles of phase diagrams in materials systems*. McGraw-Hill, 232 pp.

Graf, D. L. 1982. Chemical osmosis, reverse chemical osmosis and the origin of subsurface brines. *Geochim. Cosmochim. Acta* 46:1431–1448.

Green, E. J. 1970. Predictive thermodynamic models for mineral systems, I. Quasi-chemical analysis of the halite-sylvite subsolidus. *Am. Mineral.* 55:1692–1713.

Greenwood, H. J. 1967. Mineral equilibria in the system MgO–SiO_2–H_2O–CO_2. In *Researches in geochemistry, volume 2*, ed. P. H. Abelson, pp. 542–567. Wiley.

———. 1969. The compressibility of gaseous mixtures of carbon dioxide and water between 0 and 500 bars pressure and 450° and 800° centigrade. *Am. J. Sci.* 267A:191–208.

———. 1973. Thermodynamic properties of gaseous mixtures of H_2O and CO_2 between 450° and 800°C and 0 to 500 bars. *Am. J. Sci.* 273:561–571.

Grover, J. 1977. Chemical mixing in multicomponent solutions: an introduction to the use of Margules and other thermodynamic excess functions to represent non-ideal behavior. In *Thermodynamics in geology*, ed. D. G. Fraser, pp. 67–97. D. Reidel.

Guggenheim, E. A. 1935. The specific thermodynamic properties of aqueous solutions of strong electrolytes. *Phil. Mag.* 19:588–643.

———. 1967. *Thermodynamics*. 6th ed. North-Holland. 390 pp.

Guggenheim, E. A., and Turgeon, J. C. 1955. Specific interaction of ions. *Trans. Faraday Soc.* 51:747–761.

Güntelberg, E. 1926. Interaction of ions. *Z. Physik. Chem.* 123:199–247.

Haas, H., and Holdaway, M. J. 1973. Equilibria in the system Al_2O_3–SiO_2–H_2O involving the stability limits of pyrophyllite, and thermodynamic data of pyrophyllite. *Am. J. Sci.* 273:449–464.

Haas, J. L., Jr. 1974. PHAS20, a program for simultaneous multiple regression of a mathematical model to thermochemical data. NTIS AD-780301. 162 pp.

Haas, J. L., Jr., and Fisher, J. R. 1976. Simultaneous evaluation and correlation of thermodynamic data. *Am. J. Sci.* 276:525–545.

Haas, J. L., Jr., Robinson, G. R., Jr., and Hemingway, B. S. 1981. Thermodynamic tabulations for selected phases in the system CaO–Al_2O_3–SiO_2–H_2O at 101.325 kPa (1 atm) between 273.15 and 1800 K. *J. Phys. Chem. Ref. Data* 10:575–669.

Haggerty, S. E. 1976. Opaque mineral oxides in terrestrial igneous rocks. *Rev. Mineralogy* 3:Hg101–Hg175.

Hamer, W. J., and Wu, Y.-C. 1972. Osmotic coefficients and mean activity coefficients of uniunivalent electrolytes in water at 25°C. *J. Phys. Chem. Ref. Data* 1:1047–1099.

Harned, H. S., and Robinson, R. A. 1940. Temperature variation of the ionization constants of weak electrolytes. *Trans. Faraday Soc.* 36:973–978.

Harvie, C. E., and Weare, J. H. 1980. The prediction of mineral solubilities in natural waters: the Na–K–Mg–Ca–Cl–SO$_4$–H$_2$O system from zero to high concentration at 25°C. *Geochim. Cosmochim. Acta* 44:981–997.

Hathaway, J. C., and Sachs, P. L. 1965. Sepiolite and clinoptilite from the Mid-Atlantic Ridge. *Am. Mineral.* 50:852–867.

Head, J. H. 1970. FITIT, a computer program to least squares fit non-linear theories. *U.S. Air Force Academy Tech. Report* 70-5.

Helgeson, H. C. 1967. Thermodynamics of complex dissociation in aqueous solution at elevated temperatures. *J. Phys. Chem.* 71:3121–3136.

———. 1969. Thermodynamics of hydrothermal systems at elevated temperatures and pressures. *Am. J. Sci.* 167:729–804.

Helgeson, H. C., Brown, T. H., and Leeper, R. H. 1969. *Handbook of theoretical activity diagrams.* Freeman, Cooper. 253 pp.

Helgeson, H. C., Brown, T. H., Nigrini, A., and Jones, T. A. 1970. Calculation of mass transfer in geochemical processes involving aqueous solutions. *Geochim. Cosmochim. Acta* 34:569–592.

Helgeson, H. C., Delany, J. M., Nesbitt, H. W., and Bird, D. K. 1978. Summary and critique of the thermodynamic properties of rock-forming minerals. *Am. J. Sci.* 278A. 229 pp.

Helgeson, H. C., Garrels, R. M., and Mackenzie, F. T. 1969. Evaluation of irreversible reactions in geochemical processes involving minerals and aqueous solutions: II. Applications. *Geochim. Cosmochim. Acta* 33:455–481.

Helgeson, H. C., and Kirkham, D. H. 1974. Theoretical prediction of the thermodynamic behavior of aqueous electrolytes at high pressures and temperatures. II. Debye-Hückel parameters for activity coefficients and relative partial molal properties. *Am. J. Sci.* 274:1199–1261.

Helgeson, H. C., Kirkham, D. H., and Flowers, G. C. 1981. Theoretical prediction of the thermodynamic behavior of aqueous electrolytes at high pressures and temperatures. IV. Calculation of activity coefficients, osmotic coefficients, and apparent molal and standard and relative partial molal properties to 600°C and 5 kb. *Am. J. Sci.* 281:1249–1516.

Hemingway, B. S., Haas, J. L., Jr., and Robinson, G. R., Jr. 1982. Thermodynamic properties of selected minerals in the system Al$_2$O$_3$–CaO–SiO$_2$–H$_2$O at 298.15 K and 1 bar (10^5 pascals) pressure and at higher temperatures. *U.S. Geol. Surv. Bull.* 1544. 70 pp.

Hemingway, B. S., Krupka, K. M., and Robie, R. A. 1981. Heat capacities of the alkali feldspars between 350 and 1000 K from differential scanning calorimetry, the thermodynamic functions of the alkali feldspars from 298.15 to 1400 K, and the reaction quartz + jadeite = analbite. *Am. Mineral.* 66:1202–1215.

Hemingway, B. S., and Robie, R. A. 1977a. Enthalpies of formation of low albite (NaAlSi$_3$O$_8$), gibbsite (Al(OH)$_3$), and NaAlO$_2$; revised values for ΔH_f°, 298 and ΔG_f°, 298 of some aluminosilicate minerals. *J. Res. U.S. Geol. Surv.* 5:413–429.

————. 1977*b*. The entropy and Gibbs free energy of formation of the aluminum ion. *Geochim. Cosmochim. Acta* 41:1402–1404.

Hemingway, B. S., Robie, R. A., Fisher, J. R., and Wilson, W. H. 1977. Heat capacities of gibbsite, $Al(OH)_3$, between 13 and 480 K and magnesite, $MgCO_3$, between 13 and 380 K and their standard entropies at 298.15 K, and the heat capacities of calorimetry conference benzoic acid between 12 and 316 K. *J. Res. U.S. Geol. Surv.* 5:797–806.

Hemley, J. J., Montoya, J. W., Christ, C. L., and Hostetler, P. B. 1977*a*. Mineral equilibria in the MgO–SiO_2–H_2O system. I. Talc–chrysotile–forsterite–brucite stability relations. *Am. J. Sci.* 277:322–351.

Hemley, J. J., Montoya, J. W., Shaw, D. R., and Luce, R. W. 1977*b*. Mineral equilibria in the MgO–SiO_2–H_2O system. *Am. J. Sci.* 277:353–383.

Högfeldt, E. 1982. *Stability constants of metal ion complexes, part A: Inorganic ligands.* IUPAC Chemical Data Series, No. 21. Pergamon Press. 310 pp.

Holland, T. J. B. 1980. The reaction albite = jadeite + quartz determined experimentally in the range 600–1200°C. *Am. Mineral.* 65:129–134.

Holloway, J. R. 1977. Fugacity and activity of molecular species in supercritical fluids. In *Thermodynamics in geology*, ed. D. G. Fraser, pp. 161–182. D. Reidel.

Holm, J. L., and Kleppa, O. J. 1966. The thermodynamic properties of the aluminum silicates. *Am. Mineral.* 51:1608–1622.

Hückel, E. 1925. The theory of concentrated aqueous solutions of strong electrolytes. *Physik. Z.* 26:93–147.

Huebner, J. S. 1971. Buffering techniques for hydrostatic systems at elevated pressures. In *Research techniques for high pressure and high temperature*, ed. G. C. Ulmer, pp. 123–178. Springer-Verlag.

Hultgren, R., Desai, P. D., Hawkins, D. T., Gleiser, M., and Kelley, K. K. 1973. *Selected values of the thermodynamic properties of binary alloys.* American Society of Metals. 1435 pp.

Ingle, S. D., Keniston, J. A., and Schults, D. W. 1979. *REDEQL.EPAK, aqueous chemical equilibrium program.* Corvallis Environmental Research Laboratory, Corvallis, Oregon.

Ingri, N., Kakolowicz, W., Sillen, L. G., and Warnquist, B. 1967. High-speed computers as a supplement of graphical methods. V. HALTAFALL, a general program for calculating the composition of equilibrium mixtures. *Talanta* 14:1261–1286.

Jacobs, G. K., and Kerrick, D. M. 1981. Methane: an equation of state with application to the ternary system H_2O–CO_2–CH_4. *Geochim. Cosmochim. Acta* 45:607–614.

Jacobs, G. K., Kerrick, D. M., and Krupka, K. M. 1981. The high-temperature heat capacity of natural calcite ($CaCO_3$). *Phys. Chem. Minerals* 7:55–59.

Jacobson, R. L., and Langmuir, D. 1974. Dissociation constants of calcite and $CaHCO_3^+$ from 0° to 50°C. *Geochim. Cosmochim. Acta* 38:301–318.

James, R. V., and Rubin, J. 1979. Applicability of the local equilibrium assumption to transport through soils of solutes affected by ion exchange. In *Chemical modeling in aqueous systems*, ed. E. A. Jenne, pp. 225–235.

Jenne, E. A., ed. 1979. *Chemical modeling in aqueous systems: speciation, sorption, solubility, and kinetics.* American Chemical Society (Symposium Series 93). 914 pp.

———. 1981. Geochemical modeling: a review. *Battelle Pacific Northwest Laboratory Report* PNL-3574. 47 pp.

Johannes, W., Bell, P. M., Boettcher, A. L., Chipman, D. W., Hays, J. F., Mao, H. K., Newton, R. C., and Seifert, F. 1971. An inter-laboratory comparison of piston-cylinder pressure calibration using the albite breakdown reaction. *Contrib. Mineral. Petrol.* 32:24–38.

Johansen, E. S. 1967. Chemical equilibrium and linear programming. *Acta Chem. Scand.* 21:2273–2279.

Johnson, G. K., Flowtow, H. E., O'Hare, P. A. G., and Wise, W. S. 1982. Thermodynamic studies of zeolites: analcime and dehydrated analcime. *Am. Mineral.* 67:736–748.

Johnson, N. M. 1971. Mineral equilibrium in ecosystem geochemistry. *Ecology* 52:529–531.

Karpov, I. K. and Kashik, S. A. 1968. Computer calculation of standard isobaric-isothermal potentials of silicates by multiple regression from a crystallochemical classification. *Geokhimiya* 7:806–814.

Kern, R., and Weisbrod, A. 1967. Thermodynamics for geologists. Freeman, Cooper. 304 pp.

Kerrick, D. M. 1974. Review of metamorphic mixed volatile (H_2O–CO_2) equilibria. *Am. Mineral.* 59:729–762.

Kerrick, D. M., and Jacobs, G. K. 1981. A modified Redlich-Kwong equation for H_2O, CO_2, and H_2O–CO_2 mixtures at elevated temperatures and pressures. *Am. J. Sci.* 281:735–767.

Kerrick, D. M., and Slaughter, J. 1976. Comparison of methods for calculating and extrapolating equilibria in P–T–X_{CO_2} space. *Am. J. Sci.* 276:883–916.

Kharaka, Y. K., and Barnes, I. 1973. SOLMNEQ: solution-mineral equilibrium computations. *National Tech. Infor. Serv. Tech. Rept.* PB214-899. 82 pp.

Kieffer, S. W. 1979a. Thermodynamics and lattice vibration of minerals: 1. Mineral heat capacities and their relationships to simple lattice vibrational models. *Rev. Geophys. Space Phys.* 17:1–19.

———. 1979b. Thermodynamics and lattice vibrations of minerals: 2. Vibrational characteristics of silicates. *Rev. Geophys. Space Phys.* 17:20–34.

———. 1979c. Thermodynamics and lattice vibrations of minerals: 3. Lattice dynamics and an approximation for minerals with application to simple substances and framework silicates. *Rev. Geophys. Space Phys.* 17:35–59.

———. 1980. Thermodynamics and lattice vibrations of minerals: 4. Application to chain and sheet silicates and orthosilicates. *Rev. Geophys. Space Phys.* 18:862–886.

———. 1982. Thermodynamics and lattice vibrations of minerals: 5. Applications to phase equilibria, isotopic fractionation, and high-pressure thermodynamic properties. *Rev. Geophys. Space Phys.* 20:827–849.

Kielland, J. 1937. Individual activity coefficients of ions in aqueous solutions. *J. Am. Chem. Soc.* 59:1675–1678.

King, M. B., and Queen, N. B. 1979. Use of rational functions for representing data. *J. Chem. Eng. Data* 24:178–181.

Kinsman, D. J. J., and Holland, H. D. 1969. Coprecipitation of cations with calcium carbonate. Coprecipitation of strontium (II) with aragonite between 16 and 96°. *Geochim. Cosmochim. Acta* 33:1–17.

Klotz, I. M., and Rosenberg, R. M. 1972. *Chemical thermodynamics*. Benjamin. 444 pp.

Kopp, H. 1864. On the specific heat of rigid solids. *Ann. Chem. Pharm. Suppl.* 3:289–342.

Korzhinskii, D. S. 1959. *Physiochemical basis of the analysis of the paragenesis of minerals.* Consultants Bureau.

Krauskopf, K. 1979. Introduction to geochemistry. 2d ed. McGraw-Hill, 617 pp.

Krupka, K. M., Robie, R. A., and Hemingway, B. S. 1979. High-temperature heat capacities of corundum, periclase, anorthite, $CaAl_2Si_2O_8$ glass, muscovite, pyrophyllite, $KAlSi_3O_8$ glass, grossular, and $NaAlSi_3O_8$. *Am. Mineral.* 64:86–101.

Kubaschewski, O., and Alcock, C. B. 1979. *Metallurgical thermochemistry*. 5th ed. Pergamon Press. 449 pp.

Kujawa, F. B., and Eugster, H. P. 1966. Stability sequences and stability levels in unary systems. *Am. J. Sci.* 264:620–642.

Lafon, G. M. 1978. Discussion: equilibrium criteria for two-component solids reacting with fixed composition in an aqueous phase-example: the magnesian calcites. *Am. J. Sci.* 278:1455–1468.

Langmuir, D. 1968. Stability of calcite based on aqueous solubility measurements. *Geochim. Cosmochim. Acta* 32:835–851.

———. 1969. The Gibbs free energies of substances in the system $Fe–O_2–H_2O–CO_2$. *U.S. Geol. Surv. Prof. Paper* 650-B, pp. 180–184.

———. 1979. Techniques of estimating thermodynamic properties for some aqueous complexes of geochemical interest. In *Chemical modeling in aqueous systems*, ed. E. A. Jenne, pp. 353–387. American Chemical Society (Symposium Series 93).

Langmuir, D., and Whittemore, D. O. 1971. Variations in the stability of precipitated ferric oxyhydroxides. In *Nonequilibrium systems in natural water chemistry*, ed. J. D. Hem, pp. 209–234. American Chemical Society (Advanced Chemistry Series 106).

Latimer, W. L. 1921. The mass effect in the entropy of solids and gases. *J. Am. Chem. Soc.* 43:818–826.

———. 1951. Method of estimating the entropy of solid compounds. *J. Am. Chem. Soc.* 73:1480–1482.

———. 1952. *Oxidation potentials.* 2d ed. Prentice-Hall. 392 pp.

Lewis, G. N., and Randall, M. 1961. *Thermodynamics.* Revised by K. S. Pitzer and L. Brewer. McGraw-Hill. 723 pp.

Lide, D. R., Jr. 1981. Critical data for critical needs. *Science* 212:1343–1349.

Lidén, J. 1983. Equilibrium approaches to natural water systems: a study of anoxic- and ground-waters based on in situ data acquisition. Ph.D. thesis, University of Umea, Umea. 52 pp. (plus 7 appendixes).

Lietzke, M. H., and Stoughton, R. W. 1962. The calculation of activity coefficients from osmotic coefficient data. *J. Phys. Chem.* 66:508–509.

Lietzke, M. H., Stoughton, R. W., and Young, T. F. 1961. The bisulfate acid constant from 25 to 225° as computed from solubility data. *J. Phys. Chem.* 65:2247–2249.

Lilley, T. H., and Briggs, C. C. 1976. Activity coefficients of calcium sulfate in water 25°C. *Proc. Roy. Soc. London, Ser. A* 349:355–368.

Lindberg, R. D. 1983. A geochemical appraisal of oxidation-reduction potential and interpretation of Eh measurements of ground water. Ph.D. thesis, Univ. Colorado. 363 pp.

Lindsley, D. H., Grover, J. E., and Davidson, P. M. 1981. The thermodynamics of $Mg_2Si_2O_6$–$CaMgSi_2O_6$ join: a review and an improved model. In *Thermodynamics of minerals and melts*, ed. R. C. Newton, A. Navrotsky and B. J. Wood, pp. 149–175. Springer-Verlag.

Linke, W. F., and Seidell, A. 1958. *Solubilities of inorganic and metal-organic compounds, volume 1.* 4th ed. American Chemical Society. 1487 pp.

Lippman, F. 1977. The solubility products of complex minerals, mixed crystals, and three-layer clay minerals. *N. Jb. Miner. Abh.* 3:243–263.

Ludington, S. 1978. The biotite-apatite geothermometer revisited. *Am. Mineral.* 63:551–553.

———. 1979. Thermodynamics of melting of anorthite deduced from phase equilibrium studies. *Am. Mineral.* 64:77–85.

MacInnes, D. A. 1919. The activities of the ions of strong electrolytes. *J. Am. Chem. Soc.* 41:1086–1092.

Maier, C. G., and Kelly, K. K. 1932. An equation for the representation of high temperature heat content data. *J. Am. Chem. Soc.* 54:3243–3246.

Manual of symbols and terminology for physicochemical quantities and units. 1979. Pergamon Press. 41 pp.

Margenau, H., and Murphy, G. M. 1956. *The mathematics of physics and chemistry.* 2d ed. Van Nostrand. 604 pp.

Mason, E. A., and Spurling, T. H. 1969. The virial equation of state. In *The international encyclopedia of physical chemistry and chemical physics: topic 10. The fluid state.* Pergamon Press. 297 pp.

Mattigod, S. V., and Sposito, G. 1978. An improved method for estimating the standard free energies of formation ($G_{f,298.15}$) of smectites. *Geochim. Cosmochim. Acta* 42:1753–1762.

———. 1979. Chemical modeling of trace metal equilibria in contaminated soil solutions using the computer program GEOCHEM. In *Chemical modeling in aqueous systems*, ed. E. A. Jenne, pp. 837–856. American Chemical Society (Symposium Series 93).

May, H. M., Helmke, P. A., and Jackson, M. L. 1979. Gibbsite solubility and thermodynamic properties of hydroxy-aluminum ions in aqueous solution at 25°C. *Geochim. Cosmochim. Acta* 43:861–868.

Mazo, R. M. 1977. Statistical mechanical calculation of aluminum-silicon disorder in albite. *Am. Mineral.* 62:1232–1237.

McDuff, R. E., and Morel, P. M. 1973. Description and use of the chemical equilibrium program REDEQL2. *Keck Lab. Tech. Rept.* EQ-73-02 (California Institute of Technology). 75 pp.

McGlashan, M. L. 1966. The use and misuse of the laws of thermodynamics. *J. Chem. Educ.* 43:226–232.

———. 1973. Internationally recommended names and symbols for physicochemical quantities and units. *Ann. Rev. Phys. Chem.* 24:51–76.

———. 1979. *Chemical thermodynamics.* Academic Press. 345 pp.

Miller, C., and Benson, L. 1983. Simulation of solute transport in a chemically reactive heterogeneous system: model development and application. *Water Resour. Res.* 19:381–391.

Miller, S. D. 1979. Chemistry of a pyritic strip-mine spoil. Ph.D. thesis, Yale University. 201 pp.

———. 1980. Sulfur and hydrogen ion buffering in pyritic strip-mine spoil. In *Biogeochemistry of ancient and modern environments,* ed. P. A. Trudinger, M. R. Walter and B. J. Ralph, pp. 537–543. Australian Academy of Sciences.

Millero, F. J. 1967. High precision magnetic float densimeter. *Rev. Sci. Inst.* 38:1441–1444.

———. 1972. The partial molal volumes of electrolytes in aqueous solutions. In *Water and aqueous solutions,* ed. R. A. Horne, pp. 519–564. Wiley.

Morel, F., and Morgan, J. J. 1972. A numerical method for computing equilibria in aqueous systems. *Environ. Sci. Tech.* 6:58–67.

Morse, S. A. 1980. *Basalts and phase diagrams, an introduction to the quantitative use of phase diagrams in igneous petrology.* Springer-Verlag. 493 pp.

Myers, J., and Eugster, H. P. 1983. The system Fe–Si–O: oxygen buffer calibrations to 1500 K. *Contr. Mineral. Petrol.* 82:75–90.

Nafziger, R. H., Ulmer, G. C., and Woermann, E. 1971. Gaseous buffering for the control of oxygen fugacity at one atmosphere. In *Research techniques for high pressure and high temperature,* ed. G. C. Ulmer, pp. 9–42. Springer-Verlag.

Navrotsky, A. 1977. Progress and new directions in high temperature calorimetry. *Phys. Chem. Minerals* 2:89–104.

———. 1981. Thermodynamics of mixing in silicate glasses and melts. In *Thermodynamics of minerals and melts,* ed. R. C. Newton, A. Navrotsky, and B. J. Wood, pp. 189–206. Springer-Verlag.

Navrotsky, A., and Coons, W. E. 1976. Thermochemistry of some pyroxenes and related compounds. *Geochim. Cosmochim. Acta* 40:1281–1288.

Navrotsky, A., Hon, R., Weill, D. F., and Henry, D. J. 1980. Thermochemistry of glasses and liquids in the systems $CaMgSi_2O_6$–$CaAl_2Si_2O_8$–$NaAlSi_3O_8$, SiO_2–$CaAl_2Si_2O_8$–$NaAlSi_3O_8$, and SiO_2–Al_2O_3–CaO–Na_2O. *Geochim. Cosmochim. Acta* 44:1409–1423.

Nelson, L. C., and Obert, E. F. 1954. Generalized *pvT* properties of gases. *Trans. ASME* 76:1057–1063.

Neumann, F. W. 1933. Solubility relations of barium sulfate in aqueous solutions of strong electrolytes. *J. Am. Chem. Soc.* 55:879–884.

Nicholls, J. 1977. The activities of components in natural silicate melts. In *Thermodynamics in geology,* ed. D. G. Fraser, pp. 327–348. D. Reidel.

Nordstrom, D. K. 1982. The effect of sulfate on aluminum concentrations in natural waters: some stability relations in the system $Al_2O_3-SO_3-H_2O$ at 298 K. *Geochim. Cosmochim. Acta* 46:681–692.

Nordstrom, D. K., and Ball, J. W. 1984. Chemical models, computer programs and metal complexation in natural waters. In *International symposium on trace metal complexation in natural waters*. ed. C. J. M. Kramer and J. C. Duinker. pp. 149–169. Martinus Nijhoff/Dr. J. W. Junk Publishing Co.

Nordstrom, D. K., Jenne, E. A., and Ball, J. W. 1979. Redox equilibria of iron in acid mine waters. In *Chemical modeling in aqueous systems*, ed. E. A. Jenne, pp. 51–79. American Chemical Society (Symposium Series 93).

Nordstrom, D. K., Plummer, L. N., Wigley, T. M. L., Wolery, T. J., Ball, J. W., Jenne, E. A., Bassett, R. L., Crerar, D. A., Florence, T. M., Fritz, B., Hoffman, M., Holdren, G. R., Jr., Lafon, G. M., Mattigod, S. V., McDuff, R. E., Morel, F., Reddy, M. M., Sposito, G., and Thrailkill, J. 1979. A comparison of computerized chemical models for equilibrium calculations in aqueous systems. In *Chemical modeling in aqueous systems*, ed. E. A. Jenne, pp. 857–892. American Chemical Society (Symposium Series 93).

Norvell, W. A., and Lindsay, W. L. 1982. Estimation of the concentration of Fe^{3+} and the $(Fe^{3+})(OH^-)^3$ ion product from equilibria of EDTA in soil. *Soil Sci. Soc. Am. J.* 46:710–715.

Nriagu, J. O. 1975. Thermochemical approximations for clay minerals. *Am. Mineral.* 60:834–839.

Öhman, L.-O. 1983. Equilibrium studies of ternary aluminum (III) complexes. Ph.D. thesis, University of Umea, Umea. 74 pp. (plus 8 appendixes).

Openshaw, R. E., Hemingway, B. S., Robie, R. A., Waldbaum, D. R., and Krupka, K. M. 1976. The heat capacities at low temperatures and the heat capacities at 298.15 K of low albite, analbite, microcline and high sanidine. *J. Res. U.S. Geol. Survey* 4:195–204.

Parkhurst, D. L., Plummer, L. N., and Thorstenson, D. C. 1982. BALANCE—a computer program for calculating mass transfer for geochemical reactions in ground water. *U.S. Geol. Surv. Water Resour. Invest. Rept. 82-14*. 29 pp.

Parkhurst, D. L., Thorstenson, D. C., and Plummer, L. N. 1980. PHREEQE—a computer program for geochemical calculations. *U.S. Geol. Surv. Water Resour. Invest. Rept. 80-96*. 210 pp.

Parks, G. A., and Nordstrom, D. K. 1979. Estimated free energies of formation, water solubilities, and stability fields for schuetteite $(Hg_3O_2SO_4)$ and corderoite $(Hg_3S_2Cl_2)$ at 298 K. In *Chemical modeling in aqueous systems*, ed. E. A. Jenne, pp. 339–352. American Chemical Society (Symposium Series 93).

Pearson, F. J., and Rettman, P. L. 1976. Geochemical and isotopic analyses of waters associated with the Edwards limestone aquifer, central Texas. *U.S. Geol. Surv. Open-File Rept.* 35 pp.

Pedley, J. B., and Rylance, J. 1977. Computer analysis of thermochemical data. In *Proceedings, Fifth Biennial CODATA Conference*, ed. B. Dreyfus, pp. 557–580. Pergamon Press.

Pitzer, K. S. 1977. Electrolyte theory—improvements since Debye-Hückel. *Acc. Chem. Res.* 10:371–377.

———. 1979. Theory: ion interaction approach. In *Activity coefficients in electrolyte solutions, volume 1,* ed. R. M. Pytkowicz, pp. 157–208. CRC Press.

———. 1980. Electrolytes. From dilute solutions to fused salts. *J. Am. Chem. Soc.* 102:2902–2906.

Planck, M. 1926. *Treatise on thermodynamics,* 3d ed. Longmans, Green. 297 pp. (Reprinted by Dover, 1945).

Plummer, L. N., and Busenberg, E. 1982. The solubilities of calcite, aragonite and vaterite in CO_2–H_2O solutions between 0 and 90°C, and an evaluation of the aqueous model for the system $CaCO_3$–CO_2–H_2O. *Geochim. Cosmochim. Acta* 46:1011–1040.

Plummer, L. N., Jones, B. F., and Truesdell, A. H. 1976. WATEQF—a FORTRAN IV version of WATEQ, a computer program for calculating chemical equilibrium of natural waters. *U.S. Geol. Surv. Water Resour. Invest.* 76-13. 61 pp.

Plummer, L. N., and Mackenzie, F. T. 1974. Predicting mineral solubility from rate data: application to the dissolution of magnesian calcites. *Am. J. Sci.* 274:61–83.

Plummer, L. N., Parkhurst, D. L., and Thorstenson, D. C. 1983. Development of reaction models for groundwater systems. *Geochim. Cosmochim. Acta* 47:665–685.

Potter, R. W., II. 1979. Computer modelling of low temperature geochemistry. *Rev. Geophys. Space Phys.* 17:850–860.

Powell, R. 1978. *Equilibrium thermodynamics in petrology. An introduction.* Harper & Row, 287 pp.

Powell, R. E., and Latimer, W. M. 1951. The entropy of aqueous solutes. *J. Chem. Phys.* 19:1139–1141.

Prausnitz, J. M. 1969. *Molecular thermodynamics of fluid-phase equilibria.* Prentice-Hall. 523 pp.

Presnall, D. C. 1969. Pressure-volume-temperature measurements on hydrogen from 200°C to 600°C and up to 1800 atmospheres. *J. Geophys. Res.* 74:6026–6033.

———. 1971. Compressibility measurements of gases using externally heated pressure vessels. In *Research techniques for high pressure and high temperature,* ed. G. C. Ulmer, pp. 259–278. Springer Verlag.

Prigogine, I., and Defay, R. 1954. *Chemical thermodynamics.* Longmans, Green. 543 pp.

Quist, A. S., and Marshall, W. L. 1968. Electrical conductances of aqueous sodium chloride solutions from 0° to 800°C and at pressures to 4000 bars. *J. Phys. Chem.* 73:684–703.

Ramette, R. W. 1977. Deducing the pK-temperature equation. *J. Chem. Ed.* 54:280–283.

Reardon, E. J. 1975. Dissociation constants of some monovalent sulfate ion pairs at 25° from a stoichiometric activity coefficients. *J. Phys. Chem.* 79:422–425.

Redlich, O., and Kwong, J. N. S. 1949. The thermodynamics of solutions. V. An equation of state. Fugacities of gaseous solutions. *Chem. Rev.* 44:233–244.

Reed, M. H. 1982. Calculation of multicomponent chemical equilibria and reaction processes involving minerals, gases and an aqueous phase. *Geochim. Cosmochim. Acta* 46:513–528.

Reid, R. C., Prausnitz, J. M., and Sherwood, T. K. 1977. *The properties of gases and liquids.* McGraw-Hill. 688 pp.

Reisman, A. 1970. *Phase equilibria.* Academic Press. 541 pp.

Reiss, H. 1965. *Methods of thermodynamics.* Blaisdell. 217 pp.

Rice, J. M., and Ferry, J. M. 1982. Buffering, infiltration, and the control of intensive variables during metamorphism. *Rev. Mineralogy* 10:263–326.

Richardson, S. W., Gilbert, M. C., and Bell, P. M. 1969. Experimental determination of kyanite-andalusite and andalusite-sillimanite equilibria; the aluminum silicate triple point. *Am. J. Sci.* 267:259–272.

Rickard, D. T., and Wickman, F. E., eds. 1981. *Chemistry and geochemistry of solutions at high temperatures and pressures.* Pergamon Press. 563 pp.

Robie, R. A., Finch, C. B., and Hemingway, B. S. 1982. Heat capacity and entropy of fayalite (Fe_2SiO_4) between 5.1 and 383 K: comparison of calorimetric and equilibrium values for the QFM buffer reaction. *Am. Mineral.* 67:463–469.

Robie, R. A., and Hemingway, B. S. 1972. Calorimeters for heat of solution and low-temperature heat capacity measurements. *U.S. Geol. Surv. Prof. Paper 755.* 32 pp.

Robie, R. A., Hemingway, B. S., and Fisher, J. R. 1978. Thermodynamic properties of minerals and related substances at 298.15 K and 1 bar (10^5 pascals) pressure and at higher temperatures. *U.S. Geol. Surv. Bull.* 1452. 456 pp. (Reprinted with corrections, 1979)

Robie, R. A., Hemingway, B. S., and Wilson, W. H. 1976. The heat capacities of calorimetry conference copper and of muscovite, $KAl_2(AlSi_3)O_{10}(OH)_2$, pyrophyllite, $Al_2Si_4O_{10}(OH)_2$ and illite, $K_3(Al_7Mg)(Si_{14}Al_2)O_{40}(OH)_8$ between 15 and 375 K and their standard entropies at 298.15 K. *J. Res. U.S. Geol. Surv.* 4:631–644.

Robinson, G. R., Jr., and Haas, J. L. 1983. Heat capacity, relative enthalpy, and calorimetric entropy of silicate minerals: an empirical method of prediction. *Am. Mineral.* 68:541–553.

Robinson, G. R., Haas, J. L., Jr., Schafer, C. M., and Hazelton, H. T., Jr. 1982. Thermodynamic and thermophysical properties of selected phases in the MgO–SiO_2–H_2O–CO_2, CaO–Al_2O_3–SiO_2–H_2O–CO_2, and Fe–FeO–Fe_2O_3–SiO_2 chemical systems, with special emphasis on the properties of basalts and their mineral components. *U.S. Geol. Surv. Open-File Rept.* 83-79. 429 pp.

Robinson, R. A., and Stokes, R. H. 1955. *Electrolyte solutions.* Butterworths. 559 pp.

Rock, P. A. 1967. Fixed pressure standard states in thermodynamic and kinetics. *J. Chem. Educ.* 44:104–108.

Rogers, P. S. Z. 1981. Thermodynamics of geothermal fluids. Ph.D. thesis, University of California at Berkeley and Livermore Berkeley Laboratory, LBL-12356. 253 pp.

Rubin, J. 1983. Transport of reacting solutes in porous media: relation between mathematical nature of problem formulation and chemical nature of reactions. *Water Resour. Res.* 19:1231–1252.

Rubin, J., and James, R. V. 1973. Dispersion-affected transport of reacting solutes in saturated porous media: Galerkian method applied to equilibrium-controlled exchange in unidirectional steady water flow. *Water Resour. Res.* 9:1332–1336.

Rumble, D., III. 1973. Fe-Ti oxide minerals from regionally metamorphosed quartzites of western New Hampshire. *Contrib. Mineral. and Petrol.* 42:181–195.

———. 1976*a*. Oxide minerals in metamorphic rocks. *Rev. Mineralogy* 3:R1–R24.

———. 1976*b*. The use of mineral solid solutions to measure chemical potential gradients in rocks. *Am. Mineral.* 61:1167–1174.

Sato, M. 1960. Oxidation of sulfide ore bodies, 1. Geochemical environments in terms of Eh and pH. *Econ. Geol.* 55:928–961.

———. 1971. Electrochemical measurements and control of oxygen fugacity and other gaseous fugacities with solid electrolyte systems. In *Research techniques for high pressure and high temperature*, ed. G. C. Ulmer, pp. 43–100. Springer-Verlag.

———. 1972. Intrinsic oxygen fugacities of iron-bearing oxide and silicate minerals under low pressure. *Geol. Soc. Am. Mem.* 135:289–308.

Sato, M., and Wright, T. L. 1966. Oxygen fugacities directly measured in magmatic gases. *Science* 153:1103–1105.

Scatchard, G. 1936. Concentrated solutions of electrolytes. *Chem. Rev.* 19:309–327.

Sharma, S. K., Virgo, D., and Mysen, B. 1978. Structure of melts along the join SiO_2–$NaAlSiO_4$ by Raman spectroscopy. *Carnegie Inst. Washington Yearbook* 77:652–658.

Shedlovsky, T., and MacInnes, D. A. 1935. The first ionization constant of carbonic acid, 0 to 38°, from conductance measurements. *J. Am. Chem. Soc.* 57:1705–1710.

Sillén, L. G. 1967. Master variables and activity scales. In *Equilibrium concepts in natural water systems*, ed. W. Stumm, pp. 45–69. American Chemical Society (Advanced Chemistry Series 67).

Sillén, L. G., and Martell, A. E. 1964. Stability constants of metal-ion complexes, *Chem. Soc. Spec. Publ.* 17. 754 pp. (Supplement no. 1, *Chem. Soc. Spec. Publ.* 25. 1971. 865 pp.)

Slaughter, M. 1966*a*. Chemical binding in silicate minerals—I. Model for determining crystal-chemical properties. *Geochim. Cosmochim. Acta* 30:299–313.

———. 1966*b*. Chemical binding in silicate minerals—II. Computational methods and approximations for the binding energy of complex silicates. *Geochim. Cosmochim. Acta* 30:315–322.

———. 1966*c*. Chemical binding in silicate minerals—III. Application of energy calculations to the prediction of silicate mineral stability. *Geochim. Cosmochim. Acta* 30:323–339.

Smith, R. M., and Martell, A. E. 1976. *Critical stability constants, Volume 4: Inorganic complexes.* Plenum Press. 257 pp.

Smith, W. R., and Missen, R. W. 1982. *Chemical reaction equilibrium analysis.* Wiley-Interscience. 364 pp.

Sokolnikoff, I. S., and Redheffer, R. M. 1966. *Mathematics of physics and modern engineering.* McGraw-Hill. 752 pp.

Spear, F. S., Ferry, J. M., and Rumble, D., III. 1982. Analytical formulation of phase equilibria: the Gibbs' method. *Rev. Mineralogy* 10:105–152.

Spear, F. S., Rumble, D., III, and Ferry, J. M. 1982. Linear algebraic manipulation of *n*-dimensional composition space. *Rev. Mineralogy* 10:53–104.

Spencer, K. J., and Lindsley, D. H. 1981. A solution model for coexisting iron-titanium oxides. *Am. Mineral.* 66:1189–1201.

Sposito, G., and Mattigod, S. V. 1980. *GEOCHEM: a computer program for the calculation of chemical equilibria in soil solutions and other natural water systems.* Dept. of Soil and Environmental Sciences, University of California, Riverside. 92 pp.

Staples, B. R., and Nuttall, R. L. 1977. The activity and osmotic coefficients of aqueous calcium chloride at 298.15 K. *J. Phys. Chem. Ref. Data* 6:385–407.

Stockmayer, W. H. 1978. Data evaluation: a critical activity. *Science* 201:577.

Stull, D. R., and Prophet, H. 1971. JANAF Thermochemical Tables, 2nd ed. *NSRDS-NBS* 37. 1141pp. (1974 supplement by Chase, M. W., Curnutt, J. L., Hu, A. T., Prophet, H., and Walker, L. C., *J. Phys. Chem. Ref. Data* 3:311–480. 1975 Supplement by Chase, M. W., Curnutt, J. L., Prophet, H., McDonald, R. A., and Syverud, A. N., *J. Phys. Chem. Ref. Data* 4:1–185.)

Stumm, W., and Morgan, J. J. 1981. *Aquatic chemistry.* Wiley-Interscience. 780 pp.

Swalin, R. A. 1972. *Thermodynamics of solids.* 2d ed. Wiley. 387 pp.

Tardy, Y. 1979. Relationships among Gibbs free energies of formation of compounds. *Am. J. Sci.* 279:217–224.

Tardy, Y., and Garrels, R. M. 1974. A method of estimating the Gibbs energies of formation of layer silicates. *Geochim. Cosmochim. Acta* 38:1101–1116.

———. 1976. Prediction of Gibbs energies of formation—I. Relationships among Gibbs energies of formation of hydroxides, oxides and aqueous ions. *Geochim. Cosmochim. Acta* 40:1051–1056.

———. 1977. Prediction of Gibbs energies of formation from the elements—II. Monovalent and divalent metal silicates. *Geochim. Cosmochim. Acta* 41:87–92.

Tardy, Y., and Viellard, P. 1977. Relationships among Gibbs free energies and enthalpies of sulfates, nitrates, carbonates, oxides, and aqueous ions. *Contrib. Mineral. Petrol.* 63:89–102.

Tardy, Y., and Viellard, P. 1977. Relationships among Gibbs free energies and enthalpies of formation of phosphates, oxides and aqueous ions. *Contrib. Mineral. Petrol.* 63:75–88.

Taylor, M., and Brown, G. E. 1979. Structure of silicate mineral glasses, I. *Geochim. Cosmochim. Acta* 43:61–75.

Thompson, A. B., and Perkins, E. H. 1981. Lambda transitions in minerals. In *Thermodynamics of minerals and melts*, ed. R. C. Newton, A. Navrotsky, and B. J. Wood, pp. 35–62. Springer-Verlag.

Thompson, J. B., Jr. 1959. Local equilibrium in metasomatic processes. In *Researches in geochemistry*, ed. P. H. Abelson, pp. 427–457. Wiley.

———. 1967. Thermodynamic properties of simple solutions. In *Researches in geochemistry, volume 2*, ed. P. H. Abelson, pp. 340–361. Wiley.

———. 1970. Geochemical reaction and open systems. *Geochim. Cosmochim. Acta* 34:529–551.

———. 1972. Oxides and sulfides in regional metamorphism of pelitic schists. *24th Inter. Geol. Cong., Section 10*, pp. 27–35.

―――. 1982. Composition space: an algebraic and geometric approach. *Rev. Mineralogy*, 10:1–32.

Thompson, J. B., Jr., and Waldbaum, D. R. 1969. Mixing properties of sanidine crystalline solutions: III. Calculations based on two-phase data. *Am. Mineral.* 54:811–838.

Thorstenson, D. C. 1984. The concept of electron activity and its relation to redox potentials in aqueous geochemical systems. *U.S. Geol. Surv. Water-Supply Paper* (in press).

Thorstenson, D. C., and Plummer, L. N. 1977. Equilibrium criteria for two-component solids reacting with fixed composition in an aqueous phase—example: the magnesian calcites. *Am. J. Sci.* 277:1203–1223.

Truesdell, A. H. 1968. The advantage of using pE rather than Eh in redox equilibrium calculations. *J. Geol. Educ.* 16:17–20.

Truesdell, A. H., and Jones, B. F. 1974. WATEQ, a computer program for calculating chemical equilibria of natural waters. *J. Res. U.S. Geol. Surv.* 2:233–248.

Tunnell, G. 1977. Thermodynamic relations in open systems. *Carnegie Inst. Washington Publication* 408A. 69 pp.

Ulbrich, H. H., and Waldbaum, D. R. 1976. Structural and other contributions to the third-law entropies of silicates. *Geochim. Cosmochim. Acta* 40:1–24.

Ulmer, G. C., ed., 1971. *Research techniques for high pressure and high temperature.* Springer-Verlag. 367 pp.

Vallochi, A. J., Street, R. L., and Roberts, P. V. 1981. Transport of ion-exchanging solutes in groundwater: chromatographic theory and field simulation. *Water Resour. Res.* 17:1517–1527.

van Breemen, N. 1976. Genesis and solution chemistry of acid sulfate soils in Thailand. Ph.D. thesis, Wageningen. 263 pp.

Van Luik, A. E., and Jurinak, J. J. 1978. A chemical model of heavy metals in the Great Salt Lake. *Utah State University Res. Rept.* 34. 155 pp.

―――. 1979. Equilibrium chemistry of heavy metals in concentrated electrolyte solution. In *Chemical modeling in aqueous systems*, ed. E. A. Jenne, pp. 683–710. American Chemical Society (Symposium Series 93).

Van Zeggeren, F., and Storey, S. H. 1970. *The computation of chemical equilibria.* Cambridge Univ. Press. 176 pp.

Wagman, D. D., Evans, W. H., Parker, V. B., Harlow, I., Bailey, S. W., and Schumm, R. H. 1968. Selected values of chemical properties. Tables of the first thirty-four elements in the standard order of arrangement. *U.S. Natl. Bur. Standards Tech. Note* 270-3. 264 pp.

Wagman, D. D., Evans, W. H., Parker, V. B., and Schumm, R. H. 1976. Chemical thermodynamic properties of compounds of sodium, potassium, and rubidium: an interim tabulation of selected values. *U.S. Natl. Bur. Standards NBSIR* 76-1034. 73 pp.

Wagman, D. D., Evans, W. H., Parker, V. B., Schumm, R. H., Harlow, I., Bailey, S. M., Churney, K. L., and Nuttall, R. L. 1982. The NBS tables of chemical thermodynamic properties: selected values for inorganic and C_1 and C_2 organic substances in SI units. *J. Phys. Chem. Ref. Data* 11 (Suppl. 2):1–392.

Waldbaum, D. R., and Robie, R. A. 1971. Calorimetric investigation of Na-K mixing and polymorphism in the alkali feldspars. *Zeit. Krist.* 134:381–420.

Walther, J. V., and Helgeson, H. C. 1977. Calculation of the thermodynamic properties of aqueous silica and the solubility of quartz and its polymorphs at high pressures and temperatures. *Am. J. Sci.* 277:1315–1351.

Walther, J. V., and Orville, P. M. 1983. The extraction-quench technique for determination of the thermodynamic properties of solute complexes: application to quartz solubility in fluid mixtures. *Am. Mineral.* 68:731–741.

Weller, W. W., and Kelley, K. K. 1963. Low-temperature heat capacities and entropies at 298.15 K of akermanite, cordierite, gehlenite, and merwinite. *U.S. Bur. Mines Rept. Inv.* 6343. 7 pp.

Westall, J. 1979. MICROQL: I. A chemical equilibrium program in BASIC. II. Computation of adsorption equilibria in BASIC. *Swiss Inst. Tech. EAWAG.* 77 pp.

Westall, J., Zachary, J. L., and Morel, F. M. M. 1976. MINEQL, a computer program for the calculation of chemical equilibrium composition of aqueous systems. *Mass. Inst. Tech. Dept. Civil Eng. Tech. Note* 18. 91 pp.

White, W. B., Johnson, S. M., and Dantzig, G. B. 1958. Chemical equilibrium in complex mixtures. *J. Chem. Phys.* 28:751–755.

Whitfield, M. 1975a. An improved specific interaction model for seawater at 25°C and 1 atmosphere pressure. *Mar. Chem.* 3:197–205.

———. 1975b. Sea water as an electrolyte solution. In *Chemical oceanography, volume 1*, ed. J. P. Riley and G. Skirrow, pp. 43–171. Academic Press.

———. 1975c. The extension of chemical models for sea water to include trace components at 25°C and 1 atm. pressure. *Geochim. Cosmochim. Acta* 39:1545–1557.

———. 1979. Activity coefficients in natural waters. In *Activity coefficients in electrolyte solutions, volume 2*, ed. R. M. Pytkowicz, pp. 153–300. CRC Press.

Whittemore, D. O., and Langmuir, D. 1972. Standard electrode potential of $Fe^{3+} + e^- = Fe^{2+}$ from 5–35°C. *J. Chem. Eng. Data* 17:288–290.

Wigley, T. M. L. 1977. WATSPEC: a computer program for determining the equilibrium speciation of aqueous solutions. *Brit. Geomorph. Res. Group Tech. Bull.* 20. 48 pp.

Wilcox, D. E., and Bromley, R. A. 1963. Computer estimation of heat and free energy of formation for simple inorganic compounds. *Ind. Eng. Chem.* 55:32–39.

Wise, S. S., Margrove, J. L., Feder, H. M., and Hubbard, W. N. 1963. Fluorine bomb calorimetry. V. The heats of formation of silicon tetrafluoride and silica. *J. Phys. Chem.* 67:815–821.

Wolery, T. J. 1979. *Calculation of chemical equilibrium between aqueous solution and minerals: the EQ 3/6 software package.* Lawrence Livermore Laboratory (UCRL-52658). 41 pp.

———. 1983. *EQ3NR, a computer program for geochemical aqueous speciation-solubility calculations, user's guide and documentation.* Lawrence Livermore Laboratory (UCRL-53414). 191 pp.

————. 1984. *EQ6, a computer program for reaction-path modelling of aqueous geochemical systems, user's guide and documentation.* Lawrence Livermore Laboratory (in press).

Wollast, R., Mackenzie, F. T., and Bricker, O. P. 1968. Experimental precipitation and genesis of sepiolite at earth-surface conditions. *Am. Mineral.* 53:1645–1662.

Wood, J. R. 1975. Thermodynamics of brine-salt equilibria—I. The systems NaCl–KCl–MgCl$_2$–CaCl$_2$H$_2$O and NaCl–MgSO$_4$–H$_2$O at 25°C. *Geochim. Cosmochim. Acta* 39:1147–1163.

————. 1976. Thermodynamics of brine-salt equilibria—II. The system NaCl–KCl–H$_2$O from 0 to 200°C. *Geochim. Cosmochim. Acta* 40:1211–1220.

Woods, L. C. 1975. *The thermodynamics of fluid systems.* Oxford Univ. Press. 359 pp.

Zeleznik, F. J., and Gordon, S. 1968. Calculation of complex chemical equilibria. *Ind. Eng. Chem.* 60:27–57.

Zen, E-an. 1966. Construction of pressure-temperature diagrams for multi-component systems after the methods of Schreinemakers—a geometric approach. *U.S. Geol. Surv. Bull.* 1225. 56 pp.

————. 1969. Free energy of formation of pyrophyllite from hydrothermal data—values, discrepancies, and implications. *Am. Mineral.* 54:1592–1606.

————. 1977. The phase equilibrium calorimeter, the petrogenetic grid, and a tyranny of numbers: presidential address. *Am. Mineral.* 62: 189–204.

INDEX